Hardwood Reforestation and Restoration

Hardwood Reforestation and Restoration

Special Issue Editors

Daniel Gagnon
Benoit Truax

MDPI • Basel • Beijing • Wuhan • Barcelona • Belgrade

MDPI

Special Issue Editors

Daniel Gagnon
University of Regina
Canada

Benoit Truax
Eastern Townships Forest Research Trust
Canada

Editorial Office
MDPI
St. Alban-Anlage 66
4052 Basel, Switzerland

This is a reprint of articles from the Special Issue published online in the open access journal *Forests* (ISSN 1999-4907) in 2018 (available at: https://www.mdpi.com/journal/forests/special_issues/hardwood)

For citation purposes, cite each article independently as indicated on the article page online and as indicated below:

LastName, A.A.; LastName, B.B.; LastName, C.C. Article Title. *Journal Name* **Year**, *Article Number*, Page Range.

ISBN 978-3-03897-730-8 (Pbk)
ISBN 978-3-03897-731-5 (PDF)

Cover image courtesy of Danial Gagon.

Contents

About the Special Issue Editors

Daniel Gagnon is a Professor in the Department of Biology at the University of Regina (Saskatchewan, Canada), where he teaches forest ecology, terrestrial ecosystems ecology and plant taxonomy. Dr. Gagnon is also a trustee of the Eastern Township Forest Research Trust.

Benoit Truax is the General Director, a researcher and a trustee of the Eastern Township Forest Research Trust, a charitable organization whose objectives are to promote the conservation of forests, their restoration, as well as developing means of sustainable use of its resources.

Preface to "Hardwood Reforestation and Restoration"

Northern Hemisphere hardwood-dominated forests (North America, Western Europe, Eastern Asia) provide valuable renewable timber and numerous ecosystem services. Many of these forests have been subjected to harvesting or conversion to agriculture, sometimes over centuries, and it has greatly reduced their former extent and diversity. Natural regeneration following harvesting or during post-agricultural succession has often failed to restore these forests adequately. Past harvesting practices and the valuable timber of some species have led to a reduction in their abundance. The loss of apex predators has caused large ungulate herbivore populations to increase and exert intense browsing pressure on hardwood regeneration, often preventing it. Particularly important are fruit, nut and acorn bearing species, because of their vital role in forest food webs and biodiversity. Restoring hardwood species to natural forests in which they were formerly more abundant will require a number of forest management actions (e.g., disease or insect resistant/tolerant hybrids, ungulate exclosures/protectors, enrichment planting, underplanting, etc.). Similarly, reforesting areas that were once natural forests will also require new silvicultural knowledge. Global warming trends will intensify the need for interventions to restore and maintain the diversity and function of temperate hardwood forests, as well as for the need for increased hardwood reforestation.

Northern hemisphere hardwood forests are composed of plant genera of the Arcto-Tertiary flora that occurred in the past when all Northern Hemisphere continents were still connected to each other (Laurasia). All hardwood tree genera originate from this flora and have different species in two or three of the northern hemisphere continents. These species diverged from each other after the continents became separated by tectonic movements, mountain orogenies, and other environmental changes. Oak (*Quercus*), beech (*Fagus*), walnut (*Juglans*), ash (*Fraxinus*) and maple (*Acer*) are some of the most common and valuable hardwood genera. Oaks, and other genera in their family, are particularly valued for their timber as well as for the ecosystem services they provide (e.g. nut bearing trees for fauna). Nine out of 10 articles of this Special Issue include oaks as study species. They present research results for 15 species of *Quercus*, as well as for a few other Fagaceae genera (*Pasania, Lithocarpus, Notholithocarpus*).

The ancient flora connection of northern hemisphere hardwood tree species has advantages and disadvantages. Major problems for North American hardwoods have occurred when pathogens and insects introduced from Asia encounter "naive" North American species closely related to Asian species. The Asian hardwood species have had a long co-evolution with these pathogens and insects and have developed resistance or some high level of tolerance to them. There are many examples of formerly abundant North American hardwood species that have been, or are currently being decimated by an introduced pathogen or insect (American chestnut blight, emerald ash borer, Dutch elm disease, butternut canker, etc.). However, one of the advantages of this ancient genetic connection is the possibility of creating hybrids, with just enough genes from Asian species, to provide resistance to American species. Such is the case of the successful hybrid between the North American *Castanea dentata* and an Asian *Castanea* species, allowing American chestnut to currently being replanted with success. A similar solution is being sought for a resistant hybrid of *Juglans cinerea* with an Asian *Juglans* species, in order to stem the devastation caused by the butternut canker. For insect pests, such as the Asian emerald ash borer, hybrids may also present a viable solution. In addition to Asian pathogens and insect pests introduced to North America, there are also introduced invasive plant species from Asia that are causing major problems in North America by competing with native hardwood regeneration in forest understories (e.g. *Frangula alnus, Lonicera mackii*).

This Special Issue presents research that has been conducted in all of the continents of the Northern Hemisphere. Several articles address the effect of overabundant herbivore populations on planted hardwoods, and how to protect them, but also on how heavy browsing in a hardwood forest understory has an indirect negative effect on understory bird species populations. Other articles illustrate the negative effects of exotic invasive understory plant species on hardwood regeneration through removal experiments. Changes in land use and climate change effects are also addressed as other threats to hardwood forests.

The articles within this Special Issue address many of the problems and issues that hardwood tree species are facing throughout the Northern Hemisphere. Some authors have identified some problems as a first step, but many have also proposed management methods for mitigating these problems.

Daniel Gagnon, Benoit Truax
Special Issue Editors

forests

MDPI

Article

Bringing the Natives Back: Identifying and Alleviating Establishment Limitations of Native Hardwood Species in a Conifer Plantation

Yu-Tsen Li [1,2], Yueh-Hsin Lo [1,3], Yi-Ching Lin [4], Biing T. Guan [1,*], Juan A. Blanco [3] and Chi-How You [5]

1   School of Forestry and Resource Conservation, National Taiwan University, Taipei 10617, Taiwan; b95605089@ntu.edu.tw (Y.-T.L.); yuehhsin.lo@gmail.com (Y.-H.L.)
2   Experimental Forest, National Taiwan University, Zhushan 55750, Taiwan
3   Departmento de Ciencias del Medio Natural, Universidad Pública de Navarra, Pamplona, 31006 Navarra, Spain; juan.blanco@unavarra.es
4   Department of Life Science, Tunghai University, Taichung 40704, Taiwan; yichingtree@gmail.com
5   Hsinchu Forest District Office, Taiwan Forest Bureau, Hsinchu 30046, Taiwan; chihao1112@gmail.com
*   Correspondence: btguan@ntu.edu.tw; Tel.: +886-02-33664628

Received: 13 November 2017; Accepted: 6 December 2017; Published: 1 January 2018

Abstract: To facilitate the reintroduction of five native late-successional Taiwanese Fagaceae species into Japanese cedar (*Cryptomeria japonica* (D.) Don) plantations, we experimented with methods to alleviate their establishment limitations. We tested different combinations of tree species, seedling development stages, and site preparation techniques. First, we directly sowed both fresh and germinated acorns under both closed and opened (thinned) canopies. Both fresh and germinated acorns survived only six months at most. Wildlife consumption was the most critical factor hindering their survival. We subsequently experimented with different methods for increasing establishment rates, such as thinning in combination with understory control, applying chemical animal repellents to seeds, using physical barriers against seed predators, and using seedlings of different ages. Among the methods experimented, none was effective. The effects of silvicultural treatments to deter seed consumption lasted only the first few weeks after sowing, whereas the effects of physical barriers were inconsistent. We also tested planting 3-month and 1-year-old seedlings. Seedling survival after 9 months was about 20% on average for 3-month-old seedlings but reached 80% for 1-year-old seedlings. Our results suggest that planting seedlings older than six months or establishing physical obstacles to prevent seed predation will be the most effective strategies to reintroduce late-successional hardwood Fagaceae species into Japanese cedar plantations.

Keywords: forest restoration; Fagaceae species; seed predation; seedling establishment; sub-tropical hardwoods; native mixed forests

1. Introduction

Forest biodiversity has been declining worldwide at an alarming rate over the past decades due to deforestation and forest fragmentation. Thus, developing effective strategies to restore forest biodiversity has been recognized as an essential element of biodiversity conservation [1]. It is suggested that plantation forests are excellent ecosystems for forest restoration, because of the microhabitat similarities between planted and natural forests [2]. Particularly, substituting planted monocultures with native mixed forests can also be a tool for increasing forest resilience to uncertain forest conditions [3]. Many plantation forests are no longer serving a timber production purpose, as in the case of Taiwan, where due to the rising awareness of conservation, management practices cannot be executed in conifer plantations, allowing natural succession to slowly take place [4].

Returning those plantation forests to their natural states provides a unique opportunity for forest restoration, and developing effective strategies to achieve that objective becomes an essential task for the conservation of forest biodiversity. Restoration guidelines for such a purpose, however, are not well established [2]. Given the unique ecological features of each plantation, and the lack of ecological knowledge on many non-commercial, subtropical native tree species, more empirical studies are required for establishing such guidelines.

Native hardwood species usually fail to return to plantation forests because of recruitment limitation (e.g., failure in seed dispersal, seed survival, or seedling survival). Planting nursery-grown seedlings and direct seeding are two common tools used to alleviate recruitment limitations of native species in forest restoration. The former has the advantage of a high success rate, but it also has the disadvantage of high costs. The latter has some biological benefits and the advantage of low cost, but its success rate is generally low [5–9]. Both approaches use plant materials that are thought to represent the most vulnerable stages during trees' life cycle [10,11].

Seed and seedling survival are limited by multiple biotic and abiotic factors [10,12,13]. Herbivory, light availability, and drought are the three most common and important causes of mortality [11]. Previous studies, however, have indicated that the relative importance of factors limiting seed and seedling establishment is highly context-dependent, and may vary among different ecosystems and sites [10]. Therefore, the success of restoration strategies depends on effectively identifying the critical stages of recruitment limitation. Thus, systematic empirical studies including seed and seedling stages are necessary to identify the critical stage of recruitment limitation.

Japanese cedar (*Cryptomeria japonica* (D.) Don, also known as *sugi*) plantations are a good case study for testing such empirical approaches. Introduced into Taiwan from Japan more than a century ago, Japanese cedar is the most widely planted tree species in Taiwan, covering approximately 1.1 percent (41,390 ha) of the island's total land area [4]. Due to the increasing production costs and declining timber prices, most of Taiwan's Japanese cedar plantations are either approaching or have passed the prescribed rotation age. Moreover, some existing Japanese cedar plantations were established on sites that are now considered unsuitable for timber production, primarily owing to concerns about watershed protection. To restore and promote biodiversity, the current local management plans mandate the restoration of these plantations by gradually reintroducing native species, particularly hardwoods. Under this mandate, the traditional harvesting-planting approach is no longer viable and an alternative approach is needed. Restoring these plantations can also serve as a model system to explore the effectiveness of different restoration practices in a subtropical island environment such as Taiwan's, for which little empirical experience exists.

Thinning (or selective cutting) and understory vegetation control are regularly prescribed in plantation forest management, and such practices may also facilitate forest restoration (e.g., [12]). The partial removal of trees can enhance local light availability and create physical environments similar to canopy gaps, which are essential for seedling survival [14–16]. Because only a portion of the trees are removed, the overall abiotic environment of the stand is usually not altered substantially [17].

The presence of understory vegetation may have both positive and negative effects on the survival of seeds and seedlings [10–12]. The presence of understory vegetation may reduce seedling survival by reducing light availability, by increasing competition between seedlings and understory vegetation, or both [6,8,11]. However, understory vegetation may also reduce seed and seedling predation by providing protection [8,18].

A preliminary feasibility study conducted in 2006 by randomly planting 4-year-old saplings of *Quercus glauca* (Thunberg) Oersted, *Q. longinux* (Hayata) Schottky, *Q. gilva* (Blume) Oersted, *Pasania hancei* (Benth.) Schottky var. *ternaticupula* (Hayata) Liao, and *P. harlandii* (Hance ex Walp.) Oerst. in a Japanese cedar plantation showed encouraging results, indicating that saplings of native, late-successional Fagaceae species can successfully establish [19]. However, compared to planting seedlings or directly sowing seeds, such a practice is clearly more expensive. Therefore, empirical studies are still needed to devise less expensive methods for restoring monospecific Japanese cedar plantations to native mixed hardwood forests.

To fill this knowledge gap, in this study we used a systematic approach to identify critical establishment stages of native broad-leaved species in a Japanese cedar plantation in central Taiwan. We addressed the following two questions: (1) Is seed germination a bottleneck for tree establishment? If so, which silvicultural techniques can improve seed germination? (2) Is seedling survival a bottleneck for tree establishment? If so, which silvicultural techniques can increase survival rates?

We designed three field experiments to answer each of the questions. First, direct seeding was carried out to evaluate seed germination. We sowed both fresh and germinating acorns on the forest floor surface under both closed and opened canopy. Second, given the lack of success in direct seed sowing, we experimented with different methods to prevent seed predation, including seed concealment and chemical repellents, to evaluate if seed predation could be deterred. Third, we planted seedlings of various ages under both closed and opened canopy. In combination with the previous questions, the impact of thinning on hardwood seedlings' recruitment was also evaluated.

2. Materials and Methods

2.1. Study Site

This study was conducted in a 10-ha Japanese cedar plantation in the Heshe District of the National Taiwan University Experimental Forest, central Taiwan (120°52′ E, 23°37′ N, 1442–1602 m.a.s.l.). Based on the information obtained from the nearest weather station (approximately 5 km away), mean annual temperature is 19.8 °C, with a mean annual rainfall of 1500 mm, indicating a warm-humid temperate climate regime.

Originally an evergreen broad-leaved late-successional stage forest dominated by Fagaceae and Lauraceae species, the site was clearcut in 1958 and planted with Chinese fir (*Cunninghamia lanceolata* Hook.). Due to extensive typhoon damages in 1969, the stand was salvaged and replanted with Japanese cedar in 1971. In 2005, the plantation was selected as a demonstration site to study the gradual restoration of Japanese cedar plantations to native forests. Remnants of the native forest can still be found within a 500-m radius from the edges of the plantation. We considered those remnants to represent the reference condition for the restoration project. They set the initial goal for a successful restoration. A preliminary inventory revealed that saplings of the late-successional Lauraceae species (mainly dispersed by birds) were relatively abundant at the study site. However, only a few saplings of Fagaceae species were present. Thus, we decided to focus our efforts only on understanding the bottlenecks for a successful reintroduction of native Fagaceae species. We used acorns collected from the surrounding areas. Because many Fagaceae species display a masting behavior, the study species used every year were determined by the availability of acorns at the time of study.

2.2. Thinning Treatments and Establishment of Transects

In 2005, as part of the preliminary study, 20 percent of the standing volume was thinned to create gaps of various sizes. Then, in 2009, we established four research plots, two in thinned gaps and two in unthinned areas, within the plantation (see Supplementary Information). The canopy openness of the two thinned plots was 27 and 29 percent, whereas the openness of the unthinned plots was 13 and 11 percent. For each plot, a 15-m transect was set at each of the eight cardinal and inter-cardinal directions. We then randomly selected two transects at cardinal directions and two transects at inter-cardinal directions in each plot for manual removal of understory vegetation in a 1-m-wide strip along the entire transect (referred henceforth as devegetated transects). The understory of the remaining two transects was left untouched (referred henceforth as vegetated transects). The experimental schedule for the four consecutive research campaigns is described in Table 1. Detailed diagrams of the experimental spatial design for each plot can be found in the Supplementary Information.

Table 1. Silvicultural treatments (listed by year) tested to enhance the establishment of different Fagaceae species in different annual research campaigns.

Treatment	*Lithocarpus lepidocarpus* (Hayata) Hayata	*Quercus glauca* (Thunberg) Oersted	*Pasania kawakami* (Hayata) Hayata	*Pasania hancei* (Benth.) Schottky	*Pasania harlandii* (Hance ex Walp.) Oerst.
Direct seeding					
Fresh seeds	2009, 2011		2009	2011, 2012	
Germinated seeds	2011	2010		2010, 2011, 2012	
Controlling seed predation					
Fencing		2010	2011 [1]		
Seed concealment	2011			2012	
Chemical repellent				2012	2011
Planting seedlings					
3-month-old				2011	2011
1-year-old		2011		2011	

[1] No data were obtained in 2011 from the fencing experiment due to the breakage of all fences.

2.3. Direct Seeding of Fresh and Germinated Seeds

To detect if seed germination and survival were bottlenecks during the establishment process, we placed acorns of different species every year from 2009 to 2012. We used acorns from different late-successional species relatively abundant in the surrounding areas in each specific year. Eight infrared automatic cameras were installed in the research site, one for each species-treatment combination, to record how and by which animal species the seeds were removed or consumed.

In spring 2009, fresh acorns of *Lithocarpus lepidocarpus* (Hayata) Hayata and acorns of *P. kawakamii* (Hayata) Hayata, each totaling 1920 seeds, were placed on top of the forest floor in thinned and unthinned plots, with and without understory removal (see Supplementary Information). Acorns were checked and the number of remaining seeds was recorded every two days. About a month later, none of the seeds were left [20].

Due to the complete establishment failure in 2009 of fresh acorns, we hypothesized that germinated seeds would be less palatable and, therefore, less attractive to seed predators. Using germinated seeds also ensured that seed quality or viability would not be a factor influencing our results. To get germinated seeds, acorns were gathered in October 2009. In the lab, water was used to separate low quality (floating) seeds from high quality (sinking) seeds. Selected seeds were stored stratified in wet moss at 4 °C until February 2010. Then, pots at the lab were used to bury seeds in soil for one month until March 2010, when they were extracted. Therefore, in spring 2010, germinated seeds (i.e., with radicles just emerging) of *Q. glauca* and *P. hancei* var. *ternaticupula*, each totaling 560 acorns, were placed in thinned and unthinned plots, with and without understory removal (see Supplementary Information). After placing the seeds on top of the forest floor, the number of seeds remaining was counted every other day during the first month. Thereafter, remaining seeds were counted on a monthly basis. After nine months we measured the height and basal diameter of the remaining established seedlings.

To account for potential inter-annual differences, we repeated our experiments in 2011 and 2012. In spring 2011, 560 fresh and 560 germinating seeds from each of two species, *L. lepidocarpus* and *P. hancei* var. *ternaticupula*, were placed in thinned and unthinned plots, with and without understory removal (see Supplementary Information). We placed both germinating and fresh seeds in the same transect line but at different distances. Seeds were checked and the number of remaining seeds was recorded every Monday, Wednesday, and Friday for two months.

In spring 2012, we established 80 monitoring points in an unthinned plot. Among them, we randomly selected 40 points and manually cleared the understory of half of these points. We then put 200 fresh and 200 germinated seeds of *P. hancei* (10 per acorn condition per point) on top of the forest floor (see the Supplementary Information). We monitored the number of remaining seeds on a weekly basis for one month.

2.4. Controlling Seed Predation

Guided by the results of our direct sowing trial, we tested different approaches to reduce acorn consumption. These approaches included using physical barriers, burying seeds with soil and litter, and using chemical repellents.

2.4.1. Physical Barriers

In spring 2010, we built three $1 \times 1 \times 1$ m^3 complete enclosures with polyethylene (PE) mesh. These physical barriers were set up at each transect (combining different canopy and understory types, see above). A hundred fresh *Q. glauca* seeds were placed inside each fenced area. Survival rates were calculated after 3 and 8 months. In spring 2011, a similar experiment was carried out for *P. kawakamii*.

2.4.2. Seed Concealment

In July 2011, 150 fresh *L. lepidocarpus* acorns were placed in three unthinned transects (50 seeds in each) and buried 5 cm below surface, and 150 seeds were also placed on the surface of another three unthinned transects. Seeds were monitored every Monday, Wednesday, and Friday for two months. The experiment was repeated in April 2012. First, 100 fresh and 100 germinated seeds of *P. hancei* were buried 3 cm below surface in unthinned transects. Second, 100 fresh and 100 germinated seeds of the same species were covered with Japanese cedar's litter. Seed status (alive, missing, or dead) was monitored weekly after burial.

2.4.3. Chemical Repellents

Chemical repellents were used in 2011 and 2012 in locations alongside, but outside of, the transects. In April 2011, 1200 seeds of *P. harlandii* were treated with two repellents (Cinnamamide 97% and trans-Cinnamaldehyde 98%) in four concentrations (0.0, 0.4, 0.8, and 1.2% w/w) with five replicates (30 seeds per treatment). Treated seeds were monitored every Monday, Wednesday, and Friday for two months after field placement. In May 2012, 1200 seeds of *P. hancei* were treated with trans-Cinnamaldehyde 98% in four concentrations (0.0, 5.0, 10.0, and 15.0% w/w) with 10 replicates (15 seeds per treatment). Seed status (alive, missing, or dead) was monitored weekly after the treated seeds were placed on the forest floor surface.

2.5. Planting Seedlings of Various Ages

To further investigate potential bottlenecks in Fagaceae species establishment, in April 2011 we planted 1-year-old nursery grown seedlings (less than 1.30 m tall) of *Q. glauca* and *P. hancei*, as well as 3-month-old nursery grown seedlings of *P. hancei* and *P. harlandii* in unthinned and thinned plots. Seven seedlings were planted along each transect of the plots after clearing the understory. The status of seedlings (alive or dead) was checked weekly after planting. Nine months later, we closed the experiment and measured the average basal diameter and height of the surviving seedlings.

2.6. Data Analysis

A semi-parametric approach based on a generalized additive mixed model [21] was used to analyze the seed and seedling survival probabilities of each experiment. We limited our analysis to the first 30 days because most of the surviving seeds of all species used in this experiment fully germinated after that time. In the model, treatment effects were modeled parametrically, whereas the time effect was modeled non-parametrically using a thin-plate regression spline with a smoothing term for each treatment. For all species, a first-order autoregressive model was used to account for the autocorrelation present in the data. We used the R package *mgcv* to perform the analyses [22].

3. Results

3.1. Seed Survival

Although the final results showed that all the acorns were eaten or removed, survival analyses indicated that canopy openness and understory presence had different effects during the removal/consumption processes depending on the species. Silvicultural treatment did not have significant effects on fresh *L. lepidocarpus* and *P. kawakamii* seeds in 2009 (Table 2, Figure 1). However, seeds had different removal rates in 2010 depending on the silvicultural treatment for *P. hancei*, but no differences were found for *Q. glauca* (Figure 2). Seed survival rates for *Q. glauca* over the first 33 days were not significantly different among the treatments. In contrast, the absence of understory cover significantly lowered survival rates for *P. hancei* seeds, whereas canopy openness had no effect (Table 3).

Table 2. Effects of thinning and understory removal on the seed survival rates of *Lithocarpus lepidocarpus* (Hayata) Hayata and *Pasania kawakamii* (Hayata) Hayata. over the first 36 days after fresh seed placement in spring 2009 and 2011.

Treatment	L. lepidocarpus			P. kawakamii		
	df	χ^2	p	df	χ^2	p
Understory control	1	0.091	0.763	1	0.748	0.387
Thinning	1	1.176	0.278	1	0.238	0.625
Understory control × thinning	1	0.856	0.355	1	1.308	0.253

Table 3. Effects of thinning and understory removal on the seed survival rates of *Quercus glauca* (Thunberg) Oersted and *Pasania hancei* var. *ternaticupula* over the first 33 days after fresh seed placement in spring 2010 and 2011.

Treatment	Q. glauca			P. hancei		
	df	χ^2	p	df	χ^2	p
Understory control	1	1.128	0.288	1	4.001	0.0455
Thinning	1	0.303	0.582	1	0.010	0.9203
Understory control × thinning	1	0.435	0.509	1	0.109	0.7414

For the 2009 trial, all the fresh seeds were consumed within one month after field placement (Figure 1). In the 2010 trial, 33 days after the germinated seeds were placed, 19.4% to 72.1% of the seeds were still present except for *Q. glauca* seeds in the devegetated transects (Figure 2). By the end of the second month, almost all of the *Q. glauca* seeds disappeared (being either consumed or removed), while some *P. hancei* seeds were still left on the ground. In summary, *Q. glauca* seeds were removed at a faster rate than that of *P. hancei*. For both species, seeds were consumed or removed at a faster rate for transects without understory vegetation in comparison to transects with understory vegetation.

In the 2011 trial, we repeated the same approach as in 2010. Final survival rates for *L. lepidocarpus* seeds detected among the treatments or seed life stage did not significantly differ. However, canopy openness had a significant effect on the survival rates of *P. hancei* seeds (Table 4). Most of the seeds, regardless of whether they were fresh or germinated, were removed in the first week (Figures 1 and 2). In the last trial of 2012, all the seeds left unburied on the forest floor were removed within three weeks, regardless of the germination status and the placement (Figure 3).

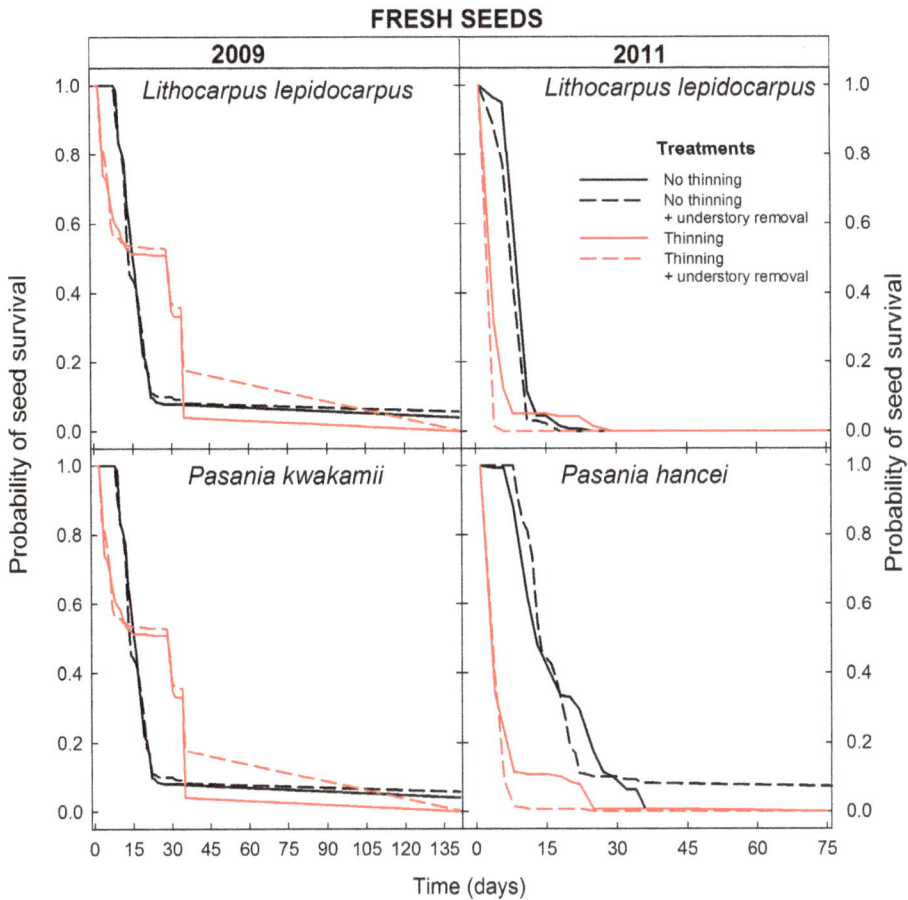

Figure 1. Average survival rates for fresh seeds of three different hardwood species under four different forms of site preparation management. The study was carried in 2009 (**left panels**) and repeated in 2011 (**right panels**).

Table 4. Effects of thinning, understory removal, and seed germination life stage treatments on the seed survival rates of *Lithocarpus lepidocarpus* (Hayata) Hayata and *Pasania hancei* var. *ternaticupula* over the first 36 days after seed placement in spring 2011.

Treatment [1]	*L. lepidocarpus*			*P. hancei* var. *ternaticupula*		
	df	χ^2	p	df	χ^2	p
Understory control	1	2.784	0.0952	1	2.975	0.0846
Thinning	1	0.057	0.8115	1	11.341	0.0008
Type of seeds (fresh or germinated)	1	0.460	0.4979	1	3.441	0.0636

[1] Non-significant interactions were removed from the analysis.

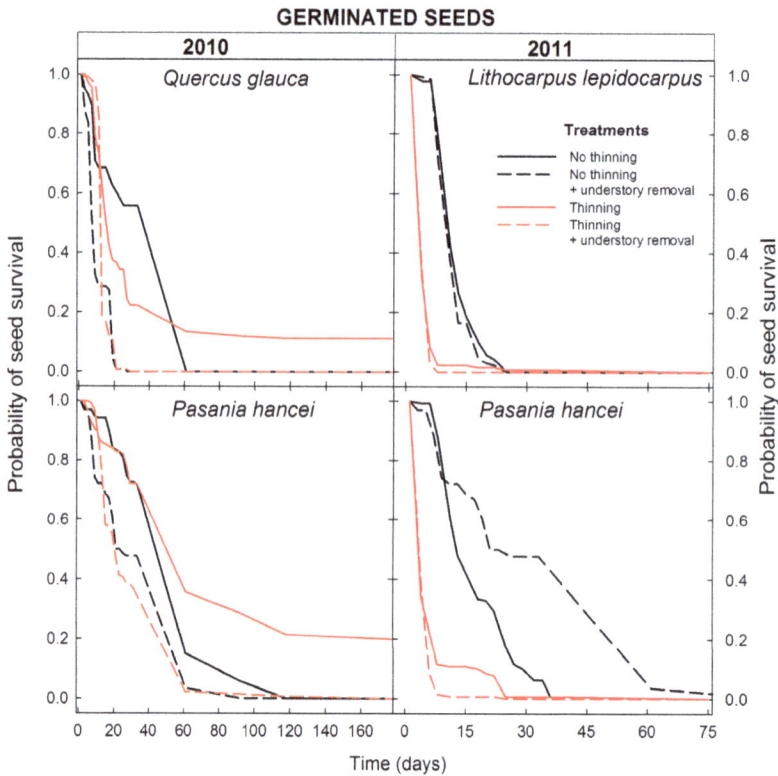

Figure 2. Average survival rates for germinated seeds of three different hardwood species under four different forms of site preparation management. The study was carried in 2010 (**left panels**) and repeated in 2011 (**right panels**).

Results from 2009 to 2011 suggested that the different removal rates were likely due more to the locations of seed placement rather than to the causes of seed removal. However, survival analyses showed that there were significant differences in seed removal rates depending on seed covers for the last year's trial (Table 5). Burying seeds in the mineral soil significantly delayed seed survival, but it did not change the outcome that after 80 days seed survival was also negligible (Figure 3). No significant interactions between canopy openness, understory presence, or seed life stage were detected in any year for any species.

Table 5. Effects of different methods to prevent seed predation (burial under soil, burial with litterfall, no ground vegetation, and control) and seed germination phases (fresh vs. germinated) on the seed survival rates of *Pasania hancei* var. *ternaticupula* over the first 39 days after seed placement in spring 2012.

Treatment [1]	P. hancei var. *ternaticupula*		
	df	χ^2	p
Method	1	11.884	0.0078
Germination phase	1	0.446	0.5043

[1] Non-significant interactions were removed from the analysis.

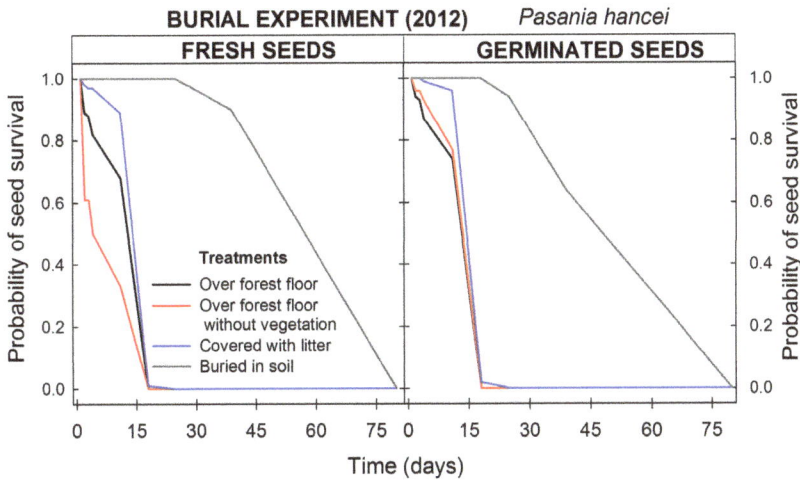

Figure 3. Survival rates of fresh and germinated seeds of *P. hancei* var. *ternaticupula* during the seed concealment experiment carried out in 2012.

3.2. Controlling Seed Predation

3.2.1. Physical Barriers

For the 2010 trial, seedling survival rates inside the fenced areas were 85.0% and 77.7% for the plots without understory (thinned and unthinned sites, respectively) eight months after the beginning of the experiment. However, survival rates were 52.3% and 61.3% for the plots with understory (thinned and unthinned sites, respectively; Figure 4). These results indicated that although understory plays a major role in seed establishment, physical barriers help to prevent acorn consumption. Unfortunately, no survival data from the 2011's trial could be obtained, as all fences were gnawed by rodents, which entered into the fenced areas. Therefore, most of the seeds were consumed, and only 11 seeds germinated.

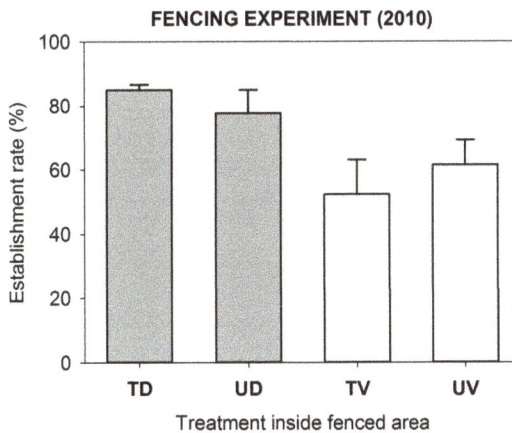

Figure 4. Seed establishment rate after eight months inside the fenced areas in 2010 for four different management regimes: T: thinned canopy; U: unthinned canopy; D: devegetated (understory removal, grey bars); V: vegetated (white bars). Bars indicate average ± standard error.

3.2.2. Seed Concealment and Chemical Repellents

In 2011, all unburied seeds were removed within 2 months, regardless of their germination stage. Four months after burial, 14.4% of acorns were able to germinate and grow to about 25 cm [19]. In 2012, most of the seeds disappeared within 20 days except the ones buried under the soil. However, 80 days after, all the seeds were removed regardless of the treatments. As for the effects of chemical repellents, all seeds disappeared within 3 weeks of the beginning of the experiment in 2011 (Figure 5), and within 15 days in 2012 (results not shown).

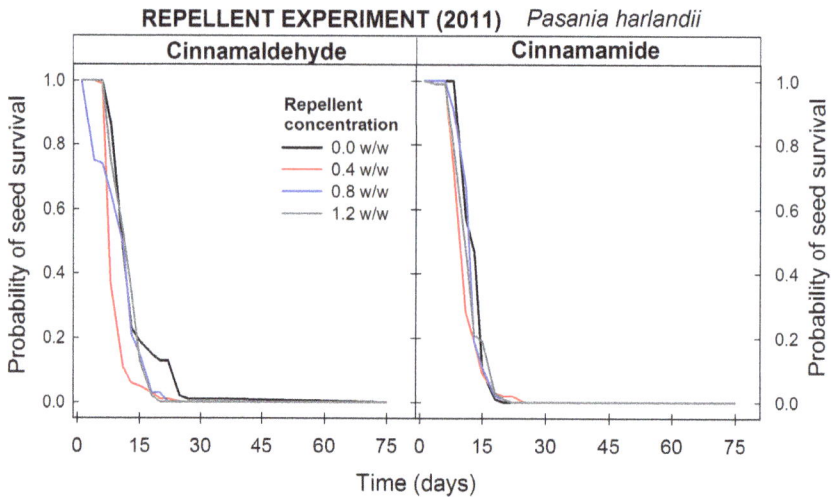

Figure 5. Seed survival rates after the applications of two different chemical repellents at five different concentrations.

3.2.3. Identification of Acorn Consumers

From the images captured by the cameras, we identified five mammal species and five bird species as potential seed consumers or removers during the observation period. The five mammal species were red-bellied squirrel (*Callosciurus erythraeus* Pallas), Formosan ferret-badger (*Melogale moschata subaurantiaca* Swinhoe), spinous country-rat (*Niviventer coxingi* Thomas), Formosan field mouse (*Apodemus semotus* Thomas), and Tada's shrew (*Crocidura tadae* Tokuda & Kano). The five bird species were Eurasian jay (*Garrulus glandarius* L.), Eurasian woodcock (*Scolopax rusticola* L.), bamboo partridge (*Bambusicola thoracica* Temminck), Steere's liocichla (*Liocichla steerii* Swinhoe), and White's ground thrush (*Turdus dauma* Latham). Among the identified faunal species, seed removal behaviors were observed for red-bellied squirrel, Formosan ferret-badger, spinous country-rat, and Eurasian jay.

3.3. Planting Seedlings of Various Ages

Survival rates differed between 1-year-old and 3-month-old seedlings. A very high proportion of 1-year-old seedlings survived until the end of the experiment. In contrast, survival rates of 3-month-old seedlings decreased to 0.2 within 80 days (Figure 6).

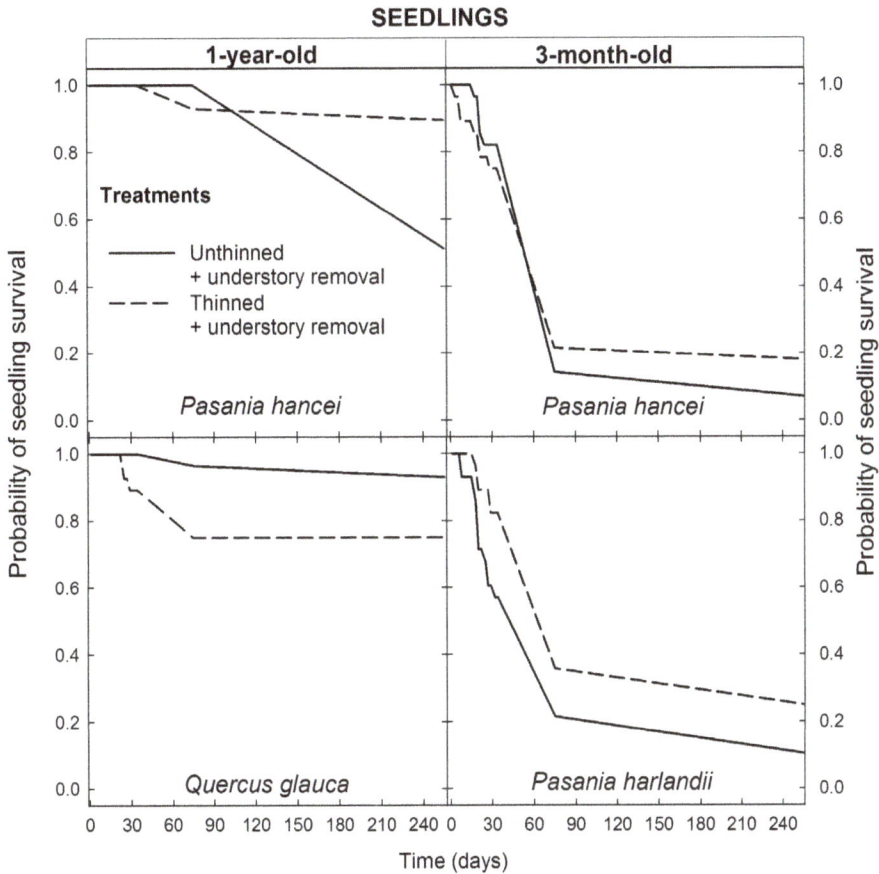

Figure 6. Survival rates of 1-year-old seedlings (**left panels**) and 3-month-old seedlings (**right panel**) of three different hardwood species during the experiment carried out in 2012.

4. Discussion

4.1. Seed Predation as the Main Cause of Establishment Failure

We used a systematic, sequential, adaptive approach to identify bottlenecks for seedling establishment and the best silvicultural treatments to overcome them. Our result from four different years showed that direct sowing (and by similitude, natural seed arrival) was unable to produce viable seedlings. Therefore, some sort of advanced regeneration or seed protection is needed to allow for the establishment of some seedlings if the establishment of native hardwood species is a management objective. Our results suggested that seedlings at least 3 months old, which survived for 3 more months in the field, have the potential to successfully establish at this site, but such survival was guaranteed only when using 1-year-old seedlings.

Seed survival and establishment rates observed in this study were low compared with those found by other studies also done with late-successional or large-seed direct sowing [5,7,13]. These low rates could result from both biotic (e.g., predation) and abiotic (e.g., drought) factors that limited seedling establishment [6,11,13].

The impact of thinning on seed predation has been demonstrated by several earlier studies (e.g., [23,24]). Previous studies have also demonstrated the significant influence of understory vegetation (e.g., [10,25]). However, the effects of canopy openness or the presence of understory vegetation on seed survival are somewhat context-dependent [12,26–28]. Understory can have both positive and negative effects on seed predation. On one hand, increased understory growth could hide the seeds better and facilitate their survival, as the predators need to spend more time finding them. On the other hand, understory could provide cover for other seed consumers that do not go into open areas but can now reach those seeds. Furthermore, some understory species also produce fleshy fruits and may provide additional food resources for seed predators. Depending on the types of seed predators present in the region, seed predation may increase or decrease in different microhabitats after silvicultural treatments [26,27,29]. For instance, seed predation by squirrels is reduced in canopy gaps, whereas predation activities by field mice are higher in canopy gaps [28].

We detected different consumption/removal patterns and different seed predators among years. At the same time, we detected initial differential survival rates for seeds of different growth stages within the first month. Different removal patterns could be due to different foraging behaviors and feeding paths of main seed predators. For example, as captured by camera, red-bellied squirrels tended to systematically remove all the seeds in one spot, and when they depleted all the acorns in that spot, they moved to the next spot. In addition, it has been reported that differences among seeds caused by seed nutrient content, physical and chemical defense mechanisms (e.g., shell hardness, presence of secondary metabolites, etc.) affect the preference of seed feeders [30]. However, our initial hypothesis that germinated seeds would be less favorable for seed consumers was rejected, as no significant differences on seed predation rates between fresh and germinated seeds were detected in the long term (more than two months after direct sowing).

In addition, the lack of a cold winter also means that there are no seasonal patterns in rodent population sizes [31]. Therefore, storing seeds for planting them at the time of lower seed predator population levels does not seem to be a viable option. At these sites, the breeding season for rodents lasts for two-thirds of the year, from January to August. During that time, energy demands are high because of pregnancy and nursing. However, the Japanese cedar plantation is short of food resources. Consequently, the seeds placed by us may have attracted those predators. These circumstances may explain why germinated seeds were also consumed or removed, even at a faster rate than fresh seeds [32]. Both factors will be important considerations in our future research on methods to improve the success of reintroduction programs. It should be highlighted that because the main purpose of this study was to develop effective silvicultural practices to improve seed survival and seedling establishment, we only placed a small number of seeds in the field, as constrained by seed availability and economic considerations. If we had used larger numbers of seeds, the saturation effect could have increased seed survival rates.

Predation is not the only mechanism by which seeds fail to establish. Scatter-hoarding behavior has also been widely observed in squirrels and jays in various forest ecosystems [33,34]. Instead of consuming the seeds, the animals remove the seeds and cache them in a different place. The captured photographs suggested that the major seed predators at the study sites were granivorous animals, including the red-bellied squirrel. Such scatter-hoarding behavior by the granivorous animals may explain the similar seedling establishment rates for *Q. glauca* and *P. hancei* after nine months. Additional species that were observed to remove the seeds were the Formosan ferret-badger, spinous country-rat, and Eurasian jay. However, seed-removal behavior was less frequent in those species than in red-bellied squirrels.

On the other hand, some of these animals may also function as seed dispersers because of their scatter-hoarding behavior. For example, in our study red-bellied squirrels seem to have removed large numbers of the seeds. The seeds that were removed may not have been eaten immediately but may have been stored for later consumption, and the cached seeds may subsequently germinate. In a follow-up expedition to the research sites in April 2017, a few of the placed acorns germinated and established. Interestingly, all those seedlings were a short distance away from their original placement locations.

Finally, climate variability may also contribute to the different observed inter-annual seed mortality patterns. Unlike other Fagaceae species in temperate regions which have other sources to supply soil moisture at the beginning of growing seasons, (e.g., snow melting), Taiwanese species depend entirely on rainfall for soil water availability during that time. The 2009–2010 El Niño event delayed the beginning of the 2010 wet season in central Taiwan for approximately six weeks, and the precipitation amount was below average during the first three months of that growing season. Hence, the first two months of the experiment were probably too dry to allow proper germination, causing the low germination rates and a germination period that was longer than expected.

4.2. Preventing Seed Predation

None of the three approaches used to discourage seed predation worked in a significant way, except physical barriers. Xiao et al. [35] and Chang et al. [36] also showed that using enclosures as physical barriers is a feasible way to increase establishment rates. Similar to the results reported by those authors, our preliminary tests also showed that seed germination rates inside the fenced areas were similar to rates obtained in the lab (data not shown). However, if we apply this method to larger areas, restoration costs will increase significantly, and could produce other undesirable ecological effects such as hindering the mobility of some wildlife species [37]. More importantly, the effect was inconsistent, as it only worked in 2010, but failed in 2011. Our 2011 results suggested that the seed-predators might have "learned" ways to overcome the barriers.

In addition to protecting the seeds from their predators with nets, we also used soil or litterfall to conceal the seeds to prevent seed predator detections. Compared to directly placing the seeds on the forest floor surface, burying seeds can slow removal rates, provide a better micro-environment for germination, and increase seedling survival and establishment rates [38–41]. In our first seed concealment trial, most of the seeds remained under the soil and had a 14.4% seedling establishment rate. These results are very similar to other previously reported results [42]. However, the protection was inconsistent, as in our second year's trial all the seeds were consumed within 80 days after burying them. Therefore, although burying seeds has the potential effect of reducing seeds from being found and consumed, such a benefit is not guaranteed.

Regarding chemical repellents, they did not have any noticeable effect on discouraging seed predators, particularly rodents, opposite to reports by some studies [43–45]. A possible explanation for this contradictory result could be that the rodents in Taiwan (e.g., red-bill squirrel or stripe squirrel) have a high tolerance to the food they eat [46,47]. Therefore, chemical repellents do not work as well as in other places. Another possible explanation is that both chemicals could have evaporated or washed out by rain a few days after application, hence lowering their concentrations to the levels where they became ineffective as repellents.

Finally, it should be taken into account that the management history of the site and surrounding region has caused the disappearances of keystone predator species that can regulate rodent populations' levels, thus creating a cascading effect on Fagaceae seeds.

4.3. Overcoming Establishment Bottlenecks: Ecosystem Management Implications

Our results indicated that the main establishment limitation is the fast removal of seeds, usually in less than a month. Similar results have been described before [48,49]. From our results, it seems clear that restoring these Japanese cedar plantations using only seeds is not effective, and therefore older seedlings should be used.

Comparing the establishment rates between 3-month-old and 1-year-old seedlings, older seedlings have higher survival and establishment rates. The effects of canopy conditions depend on seedling ages. An open canopy increased the establishment rates for young seedlings, but the situation was reversed for older seedlings. This difference could be due to a higher variety of micro-environments and understory vegetation in the thinned site, which also led to variations in seed predators' population sizes and feeding paths [50,51]. Young seedlings are less conspicuous and harder to find when the environment is more heterogeneous. In addition, thinning could increase light levels and therefore enhance the growth of shade-tolerant understory species [52]. However, understory vegetation was relatively dense at the study site, possibly because of the improved light conditions following the thinning. Although potentially beneficial for seeds, understory vegetation could be detrimental to older seedling survival, due to competition for water, nutrients, light, and growing space [53].

Seedling mortality causes also differed for seedlings with different ages. The major cause of death for 3-month-old seedlings was biting and chewing. Young seedlings still had a part of the acorn (large, more nutritious part) attached to the plants. Therefore, after seed predators finish eating all the available seeds in the area, they can turn to young seedlings to eat the remnants of seeds attached to the young plants, particularly if there is still pressure on food sources in the field. However, this situation is not applicable to 1-year-old seedlings, in which seed remnants have already fallen or decomposed, and their stems were more woodified. As a consequence, the major cause of death for older seedlings was withering (mostly by drought), similar to other studies [11].

From a management perspective, while opening the canopy and maintaining plant cover is important during the first months of the restoration effort, after seeds have been able to get established and survived their first year, understory should be controlled or removed to reduce competition for growth resources. The use of these techniques is the next natural step for developing our adaptive empirical approach to establish restoration guidelines at our study sites.

The native late-successional forests in the study area were originally dominated by hardwood species belonging to the Fagaceae and Lauraceae families. Although the Lauraceae species seem to have the capacity to establish themselves under the canopy of Japanese cedars, this is not the case for Fagaceae species. Although many Fagaceae trees with abundant seeds can be found in remnant stands near the study site, naturally regenerated seedlings and saplings are rare at the site, suggesting that recruitment is strongly limited, as our results confirm. All signs suggest that in the absence of human intervention these abandoned plantations would not be able to return to their original state, at least in any time soon. Once a forest ecosystem has been altered in a substantial way, ecosystem-level restoration measures are necessary. For example, although not tested in the current study, a potential method to reduce seed predation pressure is to re-introduce the now locally extinct natural higher-level (e.g., carnivorous) predators into the plantations to control seed predators.

5. Conclusions

This study represents the first attempt to identify establishment bottlenecks for Fagaceae species in conifer plantations in Taiwan. Our experiments indicated that without some kind of human intervention, it will take a long time (i.e., via succession) for Fagaceae species to establish in Japanese cedar plantations that are no longer serving their purposes. Among the methods tested to bring native Fagaceae species back, direct sowing was shown to have the lowest survival and seedling establishment rates, whereas seedling planting had a higher survival rate. Direct seeding with fresh seeds was observed to be virtually useless. The final survival rates from high to low were 1-year-old seedlings, 3-month-old old seedlings, germinated seeds, and fresh seeds. Seedlings older than 6 months were observed to be able to successfully establish in this site. Burying germinated seeds, in combination with thinning and a proper understory vegetation control was able to delay seed consumption but did not prevent it. Our results indicated that fencing with rodent-resistant material seems to be the only management option that is likely to result in a majority of seeds getting established. Therefore, if there are important economic or ecological limitations that prevent the use of fences, and some level of risk is accepted, we suggest burying germinated seeds (an easy-to-implement and relatively inexpensive method) as a first test to facilitate the re-introduction of native late-successional Fagaceae species in conifer plantation forests in Taiwan. If unsuccessful, more costly options should be used: either creating fenced areas or planting seedlings that are at least 6 months old.

Supplementary Materials: The following supplementary materials are available online at www.mdpi.com/1999-4907/9/1/3/s1, Figure S1: Location of the four experimental plots in the National Taiwan University Experimental Forests; Figure S2: Experimental design for the placement of seeds at the four research plots used in 2009; Figure S3: Experimental design for the placement of seeds at the four research plots used in 2010; Figure S4: Experimental design for the placement of seeds at the four research plots used in 2011; Figure S5: Experimental design for the placement of 1-year-old seedlings at the four research plots used in 2011; Figure S6: Experimental design for seed concealment experiment with seeds of *Pasania hancei* var. *ternaticupula* (Hayata) Liao in 2012.

Acknowledgments: The authors thank Po-Jen Jiang for his assistance in setting up automatic cameras. We also thank the staff of the National Taiwan University Experimental Forest for their help during fieldwork. We wish to thank the two anonymous reviewers who helped in improving this article. This study was partially supported by a grant from the Taiwan National Science Council (NSC 97-2313-B-002-041-MY3). Juan A. Blanco was funded through a Ramón y Cajal contract (ref. RYC-2011-08082) and a Marie Curie Action (ref. CIG-2012-326718-ECOPYREN3). Yueh-Hsin Lo was funded through a Marie Skłodowska-Curie Action (ref. MSCA-IF-2014-EF-656810-DENDRONUTRIENT). The FP7 post-grant Open Access publishing funds pilot funded the publication of this manuscript.

Author Contributions: Y.C.L. and B.T.G. conceived and designed the experiments; Y.C.L., Y.T.L., Y.H.L., C.H.Y. and B.T.G. performed the experiments; Y.T.L., Y.H.L., Y.C.L., and J.A.B. analyzed the data; Y.T.L., Y.H.L., B.T.G., Y.C.L., and J.A.B. wrote the paper.

Conflicts of Interest: The authors declare no conflict of interest.

References

1. Chazdon, R.L. Beyond deforestation: Restoring forests and ecosystem services on degraded lands. *Science* **2008**, *320*, 1458–1460. [CrossRef] [PubMed]

2. Brockerhoff, E.G.; Jactel, H.; Parrotta, J.A.; Quine, C.P.; Sayer, J. Plantation forests and biodiversity: Oxymoron or opportunity? *Biodivers. Conserv.* **2008**, *17*, 925–951. [CrossRef]

3. Paquette, A.; Messier, C. The role of plantations in managing the world's forests in the Anthropocene. *Front. Ecol. Environ.* **2010**, *8*, 27–34. [CrossRef]

4. Taiwan Forestry Bureau. *Summary of the Fourth Forest Resources and Land Use Inventory in Taiwan*; Taiwan Forestry Bureau: Taipei, Taiwan, 2017; p. 78.

5. Cole, R.J.; Holl, K.D.; Keene, C.L.; Zahawi, R.A. Direct seeding of late-successional trees to restore tropical montane forest. *For. Ecol. Manag.* **2011**, *261*, 1590–1597. [CrossRef]

6. Doust, S.J.; Erskine, P.D.; Lamb, D. Restoring rainforest species by direct seeding: Tree seedling establishment and growth performance on degraded land in the wet tropics of Australia. *For. Ecol. Manag.* **2008**, *256*, 1178–1188. [CrossRef]

7. Löf, M.; Birkedal, M. Direct seeding of *Quercus robur* L. for reforestation: The influence of mechanical site preparation and sowing date on early growth of seedlings. *For. Ecol. Manag.* **2009**, *258*, 704–711. [CrossRef]

8. Schmidt, L. *A Review of Direct Sowing versus Planting in Tropical Afforestation and Land Rehabilitation*; Development and Environment Series 10-2008; Forest & Landscape Denmark: Copenhagen, Denmark, 2008; p. 38. ISBN 978-87-7903-328-3.

9. Zahawi, R.A.; Holl, K.D. Comparing the performance of tree stakes and seedlings to restore abandoned tropical pastures. *Restor. Ecol.* **2009**, *17*, 854–864. [CrossRef]

10. Fenner, M.; Thompson, K. *The Ecology of Seeds*, 2nd ed.; Cambridge University Press: Cambridge, UK, 2005; p. 250. ISBN 0521653681.

11. Leck, M.A.; Parker, V.T.; Simpson, R.L. *Seedling Ecology and Evolution*, 1st ed.; Cambridge University Press: Cambridge, UK, 2008; p. 534. ISBN 9780521694667.

12. Beckage, B.; Clark, J.S.; Clinton, B.D.; Haines, B.L. A long-term study of tree seedling recruitment in southern Appalachian forests: The effects of canopy gaps and shrub understories. *Can. J. For. Res.* **2011**, *30*, 1617–1631. [CrossRef]

13. Doust, S.J.; Erskine, P.D.; Lamb, D. Direct seeding to restore rainforest species: Microsite effects on the early establishment and growth of rainforest tree seedlings on degraded land in the wet tropics of Australia. *For. Ecol. Manag.* **2006**, *234*, 333–343. [CrossRef]

14. Augspurger, C.K. Light requirements of neotropical tree seedlings—A comparative-study of growth and survival. *J. Ecol.* **1984**, *72*, 777–795. [CrossRef]

15. Brokaw, N.; Busing, R.T. Niche versus chance and tree diversity in forest gaps. *Trends Ecol. Evol.* **2000**, *15*, 183–188. [CrossRef]

16. Masaki, T.; Osumi, K.; Takahashi, K.; Hoshizaki, K.; Matsune, K.; Suzuki, W. Effects of microenvironmental heterogeneity on the seed-to-seedling process and tree coexistence in a riparian forest. *Ecol. Res.* **2007**, *22*, 724–734. [CrossRef]

17. Blanco, J.A.; Imbert, J.B.; Castillo, F.J. Effects of thinning on nutrient pools in two contrasting *Pinus sylvestris* L. forests in the western Pyrenees. *Scand. J. For. Res.* **2006**, *21*, 143–150. [CrossRef]

18. Smit, C.; Gusberti, M.; Mueller-Schaerer, H. Safe for saplings; safe for seeds? *For. Ecol. Manag.* **2006**, *237*, 471–477. [CrossRef]

19. Li, Y.-T. Restoration of a Plantation Forest to Native Broadleaved Vegetation: Indentifying and Alleviating Establishment Limitations. Master's Thesis, National Taiwan University, Taipei, Taiwan, 2013.

20. Lo, Y.-H.; Lin, Y.-C.; Blanco, J.A.; Yu, C.-H.; Guan, B.T. Moving from ecological conservation to restoration: An example from central Taiwan, Asia. In *Forest Ecosystems: More Than Just Trees*; Blanco, J.A., Lo, Y.-H., Eds.; InTech: Rijeka, Croatia, 2012; pp. 339–354. ISBN 978-953-307-667-6.

21. Wood, S.N. *Generalized Additive Models: An Introduction with R*; Chapman and Hall/CRC: Boca Raton, FL, USA, 2006; p. 476. ISBN 9781498728331.

22. R Development Core Team. *R: A Language and Environment for Statistical Computing*; R Foundation for Statistical Computing: Vienna, Austria, 2011; ISBN 3-900051-07-0. Available online: http://www.R-project. org/ (accessed on 10 November 2017).

23. Boman, J.S.; Casper, B.B. Differential postdispersal seed predation in disturbed and intact temperate forest. *Am. Midl. Nat.* **1995**, *134*, 107–116. [CrossRef]

24. Schnurr, J.L.; Canham, C.D.; Ostfeld, R.S.; Inouye, R.S. Neighborhood analyses of small-mammal dynamics: Impacts on seed predation and seedling establishment. *Ecology* **2004**, *85*, 741–755. [CrossRef]

25. Chambers, J.C.; MacMahon, J.A. A day in the life of a seed: Movements and fates of seeds and their implications for natural and managed systems. *Ann. Rev. Ecol. Syst.* **1994**, *25*, 263–292. [CrossRef]

26. Den Ouden, J.; Jansen, P.A.; Smit, R. Jays, mice and oaks: Predation and dispersal of *Quercus robur* and *Q. petraea* in north-western Europe. In *Seed Fate: Predation, Dispersal and Seedling Establishment*; Forget, P.M., Lambert, J.E., Hulme, P.E., vander Wall, S.B., Eds.; CABI Publishing: Wallingford, UK, 2005; pp. 223–240. ISBN 0851998062.

27. Hulme, P.E.; Kollmann, J. Seed predator guilds, spatial variation in post-dispersal seed predation and potential effects on plant demography—A temperate perspective. In *Seed Fate: Predation, Dispersal and Seedling Establishment*; Forget, P.M., Lambert, J.E., Hulme, P.E., vander Wall, S.B., Eds.; CABI Publishing: Wallingford, UK, 2005; pp. 9–30. ISBN 0851998062.

28. Tamura, N.; Katsuki, T. Walnut seed dispersal: Mixed effects of tree squirrels and field mice with different hoarding ability. In *Seed Fate: Predation, Dispersal and Seedling Establishment*; Forget, P.M., Lambert, J.E., Hulme, P.E., vander Wall, S.B., Eds.; CABI publishing: Wallingford, UK, 2005; pp. 241–252. ISBN 0851998062.

29. Schupp, E.W.; Milleron, T.; Russo, S.E. Dissemination limitation and the origin and maintenance of species-rich tropical forests. In *Seed Dispersal and Frugivory: Ecology, Evolution and Conservation*; Levey, D.J., Silva, W.R., Galetti, M., Eds.; CABI International: Wallingford, UK, 2002; pp. 19–34. ISBN 085199525X.

30. González-Rodríguez, V.; Villar, R. Post-dispersal seed removal in four Mediterranean oaks: Species and microhabitat selection differ depending on large herbivore activity. *Ecol. Res.* **2012**, *27*, 587–594. [CrossRef]

31. Tsai, J.-W.; Yuan, H.-W.; Tsai, P.-Y.; Lee, S.-Y.; Ding, T.-S.; Hong, C.-H. Effects of thinning on bird community and spinous country-rat (*Niviventer coxingi*) population in china-fir (*Cunninghamia lanceolata*) plantations. *Q. J. Chin. For.* **2010**, *43*, 367–382.

32. Li, L.-L. 1981 A Study on the Behavior of Red-bellied Squirrel (*Callosciurus erythraeus*). Master's Thesis, National Taiwan University, Taipei, Taiwan, 1981.

33. Zhang, Z.-B.; Xiao, Z.-S.; Li, H.-J. Impact of small rodents on tree seeds in temperate and subtropical forests, China. In *Seed Fate: Predation, Dispersal and Seedling Establishment*; Forget, P.M., Lambert, J.E., Hulme, P.E., Vander Wall, S.B., Eds.; CABI publishing: Wallingford, UK, 2005; pp. 269–282. ISBN 0851998062.

34. Vander Wall, S.B. Seed fate pathways of antelope bitterbrush: Dispersal by seed-caching yellow pine chipmunks. *Ecology* **1994**, *75*, 1911–1926. [CrossRef]

35. Xiao, Z.; Zhang, Z.; Wang, Y. Effects of enclosure protection and seed burial on direct seeding of nut-bearing trees. *Biodivers. Sci.* **2005**, *13*, 520–526. [CrossRef]

36. Chang, C.T. Microenvironmental Variation, Seed Germination and Seedling Growth under Different Canopy Openness in a Sugi (*Cryptomeria japonica*) Plantation at Shitou, Central Taiwan. Master's Thesis, National Taiwan University, Taipei, Taiwan, 2007.

37. Löf, M.; Bergquist, J.; Brunet, J.; Karlsson, M.; Welander, N.T. Conversion of Norway spruce stands to broadleaved woodland-regeneration systems, fencing and performance of planted seedlings. *Ecol. Bull.* **2010**, *53*, 165–173.

38. Xiao, Z.S.; Zhang, Z.B.; Wang, Y.S. Dispersal and germination of big and small nuts of *Quercus Serrata* in a subtropical broad-leaved evergreen forest. *For. Ecol. Manag.* **2004**, *195*, 141–150. [CrossRef]

39. Xiao, Z.S.; Zhang, Z.B.; Wang, Y.S. Effects of seed size on dispersal distance in five rodent-dispersed Fagaceous species. *Acta Oecol.* **2005**, *28*, 221–229. [CrossRef]

40. Birkedal, M.; Löf, M.; Olsson, G.E.; Bergsten, U. Effects of granivorous rodents on direct seeding of oak and beech in relation to site preparation and sowing date. *For. Ecol. Manag.* **2010**, *259*, 2382–2389. [CrossRef]

41. Savadogo, P.; Tigabu, M.; Oden, P.C. Restoration of former grazing lands in the highlands of Laos using direct seeding of four native tree species. *Mt. Res. Dev.* **2010**, *30*, 232–243. [CrossRef]

42. Huang, I.-C.; Chen, I.-Z.; Lu, S.-Y.; Chang, K.-S. Effects of chilling stratification, scarification and excised-embryo on seed germination of *Lithocarpus lepidocarpus* Hayata. *J. Chin. Soc. Hortic. Sci.* **2004**, *50*, 515–520.

43. Crocker, D.R.; Scanlon, C.B.; Perry, S.M. Repellency and Choice: Feeding responses of wild rats (*Rattus norvegicus*) to cinnamic acid derivatives. *Appl. Anim. Behav. Sci.* **1993**, *38*, 61–66. [CrossRef]

44. Gurney, J.E.; Watkins, R.W.; Gill, E.L.; Cowan, D.P. Non-lethal mouse repellents: Evaluation of cinnamamide as a repellent against commensal and field Rodents. *Appl. Anim. Behav. Sci.* **1996**, *49*, 353–363. [CrossRef]

45. Lee, H.K.; Lee, H.S.; Ahn, Y.J. Antignawing factor derived from *Cinnamomum Cassia* bark against mice. *J. Chem. Ecol.* **1999**, *25*, 1131–1139. [CrossRef]

46. Chao, J.-T.; Fang, K.-Y.; Koh, C.-N.; Chen, Y.-M.; Yeh, W.-C. Feeding on plants by the red-bellied tree squirrel *Callosciurus erythraeus* in Taipei Botanical Garden. *Bull. Taiwan For. Res. Inst.* **1993**, *8*, 39–50.

47. Liu, Y.-F. A Study on the Population and Habitat Use of the Red-bellied Squirrel (*Callosciurus erythraeus*) in Nanjenshan Area. Master's Thesis, National Pingtung University of Science and Technology, Pingtung, Taiwan, 2003.

48. Li, H.J.; Zhang, Z.B. Effect of rodents on acorn dispersal and survival of the Liaodong oak (*Quercus liaotungensis* Koidz.). *For. Ecol. Manag.* **2003**, *176*, 387–396. [CrossRef]

49. Birkedal, M.; Fischer, A.; Karlsson, M.; Löf, M.; Madsen, P. Rodent impact on establishment of direct-seeded *Fagus sylvatica*, *Quercus robur* and *Quercus petraea* on forest land. *Scand. J. For. Res.* **2009**, *24*, 298–307. [CrossRef]

50. Buckley, D.A.; Sharik, T.L. Effect of overstorey and understorey vegetation treatments on removal of planted northern red oak acorns by rodents. *North. J. Appl. For.* **2002**, *19*, 88–92.

51. Blanco, J.A.; Welham, C.; Kimmins, J.P.; Seely, B.; Mailly, D. Guidelines for modeling natural regeneration in boreal forests. *For. Chron.* **2009**, *85*, 427–439. [CrossRef]

52. Liu, T.-Y.; Lin, K.-C.; Vadeboncoeur, M.A.; Chen, M.-A.; Huang, M.-Y.; Lin, T.C. Undersotey plant community and ligth availability in conifer plantations and natural hardwood forests in Taiwan. *Appl. Veg. Sci.* **2015**, *18*, 591–602. [CrossRef]

53. Bi, J.; Blanco, J.A.; Kimmins, J.P.; Ding, Y.; Seely, B.; Welham, C. Yield decline in Chinese fir plantations: A simulation investigation with implications for model complexity. *Can. J. For. Res.* **2007**, *37*, 1615–1630. [CrossRef]

forests

MDPI

Article

Black Plastic Mulch or Herbicide to Accelerate Bur Oak, Black Walnut, and White Pine Growth in Agricultural Riparian Buffers?

Benoit Truax [1],*, Julien Fortier [1], Daniel Gagnon [1,2] and France Lambert [1]

[1] Fiducie de Recherche sur la Forêt des Cantons-de-l'Est/Eastern Townships Forest Research Trust, 1 rue Principale, Saint-Benoît-du-Lac, QC J0B 2M0, Canada; fortier.ju@gmail.com (J.F.); daniel.gagnon@uregina.ca (D.G.); france.lambert@frfce.qc.ca (F.L.)

[2] Department of Biology, University of Regina, 3737 Wascana Parkway, Regina, SK S4S 0A2, Canada

* Correspondence: btruax@frfce.qc.ca; Tel.: +1-819-821-8377

Received: 10 April 2018; Accepted: 7 May 2018; Published: 10 May 2018

Abstract: This study was conducted in a riparian buffer bordering a 1 km segment of a headwater stream crossing a pasture site located in southern Québec (Canada). Three species were planted (black walnut (*Juglans nigra* L.), bur oak (*Quercus macrocarpa* Michx.), and eastern white pine (*Pinus strobus* L.)) with three vegetation treatments (control, herbicide (one application/year for 3 years), and black plastic mulch)). The main objective was to determine to which extent herbicide and plastic mulch, used with species having different ecological characteristics, affect tree growth and soil nutrient status in riparian buffers. Survival was high (>93%) for all species in all treatments. In the control (no vegetation treatment), growth was similar among species. Black walnut had the strongest growth response to herbicide and plastic mulch, and white pine had the weakest. For all species, growth was similar in the herbicide and the plastic mulch treatments. During the fifth growing season, plastic mulch increased soil nitrate and phosphorus compared to the herbicide treatment. In the plastic mulch treatment, higher soil nitrate supply was observed for species that preferentially uptake ammonium (black walnut and white pine). Soil nutrient supplies were similar between the control and herbicide treatments. Despite the more favorable nutritional conditions it provides, permanent black plastic mulching does not provide higher growth benefits after 5 years than a 3-year herbicide treatment. The high soil nitrate supply observed in mulched black walnut and mulched white pine may indicate a limited capacity for nitrate phytoremediation by these species.

Keywords: agroforestry; riparian forest restoration; hardwoods; *Juglans nigra*; *Quercus macrocarpa*; *Pinus strobus*; vegetation management; weed control; nitrate; phosphorus

1. Introduction

During the last decades, many countries have developed policies and programs to stimulate the protection and the establishment of forest riparian buffers on farmland [1,2]. Such buffers can reduce the load of many aquatic pollutants (nitrate, phosphorus, pesticides, sediments, pathogens), while providing stream shading, streambank stabilisation, and flood protection [3–5]. Tree riparian buffers are also keystone landscape components for terrestrial, amphibian, and aquatic biodiversity because they provide habitat, shelter, food sources, and movement corridors for many species within agroecosystems [6–9].

Ecosystem services provide by riparian buffers can be improved by planting high-value hardwood species that produce edible nuts or acorns [10,11]. Many of these species have high light, water, and nutrient requirements to achieve optimal growth [12]. Hence, the large amount of deforested riparian land available along pastures and crop fields [13,14] provides tremendous opportunity to reintroduce nut producing hardwoods. The inclusion of conifers in riparian buffers can also improve their multi-functionality [15], as conifers provide a year-round shelter for bird and mammal species [16].

Although the selection of tree species has important implications for the provision of ecosystem services in riparian buffers, little information is available on the establishment success of different species in farmland riparian habitats. Most often, one or more sylvicultural treatments are needed to improve tree survival and growth in riparian zones [11,17,18]. Vegetation management (weed control) is especially needed when the site is dominated by a dense cover of grasses, which compete strongly with trees for water and nutrients [19]. On sites dominated by a tall herbaceous vegetation cover (i.e., forbs) or shrubs, vegetation management can also be used to increase light availability [20]. Among vegetation management strategies, herbicide application and black plastic (polyethylene) mulching have been widely used in planted buffers [10,15,21,22]. Herbicide and black plastic mulch are both known to enhance angiosperm and gymnosperm growth by locally creating a weed-free plantation environment where soil temperature, humidity, nitrogen (N) mineralisation, and nitrate (NO_3) availability are increased [23–27]. Because of its absorptive and transmittance properties, black plastic mulch also increases air temperature locally [28], which may positively affect photosynthetic capacity in seedlings of temperate species [29].

Despite the benefits they provide, it is unclear which of plastic mulching or herbicide is the best treatment to enhance tree growth and survival in riparian buffers and other types of agroforestry systems [30]. Sometimes, growth response to one or the other treatment differed between species, as seen in abandoned farmland and riparian afforestation trials [11,23,31,32]. Change in the magnitude of the growth response to a given vegetation management treatment is also expected when species with contrasting ecologies are compared [11]. For example, white pine (*Pinus strobus* L.), which has low nutrient requirements [16], was less responsive to plastic mulch than deciduous species [15]. Species-specific feedbacks on riparian soil NO_3 can also occur depending on the species used in combination with black plastic mulch [15]. A much higher NO_3 supply was found in the soil underneath mulched white pine and red oak (*Quercus rubra* L.) [15], which are known to preferentially uptake soil ammonium (NH_4) [33–36]. Such species feedback on soil NO_3 observed in mulched trees requires further investigation because many riparian buffers are planted to reduce NO_3 pollution, which greatly accelerates stream eutrophication [3,37].

An important distinction between herbicide and black plastic mulch is related to the duration of their effects on the growth environment. Once an herbicide treatment is stopped, the regrowth of herbaceous vegetation is generally spontaneous and rapid [20]. In riparian buffers, such vegetation cover has important functions for NO_3 pollution mitigation because ruderal herbaceous species naturally growing in riparian buffers are often nitrophilous and their capacity to immobilize N in plant biomass is relatively high (up to 150 kg N/ha) during the summer [35,38–40]. Conversely, plastic mulch generally remains on plantation sites for decades because of the lack of biodegradability of the petroleum-based commercial mulches [41]. Thus, the effects of plastic mulch on the growth environment and on vegetation cover are longer-lasting compared to an herbicide treatment that would be stopped once trees were properly established. Consequently, the use of a vegetation treatment that allows the regrowth of herbaceous vegetation underneath trees (e.g., herbicide) is expected to have a long-term effect on reducing soil NO_3 supply, thus reducing its potential leaching into streams.

This farm-scale study was conducted along a 1 km segment of a headwater stream crossing a high fertility pasture site located in southern Québec (southeastern Canada). Two nut producing hardwoods (black walnut (*Juglans nigra* L.) and bur oak (*Quercus macrocarpa* Michx.)) and eastern white pine were planted in an experimental design with three vegetation treatments (a control treatment, an herbicide treatment (one application per year for 3 years), and a permanent black plastic mulch treatment). The first objective of this study was to evaluate which vegetation treatment is the best to enhance the survival and growth of the three species in a riparian buffer. The second objective was to determine to which extent herbicide and plastic mulch used in combination with species having different nutritional requirements and ecological characteristics affect nutrient status in riparian soil once trees are well-established (fifth growing season).

In this study, four hypotheses are tested: (1) because of its longer-lasting effect on the growing environment, black plastic mulch will be more effective than the herbicide to improve soil NO_3 status and the growth of all species; (2) because of its low nutrient requirements, white pine will be less responsive to both vegetation management treatments than black walnut and bur oak; (3) in the plastic mulch treatment, the highest NO_3 supply rate will be observed for white pine and black walnut, which preferentially uptake NH_4 [33,42]; (4) similar NO_3 supply rates will be observed at 5 years in the control and herbicide treatments.

2. Materials and Methods

2.1. Site Description

The study took place at the Carocel farm (45°29'53.61" N; 71°59'1.36" W) located in the municipality of Sherbrooke (Bromptonville Borough), which is part of the Estrie administrative region of southern Québec (southeastern Canada). Gentle slopes characterize the site and the regional topography [43]. The study site is within the sugar maple–basswood ecoregion, which is part of the broader northern hardwoods forest ecosystem [43,44]. The regional climate is a continental sub-humid moderate climate, with a growing season of 180–190 days [43], mean annual precipitation of 1145 mm, and a mean annual temperature of 5.6 °C [45]. The soil at the study site has developed on glacial outwash, deposited over lacustrine clay and is generally well-drained [46]. In the 0–60 cm depth range, soil pH ranged 5.7–6.4, soil texture varied within the loam soil class (loam, silty loam, sandy clay loam), and the soil is free of large stones [47].

The riparian buffer was established in spring 2010 along a 1 km segment of a small farm stream draining into the Saint-François River. The stream has been subjected to channel reconfiguration and streambank deforestation. A mean stream width of 1.6–2.0 m and a mean depth of 0.20 m were previously recorded at the study site during summer [48]. The land use adjacent to the riparian buffer is a pasture that is annually fertilized with cattle manure. This pasture supports a cattle density of approximately 0.6 cattle/ha (M. Beauregard, pers. comm.). Prior to the installation of stream fences in 2009, the livestock had full access to the stream and riparian zone, which substantially altered channel morphology and contributed to water quality degradation (B. Truax and J. Fortier, field observations). Mostly warm water fish species, tolerant of water pollution are found in this headwater stream [48]. Prior to the buffer establishment, the vegetation cover was dominated by herbaceous species (mostly pasture grasses and other ruderal species [39]), and was repeatedly grazed by the livestock. Following buffer establishment and fencing, the height of the herbaceous cover ranged 20–50 cm at its maximum development stage during the summer.

2.2. Experimental Design

In May 2010, a randomized block design was established to test the effect of tree species and vegetation treatment on tree growth and the dynamics of soil elements. Three vegetation treatments (control, herbicide, black plastic mulch) and three species (black walnut, bur oak, white pine), replicated in six blocks, were used in this factorial experiment. The experimental design contains 54 plots (3 Species × 3 Vegetation treatments × 6 blocks). Each block contains nine contiguous experimental plots. Each plot contains a single Species/Vegetation treatment combination. Each plot measures 4.5 × 4.5 m (20.5 m²/plot) and contains nine trees planted with a square spacing of 1.5 × 1.5 m (2.25 m²/tree). Each block measured 40.5 m in length (parallel to the stream) by 4.5 m of width (perpendicular to the stream). Topographically, all blocks were placed in areas presenting relatively homogeneous soil and no signs of inadequate drainage. All blocks were located outside of the bankfull stage (above the top of the streambank), but inside the floodplain zone. For the entire experimental design, a total of 486 trees were planted along three rows parallel to the stream.

Blocks were spatially separated along a 1 km stream segment in order to maximize environmental variability at the site level. Between blocks, buffer zones were planted to create a continuous linear tree structure in the riparian zone, thereby reducing edge effects on trees located at the ends of blocks. These buffer zones contained the same tree species that were positioned in the first and last plots in a block and they were planted with the same spacing used in the experimental plots. Along all blocks, one row of hybrid willows (*Salix* × spp.) was planted outside the experimental design, between the top and the toe of the streambank (in the active stream channel zone). These willows were planted to enhance streambank stabilization and to eventually create an overhanging canopy structure above the stream, which provides stream shading and allochthonous inputs of organic matter and terrestrial insect preys [9,49,50]. Livestock was also excluded from the riparian buffer over the entire study duration with electrical fence wires that were installed in summer 2009. At the pasture edge of the experimental design, the fence was located at about 1.5–2.0 m from the first row of trees. A schematic representation of an experimental plot, where a vegetation treatment was applied, is shown in Figure 1.

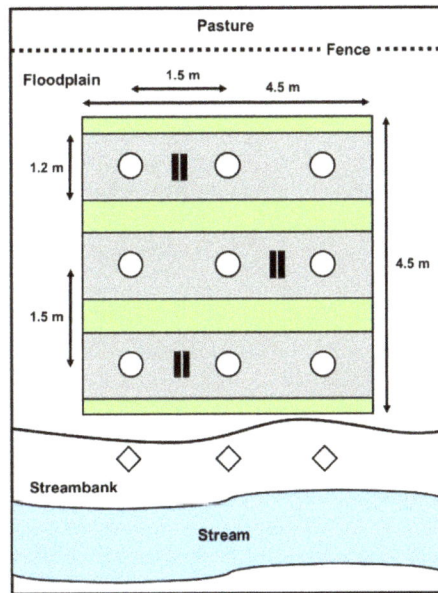

Figure 1. Schematic representation of an experimental plot for a single tree species (4.5 m × 4.5 m) that received a vegetation treatment (herbicide or plastic mulch). The plot is located in the floodplain zone and it contains nine trees (white circles) and a single vegetation treatment applied on each row on a 1.2 m wide strip (grey rectangles). Spacing between trees is 1.5 m × 1.5 m. Small black rectangles indicate the positioning of each pair of Plant Root Simulator (PRS)-probes ion exchange membranes. Hybrid willows (white diamonds) were planted between the top and toe of the streambank, outside the experimental plot.

All trees were planted manually with a shovel, directly through the herbaceous vegetation. No site preparation was done prior to planting. The herbicide (glyphosate) and the plastic mulch treatment were applied in 1.2 m wide strips on each of the three tree rows. The plastic mulch was installed shortly after tree planting (in mid-May 2010). Black plastic mulch strips (0.06 mm thick) were installed manually and pinned down with large wooden pegs. Rocks found nearby were used to maintain the mulch close to soil surface. Black plastic mulch rolls were purchased from Dubois Agrinovation Inc. (Napierville, QC, Canada). The plastic mulch strips remained on site for the entire duration of the study

(2010–2014). The herbicide application was done once in mid-June of the first three growing seasons (2010, 2011, 2012). The herbicide was manually applied with a backpack sprayer and a cardboard tube was used to protect trees against drift.

The tree species used in this study were selected for their contrasted ecological characteristics. Black walnut is a naturalized hardwood species (native to southern Ontario, Canada) that mainly grows on well-drained bottomland sites [51]. Bur oak is a native hardwood species in the study area that mainly grows in bottomlands and on riparian sites experiencing short flood events [51]. White pine is a ubiquitous native species in the study area [51]. Characteristics of planted species are presented in Table 1. One-year-old bare-root seedlings were used for black walnut (1-0), two-year-old bare-root seedlings were used for bur oak (2-0), and two-year-old container seedlings were used for white pine (2-0). Height of seedlings at planting was 40 cm for black walnut, 44 cm for bur oak, and 21 cm for white pine. Seedlings were provided by the Berthier nursery (Sainte-Geneviève-de-Berthier, QC, Canada) of the Ministère des Forêts, de la Faune et des Parcs (MFFP) of Québec.

Table 1. Ecological characteristics of the studied species.

Common Name	Habitat Range	Site Fertility Class	Early Growth	Shade Tolerance	N-Form Preference	Successional Status	Flood Tolerance
Black walnut [1]	Well-drained bottomlands	High	Moderately fast	Low	NH_4	Early	Low
Bur oak [1]	Bottomlands, riparian zones, and dry calcareous sites	High	Slow to moderate	Low to intermediate	NO_3:NH_4	Early	Intermediate
White pine [1]	All types of sites ranging from rocky hill top to peatland	Low to moderate	Slow to moderate	Intermediate	NH_4	Mid	Low

[1] References for black walnut [42,51–53], for bur oak [23,54], and for white pine [16,33,36,51].

2.3. Measurement of Soil Nutrient Dynamics

In each experimental plot (n = 54), the dynamics (supply rate) of soil nutrients (NO_3, NH_4, P, calcium (Ca), magnesium (Mg), and sulfur (S)) was determined using Plant Root Simulator (PRS[TM]-Probes) technology from Western Ag Innovations Inc. (Saskatoon, SK, Canada). The PRS-probes are a type of ion exchange membrane encapsulated in thin plastic probes. The membrane's surface exhibits sorption characteristics similar to those of a plant root. Nutrient supply rates measured with this method are strongly correlated with conventional soil extraction methods over a wide range of soil types [55], including the agricultural buffer soils of the study area [56]. This technology was also useful in understanding the effect of different tree species/vegetation treatment combinations on NO_3 dynamics in a riparian buffer located on another stream 30 km south [15].

On 21 June of the fifth growing season (2014), three pairs of probes (an anion and a cation probe in each pair) were buried in each experimental plot for a 30-day period. The PRS-probes were inserted vertically in the shallow soil (0–10 cm, A horizon) with little disturbance of soil structure. Within a plot, each pair of probes was placed on each of the three tree rows, equidistant between two trees of the same row (Figure 1). On 21 July, probes were removed from the soil, washed in the field with distilled water, and returned to Western Ag Labs for analysis. Plot-level composite samples were made by combining the three pairs of probes collected in each plot. Probe supply rates are reported as µg of nutrient per 10 cm^2 per 30 days.

2.4. Growth and Survival Measurements

At the end of the second, third, fourth, and fifth growing seasons (late October 2011–2014), total tree height, basal diameter, and, when possible, diameter at breast height (DBH at 1.3 m) were measured for each tree of the experimental design. A digital caliper was used to make diameter measurements (mean of two diameter measurements taken perpendicularly). Stem volume outside the bark was then calculated for two tree categories, those having no DBH and those having a DBH value. For trees with no DBH, the simple cone volume formula was used [57]:

$$V = \pi D_B^2 H / 12 \tag{1}$$

where V is the stem volume (cm^3), D_B is the basal diameter (cm), and H is the tree height (cm). For trees with a DBH value, the stem volume was measured by summing the volume of two stem sections: (1) from basal diameter to DBH and (2) from DBH to tree tip. For stem Section 2, Equation (1) was used, but D_B was replaced by a DBH value and H was replaced by the height of the stem section from DBH to the tree tip. For stem Section 1, the following formula was used [58]:

$$V = \pi/12(D_1^2 + D_2^2 + D_1 D_2) L \tag{2}$$

where, V is the volume (cm^3) of a stem section, D_1 is the base diameter (cm) of the stem section, D_2 is the diameter (cm) at the top of the stem section, and L is the length (or height) of the stem section. Thus, the volume of stem section 1 was measured by replacing D_1 by a basal diameter value, D_2 by a DBH value, and L by 130 cm in Equation (2).

2.5. Statistical Analyses

Main effects (Tree species and Vegetation treatment) and interaction effects (Tree species × Vegetation treatment) on measured variables were analyzed using analysis of variance (ANOVA) in a fixed factorial design [59]. Degrees of freedom were the following: Total, 53; Tree species, 2; Vegetation treatment, 2; Tree species × Vegetation treatment, 4; Block, 5; Error, 40. All of the ANOVAs were run with the complete set of data (3 species × 3 vegetation treatments × 6 blocks = 54 experimental plots). Being proportions, survival data were logit transformed prior to ANOVA [60], but survival rate results are reported in percent values. Main effects or interaction effects were declared statistically significant for four levels of significance ($p < 0.1$, $p < 0.05$, $p < 0.01$, and $p < 0.001$). The standard error of the difference (SED) was used to separate means [59]. Based on the number of degrees of freedom in this study, this mean separation method is equivalent to the least significant difference test (LSD) at an alpha level of 0.05 [59]. *A priori* contrasts were further used to test specific hypotheses between particular sets of means [59,61,62]. Contrasts are more powerful than pairwise *a posteriori* tests and can be used whether or not the F-test of the ANOVA is significant [59]. Using stem volume data collected after two, three, four, and five growing seasons, a multivariate analysis of variance (MANOVA) was used to test the Time factor and its interaction with other effects tested in this study [62]. The Pillai's trace test-statistic was used to declare significant effects (Time; Time × Tree species; Time × Vegetation treatment; Time × Tree species × Vegetation treatment). Finally, pairwise correlations were used to identify significant correlations between soil nutrient supply rate (NO$_3$, NH$_4$, P, Ca, K, and Mg) and stem volume growth after 5 years. All statistical analyses were done using JMP 11 from SAS Institute (Cary, NC, USA).

3. Results

3.1. Soil Nutrient Dynamics

A nearly significant Tree species × Vegetation treatment interaction ($p = 0.11$) was observed on NO$_3$ supply rate, with the highest values observed for black walnut and white pine in the mulch treatment (Figure 2). Contrast analysis further suggests that in the plastic mulch treatment, species that preferentially uptake NH$_4$ (black walnut and white pine) were associated with a significantly higher NO$_3$ supply rate than bur oak ($p < 0.01$) (Table 2). The Vegetation treatment effect was significant on the supply rate of several soil elements (Table 3). The strongest effect was observed for soil NO$_3$ ($p < 0.001$), with the plastic mulch treatment having significantly higher soil NO$_3$ compared to the herbicide treatment ($p < 0.01$) (Table 2). Despite that NO_3 supply rate was found to be slightly higher in the herbicide vs. the control treatment, this difference was not significant ($p = 0.37$) (Table 2). The P supply rate was 64.3% higher in the mulch treatment compared to the herbicide treatment, but this difference was nearly significant ($p = 0.11$) (Table 3). A significant Vegetation treatment effect was observed for soil supply of S ($p < 0.01$) and Ca ($p < 0.05$). For those nutrients, the observed values in the plastic mulch treatment were significantly higher than in the control treatment (Table 3). However, when the mulch treatment was compared to the herbicide treatment for those elements, only the S supply rate was found to be significantly higher in the mulch treatment. Overall, few differences in soil nutrient dynamics were observed between the herbicide and the control treatment, while highest nutrient supply values were mainly found in the plastic mulch treatment.

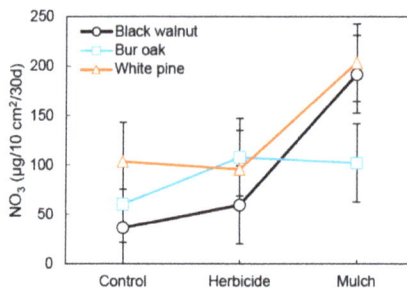

Figure 2. Species × Vegetation treatment interaction for soil NO_3 supply rate ($p = 0.11$) measured for 30 days (21 June to 21 July) of the fifth growing season in an agricultural riparian buffer. Vertical bars represent standard error of the difference (SED).

Table 2. *A priori* contrasts for the analysis of variance (ANOVA) testing the effects of Species and Vegetation treatment on soil NO_3 supply rate measured during the fifth growing season.

Contrast	Species	Treatment [2]	Black Walnut			Bur Oak			White Pine			*p*-Value
			Cont.	Herb.	Mulch	Cont.	Herb.	Mulch	Cont.	Herb.	Mulch	
(1)	NH_4 pref. [1] vs. oak	Mulch	0	0	−1	0	0	2	0	0	−1	<0.01
(2)	All	Mulch vs. Herb.	0	1	−1	0	1	−1	0	1	−1	<0.01
(3)	All	Herb. vs. Cont.	1	−1	0	1	−1	0	1	−1	0	0.37

[1] Species having preference for NH_4 uptake are white pine and black walnut. [2] Mulch = black plastic mulch; Herb. = Herbicide; Cont. = Control.

Table 3. Vegetation treatment effect on the supply rate ($\mu g/10\ cm^2/30$ days) of various soil elements measured during the fifth growing season in an agricultural riparian buffer.

Treatment	NO_3	NH_4	P	Ca	K	Mg	S
Control	66.9	3.23	9.13	2049	201.0	434.4	53.9
Herbicide	87.6	2.67	7.72	2161	149.5	380.6	52.0
Plastic mulch	165.8	2.58	12.69	2292	175.4	375.4	83.4
SED [1]	22.7	0.35	2.35	95	60.7	36.2	10.0
p-value	<0.001	0.13	0.11	<0.05	0.70	0.21	<0.01

[1] SED: standard error of the difference.

3.2. Tree Growth and Survival

After five growing seasons, survival rate was above 93% for all species across all treatments (Figure 3). A significant Tree species effect was observed ($p = 0.05$), with bur oak having a significantly higher survival rate than white pine. The MANOVA done on stem volume measured at the end of each of the last four growing seasons detected a significant Tree species × Vegetation treatment × Time interaction ($p < 0.05$) (Figure 4). In the control treatment, stem volume between the three species remained within a relatively narrow range of values at the end of each growing season (21–39 cm^3 after 2 years, 121–195 cm^3 after 3 years, 400–605 cm^3 after 4 years, and 1244–1541 cm^3 after 5 years), with the lowest volume always observed for white pine. However, in the herbicide and plastic mulch treatments, a much wider range of stem volume values was observed between the three species across the last 4 years. In the herbicide treatment, stem volume ranged 45–223 cm^3 after 2 years, 251–1386 cm^3 after 3 years, 705–3432 cm^3 after 4 years, and 1871–6557 cm^3 after 5 years, while in the plastic mulch treatment, stem volume ranged 43–178 cm^3 after 2 years, 312–1211 cm^3 after 3 years, 979–2982 cm^3 after 4 years, and 2340–6040 cm^3 after 5 years. In the herbicide and mulch treatments, species ranking for stem volume was always the same at the end of each growing season: black walnut > bur oak > white pine. Also, for a given species, the variation in stem volume between the herbicide and plastic mulch treatment was marginal from year 2 to year 5 (Figure 4).

More specifically, after five growing seasons, a significant Tree species × Vegetation treatment interaction was observed on total height ($p < 0.05$), DBH ($p < 0.05$), and stem volume growth ($p < 0.01$) (Figure 5). Furthermore, for all growth variables measured after 5 years (height, DBH, basal diameter, and stem volume), main effects (Tree species and Vegetation treatment) were always highly significant ($p < 0.001$). All of those growth indicators showed a positive response of each species to both vegetation management treatments. However, the magnitude of the growth response varied between species, with black walnut having the strongest response to vegetation treatments and white pine the weakest (Figure 5, Table 4). Indeed, although stem volume of white pine was increased by 46% in the herbicide treatment and by 88% in the plastic mulch treatment, its volume growth in those treatments was not significantly different from the control treatment ($p = 0.25$) (Table 4). After 5 years, stem volume of black walnut was 3.9–4.2 times higher in the vegetation treatments than in the control treatment, while for bur oak stem volume was 2.4–2.7 times higher in the vegetation treatments than in the control. Yet, at the species level, there were no statistical differences in any growth indicators when the herbicide and the black plastic mulch treatments where compared after 5 years (Figure 5, Table 4). Finally, in the black plastic mulch treatment, a strong negative relationship ($R^2 = 0.82$, $p < 0.05$) between soil NO$_3$ supply rate and stem volume growth was observed for white pine during the fifth growing season.

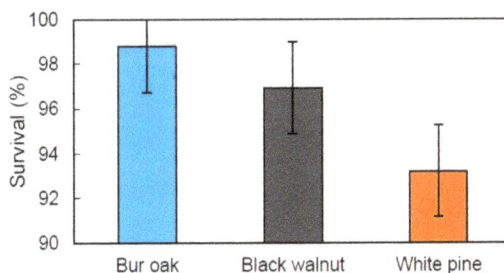

Figure 3. Species effect on survival rate ($p = 0.05$) in an agricultural riparian buffer. Vertical bars represent standard error of the difference (SED).

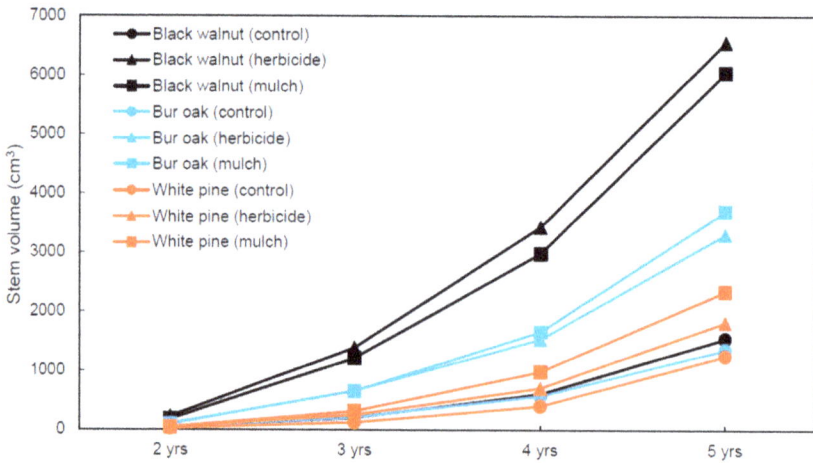

Figure 4. Stem volume per tree measured after 2, 3, 4, and 5 years of growth in an agricultural riparian buffer. The Species × Vegetation treatment × Time interaction is significant at $p < 0.05$ according to multivariate analysis of variance (MANOVA).

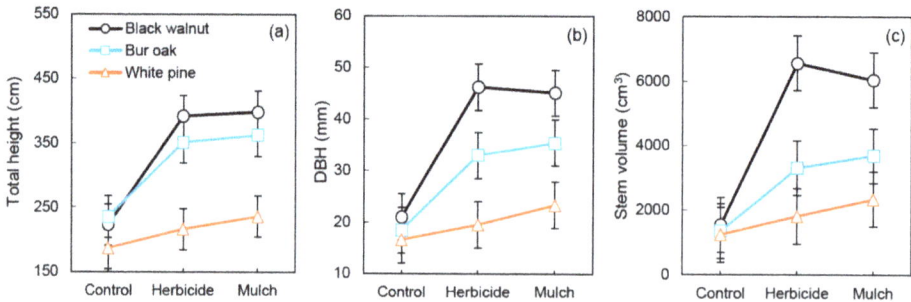

Figure 5. Species × Vegetation treatment interaction for growth variables measured at the end of the fifth growing season in an agricultural riparian buffer: (**a**) total height ($p < 0.05$), (**b**) diameter at breast height (DBH) ($p < 0.05$), and (**c**) stem volume per tree ($p < 0.01$). Vertical bars represent standard error of the difference (SED).

Table 4. *A priori* contrasts for the ANOVA testing the effects of Species and Vegetation treatment on stem volume after 5 years.

Contrast	Species	Treatment [1]	Black Walnut			Bur Oak			White Pine			*p*-Values
			Cont.	Herb.	Mulch	Cont.	Herb.	Mulch	Cont.	Herb.	Mulch	
(1)	B. walnut	Veg. vs. Cont.	2	−1	−1	0	0	0	0	0	0	<0.001
(2)	B. walnut	Herb. vs. Mulch	0	−1	1	0	0	0	0	0	0	0.55
(3)	B. oak	Veg. vs. Cont.	0	0	0	2	−1	−1	0	0	0	<0.01
(4)	B. oak	Herb. vs. Mulch	0	0	0	0	−1	1	0	0	0	0.65
(5)	W. pine	Veg. vs. Cont.	0	0	0	0	0	0	2	−1	−1	0.25
(6)	W. pine	Herb. vs. Mulch	0	0	0	0	0	0	0	−1	1	0.54

[1] Veg. = Vegetation treatments; Herb. = Herbicide; Cont. = Control.

4. Discussion

4.1. Vegetation Treatments and Tree Species Affect Riparian Buffer Growth and Soil Nutrient Dynamics

The selection of the optimal vegetation treatment can greatly accelerate tree establishment and growth in riparian buffers, and thus, reduce the amount of time needed to restore ecosystem services and functions that are linked to the buffer's structural attributes (habitats and movement corridors for forest species, carbon and nutrient storage in woody biomass, flood and wind protection, pesticide drift interception, stream shading, streambank stabilisation, etc.) [63]. We had hypothesized that superior growth would be observed for all species in the plastic mulch treatment because this treatment generally provides a longer-term increase in soil N availability compared to an herbicide treatment that would be stopped after 3 years [15,64]. As expected, plastic mulch was more effective at maintaining high soil NO_3 supply over the long term (Tables 2 and 3). Furthermore, compared to the herbicide, plastic mulch also increased the supply of P, another growth limiting nutrient [65], and the supply of S, a secondary macronutrient important for the synthesis of some amino-acids and vitamins [66] (Table 3). Although not measured in this study, permanent plastic mulching may also be superior to a 3-year herbicide treatment to increase soil and air temperatures, and soil moisture content over the long term [67]. However, despite the more favorable growth conditions provided by the mulch, the growth of all species was not statistically different between the herbicide and the mulch treatments (Figure 5, Table 4). In the Canadian Prairies, herbicide and plastic mulch also provided similar growth gains for four-year-old agroforestry trees belonging to various genera [30].

All studied species benefited from both vegetation treatments, but the magnitude of the growth response was species-specific; a trend that became more and more evident over the years (Figure 4). Hardwood species had the strongest growth response to the vegetation treatments and white pine had the weakest (Figures 4 and 5), which supports our second hypothesis and previous observations made in a nearby riparian buffer and in agroforestry systems of the Prairies [15,30]. In fact, stem volume, height, and diameter growth were not statistically different between the three vegetation treatments for white pine (Figure 5). Compared to most hardwood species, white pine has relatively low nutrient requirements to achieve optimal growth [16], but also has a very effective N retention strategy, where N losses through litter fall are minimized and N residence time in biomass is relatively long [68]. Such a N cycling strategy allows white pine to colonise sites dominated by herbaceous species, even though these species are strong competitors for mineral N in shallow soil [68]. Thus, vegetation management is not essential for white pine establishment in riparian buffers.

On the other hand, for the studied hardwoods, the use of a vegetation treatment resulted in major stem volume increases (292–325% for black walnut and 144–173% for bur oak). Hence, although all species had comparable height, DBH, and stem volume growth after 5 years in the control treatment, tree size varied considerably between species in the plastic mulch and in the herbicide treatments (Figure 5). The strong growth response of black walnut to vegetation management is consistent with previous observations within the *Juglans* genus [23,69–71]. Black walnut is especially sensitive to soil conditions as it requires deep, well-drained, nearly neutral pH, moist and fertile soil to achieve optimal growth [52]. Such favorable soil conditions were found at the study site (see Section 2.1). However, without proper management of herbaceous competition, this species remains far from reaching optimal growth conditions in agricultural riparian buffers (Figures 4 and 5). The same can be said about bur oak, although this species is known for its good tolerance to herbaceous competition [15,54]. Bur oak also has a more conservative early-growth pattern than black walnut [52,54] (Table 1), which can explain its intermediate growth response to both vegetation treatments (Figures 4 and 5).

In agreement with our third hypothesis, and with previous findings near the study area [15], results show that the use of black plastic mulch with tree species having a preference for NH_4 uptake (white pine and black walnut) [33,42] can highly increase the NO_3 supply rate in riparian soils, even on the longer term (Figure 2, Table 2). Thus, particular tree species/vegetation treatment combinations may create undesirable effects on soil N, especially when the goal is to mitigate non-point source NO_3

pollution reaching farm streams. Thus, even though black walnut grew a larger stem volume than bur oak in the plastic mulch treatment (Figure 5), the soil NO_3 supply rate in mulched walnuts was about two times higher than in mulched bur oaks (Figure 2). Such a result may appear counterintuitive given that black walnut requires high soil fertility to reach its full growth potential [52,72]. Previous studies have also reported reduced stand growth, accumulation of soil NO_3, and NO_3 leaching loss in a mature *Pinus* plantation where a high dose of N fertilizer was applied for three consecutive years [73]. Our observations are consistent with such growth decline in pine under elevated soil NO_3. Despite better growth of white pine in the plastic mulch treatment (Figure 5), a strong negative correlation between soil NO_3 supply and volume growth ($R^2 = 0.82$, $p < 0.05$) was observed in mulched pines (Figure 6). Thus, it is potentially not the higher soil NO_3 supply that enhances white pine growth in the plastic mulch treatment, but the higher soil P supply (Table 3). While N limitation is the main factor controlling photosynthesis in deciduous species, P limitation may interact with N in controlling peak photosynthetic capacity in white pine [74]. Also, under open field conditions, white pine develops many branches close to the ground where air temperature is expected to be maximal in the plastic mulch treatment. Such warmer air conditions would enhance the photosynthetic activity of white pines [29].

As hypothesized, the NO_3 supply rate measured during the fifth growing season was not statistically different between the herbicide and control treatments (Figure 2 and Tables 2 and 3). Thus, although herbicide treatments generally enhance soil NO_3 during the year of application [24], they have little residual effects on soil NO_3 dynamics in the subsequent years. The regrowth of herbaceous vegetation underneath trees in the herbicide treatment after the last application (third growing season) may explain such a result (Figure 7). Most herbaceous plants found in the agricultural riparian zone of the study site are ruderal plants [39], which are well-known for their high capacity to uptake soil NO_3 [35]. In a previous study, soil NO_3 supply rate did not differ significantly in the control treatment (no vegetation management) between five different tree species with contrasted growth rates and nutritional requirements [15], a finding corroborated by this study (Figure 2). Recent evidence also suggests that herbaceous plants are more effective than trees at reducing NO_3 leaching in agricultural riparian zones [40].

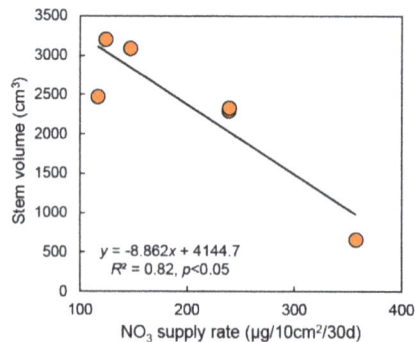

Figure 6. Negative relationship between soil NO_3 supply rate and stem volume growth after five growing seasons for white pine in the black plastic mulch treatment ($n = 6$). The solid line represents a linear least-square regression.

After 5 years, a very high survival rate (>93%) was achieved by all species (Figure 3) and the use of a vegetation treatment was not a significant factor affecting survival ($p = 0.18$), as equally observed by Sweeney and Czapka [11] in a floodplain ecosystem. Tree species selection is sometimes more important to increase survival rate than the use of vegetation treatments [75]. However, in other riparian afforestation studies, where survival rates were low, competing vegetation management was a determining factor affecting tree survival [18,31]. High survival rates were potentially linked to the high soil fertility and the good soil moisture conditions that characterised the studied riparian zone, which has a good hydrological connectivity with the stream (B. Truax and J. Fortier, field observations). The presence of electric fence wires along the stream corridor may have equally reduced herbivory by the white-tailed deer (*Odocoileus virginianus* Zimm.), which has become a serious problem for hardwood and white pine regeneration regionally [76]. Also, the species selected for this study may have contributed to obtaining high survival rates, even in the control treatment. White pine has a high capacity to invade sites dominated by herbaceous plants [68], while black walnut and bur oak, two early-successional species of bottomlands, rapidly form a deep tap root, which increases establishment success in environments dominated by shallow-rooted herbaceous plants [52,77]. However, over the years, growth stagnation related to N limitation may occur for black walnut in the control treatment, as observed in field plantations receiving only a few years of vegetation management [70].

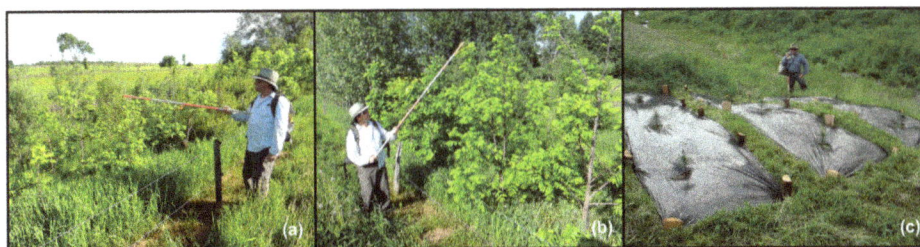

Figure 7. During the fifth growing season (June 2014), a dense herbaceous vegetation cover is observed underneath black walnut in the control treatment (**a**) and in the herbicide treatment (**b**). Permanent black plastic mulch treatment (**c**) restricts herbaceous vegetation growth over the long-term (photo taken after mulch installation around white pines in May 2010).

4.2. Selecting a Vegetation Treatment for the Establishment of Trees in Riparian Buffers

The wide-scale and intensive use of herbicides in agriculture has raised many concerns in the last decades and is still keenly debated today. A high herbicide use is known to pose serious threats to aquatic biodiversity, water quality, and human health [78–80]. Although much less intensive, the use of herbicide to establish tree riparian buffers remains controversial because herbicides are applied in zones that are hydrologically interconnected with freshwater ecosystems. The social acceptability related to the use of herbicides in tree plantations is also relatively low [81].

Because of the many limitations related to herbicide use, black plastic mulch is increasingly used along farm streams and is generally perceived as a more sustainable vegetation treatment by land managers and ecological engineers. Yet, the use of plastic mulches also leads to noticeable impacts (degradation of stable organic matter, micro-plastics and phthalates leaching, modification of soil structure and infiltration properties, impeded root growth, lack of recycling opportunity for plastics contaminated with soil particles) [41,82–84]. Moreover, plastic mulches are rarely removed from plantation sites because this operation is time consuming and expensive [41]. As a result, the impacts of plastic mulch on the soil properties will be long-lasting and should be fully considered as riparian buffer pollutant removal efficiency is tightly bound to water infiltration capacity in soil, to soil health, and to the colonization of soil by plant roots [3]. Also, the permanent barrier created by mulches restricts herbaceous vegetation growth underneath trees (Figure 7c), an important consequence because herbaceous plants have a central role in runoff and N leaching reduction [40,85]. In this study, soil NO_3 and P supplies after 5 years were the highest in the plastic mulch treatment, although species-specific trends were observed for soil NO_3 (Figure 2, Table 3). This situation may be undesirable, since many agricultural riparian buffers are designed to phytoremediate excess soil NO_3 and P generated by adjacent agricultural use. The use of plastic mulch will also reduce the area of riparian habitats that can be colonised by herbaceous plants, shrubs, and tree species, which would otherwise contribute toward creating a more diversified, multifunctional, and resilient buffer system. On the other hand, riparian zones can be important reservoirs of agricultural weeds and invasive exotic plants [6,86,87]. Therefore, the reduction in plant habitat size caused by plastic mulches could be seen positively by some farmers, land planners, and ecological engineers.

Herbicide treatments also have noticeable effects on targeted plant communities, with changes in plant species dominance being more frequent than changes in species composition and diversity [20]. In terms of impacts on soil, extensive herbicide use in tree plantations mostly affects C stocks in the organic layer, with only minor effects being observed on mineral C stocks and stability [64,88]. Moreover, extensive use of glyphosate-based herbicides in tree plantations does not represent a significant risk to human health and to terrestrial and aquatic ecosystems [89]. From an economic perspective, herbicide application is by far the more cost-effective compared to plastic mulching [32]. Technically, plastic mulch application is also more restrictive than an herbicide treatment, which can be done with a backpack sprayer, in almost all types of riparian terrains, including steep and stony streambanks. Furthermore, if a plastic mulch layer is used to install mulch strips, soil tillage may be required, while no such site preparation is needed prior to an herbicide treatment for tree buffer establishment. Finally, from the perspective of agricultural land aesthetics, the visual impact of an herbicide treatment is only temporary, while black plastic mulch remains visible for many years.

Both vegetation treatments used in this study had a similar effectiveness for enhancing hardwood growth (Figure 5, Table 4) but had different environmental impacts. Hence, their use in streamside buffers should be made with the goal of reducing their potential negative impacts, both in regard to herbaceous vegetation and non-target organisms. From that perspective, it is recommended to apply herbicide manually, outside of periods with high runoff or flood event probability, and to use herbicide formulations recommended for aquatic use, which are less toxic to stream fauna [90]. Herbicides should be applied in strips or in spots [11,21], and the number of applications kept to a minimum. Additional studies are needed to identify the most efficient extensive herbicide application strategies for different tree species planted across a gradient of riparian sites differing in resource availability and in competing vegetation characteristics. In some riparian environments, tree survival and growth rate may not be significantly affected by the use of a single or a multi-year herbicide treatment [91]. The identification of key site variables responsible for such a lack of response is essential to the future selection of sites where tree buffers could be installed at lower costs (with no vegetation management). If plastic mulch is used to establish riparian buffers, planting tree species that are nitrophilous is recommended in order to minimize the accumulation of soil NO_3 underneath mulches. Plastic mulches should also be retrieved from riparian sites once trees are well established. Finally, it should be acknowledged that the faster a riparian buffer grows, the faster it will provide key ecosystem services and functions. Thus, land managers and ecological engineers should fully consider the tradeoffs related to the use of a particular vegetation management treatment in the specific case of tree riparian buffer establishment and growth on agricultural land.

5. Conclusions

This study has shown that plastic mulch and herbicide, applied in 1.2 m wide strips, have a similar efficacy to enhance tree growth after 5 years in a fertile agricultural riparian buffer, despite the fact that plastic mulch provides longer-term improvement of soil nutrient status. Because black walnut and bur oak were highly responsive to plastic mulching and herbicide, the use of these vegetation treatments will greatly shorten the time needed to reach canopy closure with those hardwoods. On the other hand, white pine growth was less affected by both vegetation treatments, which indicates that this species can be successfully planted at low cost and with little impact along headwater streams on farmland.

The high NO_3 supply rate observed in the soil underneath plastic mulch where species that preferentially uptake NH_4 (black walnut and white pine) grew suggests caution when selecting particular tree species/vegetation treatment combinations, especially if the objective of the riparian buffer is to mitigate NO_3 pollution in waterbodies. Non-nitrophilous tree species potentially have a limited capacity to reduce NO_3 leaching from intensive cropping systems. The high soil P supply generated by the plastic mulch treatment may also be undesirable over the long term in riparian buffer soils. Conversely, after 5 years, soil nutrient supply was similar between plots receiving no vegetation treatment and plots where a 3-year herbicide treatment was applied, indicating little residual effect of the herbicide on soil nutrient status. In a context where plastic mulch is rarely removed following riparian buffer establishment, further studies are needed to evaluate the long-term impacts of such a vegetation treatment on riparian buffer soils.

Author Contributions: B.T. and D.G. conceived and designed the experiment. B.T. and F.L. were involved in sampling design and field sampling. J.F., B.T., and F.L. analyzed the data. J.F. wrote the first draft of the manuscript. B.T., D.G., and F.L. critically revised the manuscript.

Acknowledgments: We gratefully acknowledge Agriculture and Agri-Food Canada (Agricultural Greenhouse Gas Program), the Ministère des Forêts, de la Faune et des Parcs du Québec, and Tree/Arbres Canada for the funding received. We wish to thank tree planters and field assistants for their help (F. Gendron, A. Laflamme, J. Lemelin, L. Meulien, M.-A. Pétrin, A. Richard, L. Godbout). We also thank M. Beauregard and C. Vincent (Ferme Carocel) who have kindly allowed the use of part of their property for the establishment of the riparian buffer. We acknowledge the Berthier nursery of the Ministère des Forêts, de la Faune et des Parcs of Québec for providing high quality planting stock.

Conflicts of Interest: The authors declare no conflict of interest.

References

1. Muscutt, A.D.; Harris, G.L.; Bailey, S.W.; Davies, D.B. Buffer zones to improve water quality: A review of their potential use in UK agriculture. *Agric. Ecosyst. Environ.* **1993**, *45*, 59–77. [CrossRef]
2. Sweeney, B.W.; Blaine, J.G. River conservation, restoration, and preservation: Rewarding private behavior to enhance the commons. *Freshw. Sci.* **2016**, *35*, 755–763. [CrossRef]
3. Dosskey, M.G.; Vidon, P.; Gurwick, N.P.; Allan, C.J.; Duval, T.P.; Lowrance, R. The role of riparian vegetation in protecting and improving chemical water quality in streams. *JAWRA* **2010**, *46*, 261–277.
4. Mander, Ü.; Hayakawa, Y.; Kuusemets, V. Purification processes, ecological functions, planning and design of riparian buffer zones in agricultural watersheds. *Ecol. Eng.* **2005**, *24*, 421–432. [CrossRef]
5. Tabacchi, E.; Lambs, L.; Guilloy, H.; Planty-Tabacchi, A.M.; Muller, E.; Décamps, H. Impacts of riparian vegetation on hydrological processes. *Hydrol. Proc.* **2000**, *14*, 2959–2976. [CrossRef]
6. Boutin, C.; Jobin, B.; Bélanger, L. Importance of riparian habitats to flora conservation in farming landscapes of southern Québec, Canada. *Agric. Ecosyst. Environ.* **2003**, *94*, 73–87. [CrossRef]
7. Jobin, B.; Bélanger, L.; Boutin, C.; Maisonneuve, C. Conservation value of agricultural riparian strips in the Boyer River watershed, Québec (Canada). *Agric. Ecosyst. Environ.* **2004**, *103*, 413–423. [CrossRef]
8. Lovell, S.T.; Sullivan, W.C. Environmental benefits of conservation buffers in the United States: Evidence, promise, and open questions. *Agric. Ecosyst. Environ.* **2006**, *112*, 249–260. [CrossRef]
9. Baxter, C.V.; Fausch, K.D.; Saunders, W.C. Tangled webs: Reciprocal flows of invertebrate prey link streams and riparian zones. *Freshw. Biol.* **2005**, *50*, 201–220. [CrossRef]
10. Schultz, R.C.; Isenhart, T.M.; Simpkins, W.W.; Colletti, J.P. Riparian forest buffers in agroecosystems—Lessons learned from the Bear Creek Watershed, central Iowa, USA. *Agrofor. Syst.* **2004**, *61–62*, 35–50.
11. Sweeney, B.W.; Czapka, S.J. Riparian forest restoration: Why each site needs an ecological prescription. *For. Ecol. Manag.* **2004**, *192*, 361–373. [CrossRef]
12. Burns, R.M.; Honkala, B.H. *Silvics of North America*; Forest Service Agriculture: Washington, DC, USA, 1990.
13. Fortier, J.; Truax, B.; Gagnon, D.; Lambert, F. Potential for hybrid poplar riparian buffers to provide ecosystem services in three watersheds with contrasting agricultural land use. *Forests* **2016**, *7*, 37. [CrossRef]
14. Rheinhardt, R.; Brinson, M.; Meyer, G.; Miller, K. Carbon storage of headwater riparian zones in an agricultural landscape. *Carbon Bal. Manag.* **2012**, *7*, 4. [CrossRef] [PubMed]

15. Truax, B.; Gagnon, D.; Lambert, F.; Fortier, J. Riparian buffer growth and soil nitrate supply are affected by tree species selection and black plastic mulching. *Ecol. Eng.* **2017**, *106*, 82–93. [CrossRef]

16. Wendel, G.W.; Clay Smith, H. Eastern white pine. In *Silvics of North America: 1. Conifers. Agriculture Handbook 654*; Burns, R.M., Honkala, B.H., Eds.; U.S. Department of Agriculture, Forest Service: Washington, DC, USA, 1990.

17. Andrews, D.M.; Barton, C.D.; Czapka, S.J.; Kolka, R.K.; Sweeney, B.W. Influence of tree shelters on seedling success in an afforested riparian zone. *New For.* **2010**, *39*, 157–167. [CrossRef]

18. Sweeney, B.W.; Czapka, S.J.; Yerkes, T. Riparian forest restoration: Increasing success by reducing plant competition and herbivory. *Rest. Ecol.* **2002**, *10*, 392–400. [CrossRef]

19. Davies, R.J. The importance of weed control and the use of tree shelters for establishing broadleaved trees on grass-dominated sites in England. *Forestry* **1985**, *58*, 167–180. [CrossRef]

20. Balandier, P.; Collet, C.; Miller, J.H.; Reynolds, P.E.; Zedaker, S.M. Designing forest vegetation management strategies based on the mechanisms and dynamics of crop tree competition by neighbouring vegetation. *Forestry* **2006**, *79*, 3–27. [CrossRef]

21. Fortier, J.; Gagnon, D.; Truax, B.; Lambert, F. Biomass and volume yield after 6 years in multiclonal hybrid poplar riparian buffer strips. *Biomass Bioenergy* **2010**, *34*, 1028–1040. [CrossRef]

22. Borin, M.; Vianello, M.; Morari, F.; Zanin, G. Effectiveness of buffer strips in removing pollutants in runoff from a cultivated field in North-East Italy. *Agric. Ecosyst. Environ.* **2005**, *105*, 101–114. [CrossRef]

23. Lambert, F.; Truax, B.; Gagnon, D.; Chevrier, N. Growth and N nutrition, monitored by enzyme assays, in a hardwood plantation: Effects of mulching materials and glyphosate application. *For. Ecol. Manag.* **1994**, *70*, 231–244. [CrossRef]

24. Truax, B.; Gagnon, D.; Chevrier, N. Nitrate reductase activity in relation to growth and soil N-forms in red oak and red ash planted in three different environments: Forest, clear-cut and field. *For. Ecol. Manag.* **1994**, *64*, 71–82. [CrossRef]

25. Truax, B.; Gagnon, D. Effects of straw and black plastic mulching on the initial growth and nutrition of butternut, white ash and bur oak. *For. Ecol. Manag.* **1993**, *57*, 17–27. [CrossRef]

26. Munson, A.D.; Margolis, H.A.; Brand, D.G. Intensive silvicultural treatment: Impacts on soil fertility and planted conifer response. *Soil Sci. Soc. Am. J.* **1993**, *57*, 246–255. [CrossRef]

27. Black, A.; Greb, B. Nitrate accumulation in soils covered with plastic mulch. *Agron. J.* **1962**, *54*, 366. [CrossRef]

28. Kwabiah, A.B. Growth and yield of sweet corn (*Zea mays* L.) cultivars in response to planting date and plastic mulch in a short-season environment. *Sci. Hortic.* **2004**, *102*, 147–166. [CrossRef]

29. Dreyer, E.; Le Roux, X.; Montpied, P.; Daudet, F.A.; Masson, F. Temperature response of leaf photosynthetic capacity in seedlings from seven temperate tree species. *Tree Physiol.* **2001**, *21*, 223–232. [CrossRef] [PubMed]

30. Schroeder, W.R.; Naeem, H. Effect of weed control methods on growth of five temperate agroforestry tree species in Saskatchewan. *For. Chron.* **2017**, *93*, 271–281. [CrossRef]

31. Smaill, S.J.; Ledgard, N.; Langer, E.R.; Henley, D. Establishing native plants in a weedy riparian environment. *N. Z. J. Mar. Freshw. Res.* **2011**, *45*, 357–367. [CrossRef]

32. Von Althen, F.W. Effects of weed control on the survival and growth of planted black walnut, white ash and sugar maple. *For. Chron.* **1971**, *47*, 223–226. [CrossRef]

33. Bauer, G.A.; Berntson, G.M. Ammonium and nitrate acquisition by plants in response to elevated CO_2 concentration: The roles of root physiology and architecture. *Tree Physiol.* **2001**, *21*, 137–144. [CrossRef] [PubMed]

34. Beckjord, P.R.; Adams, R.E.; Smith, D.W. Effects of nitrogen fertilization on growth and ectomycorrhizal formation of red oak. *For. Sci.* **1980**, *26*, 529–536.

35. Gebauer, G.; Rehder, H.; Wollenweber, B. Nitrate, nitrate reduction and organic nitrogen in plants from different ecological and taxonomic groups of Central Europe. *Oecologia* **1988**, *75*, 371–385. [CrossRef] [PubMed]

36. Walters, M.B.; Willis, J.L.; Gottschalk, K.W. Seedling growth responses to light and mineral N form are predicted by species ecologies and can help explain tree diversity. *Can. J. For. Res.* **2014**, *44*, 1356–1368. [CrossRef]

37. Carpenter, S.R.; Caraco, N.F.; Correll, D.L.; Howarth, R.W.; Sharpley, A.N.; Smith, V.H. Nonpoint pollution of surface waters with phosphorus and nitrogen. *Ecol. Appl.* **1998**, *8*, 559–568. [CrossRef]

38. Fortier, J.; Truax, B.; Gagnon, D.; Lambert, F. Biomass carbon, nitrogen and phosphorus stocks in hybrid poplar buffers, herbaceous buffers and natural woodlots in the riparian zone on agricultural land. *J. Environ. Manag.* **2015**, *154*, 333–345. [CrossRef] [PubMed]

39. Fortier, J.; Gagnon, D.; Truax, B.; Lambert, F. Understory plant diversity and biomass in hybrid poplar riparian buffer strips in pastures. *New For.* **2011**, *42*, 241–265. [CrossRef]

40. Neilen, A.D.; Chen, C.R.; Parker, B.M.; Faggotter, S.J.; Burford, M.A. Differences in nitrate and phosphorus export between wooded and grassed riparian zones from farmland to receiving waterways under varying rainfall conditions. *Sci. Total Environ.* **2017**, *598*, 188–197. [CrossRef] [PubMed]

41. Kasirajan, S.; Ngouajio, M. Polyethylene and biodegradable mulches for agricultural applications: A review. *Agron. Sustain. Dev.* **2012**, *32*, 501–529. [CrossRef]

42. Nicodemus, M.; Salifu, K.; Jacobs, D. Nitrate reductase activity and nitrogen compounds in xylem exudate of *Juglans nigra* seedlings: Relation to nitrogen source and supply. *Tree Struct. Funct.* **2008**, *22*, 685–695. [CrossRef]

43. Robitaille, A.; Saucier, J.-P. *Paysages Régionaux du Québec Méridional*; Les Publications du Québec: Ste-Foy, QC, Canada, 1998.

44. Westveld, M. Natural forest vegetation zones of New England. *J. For.* **1956**, *54*, 332–338.

45. Government of Canada. Station Results—1981–2010 Climate Normals and Averages. Available online: http://climate.weather.gc.ca/climate_normals/station_select_1981_2010_e.html?searchType= stnProv&lstProvince=QC (accessed on 16 February 2017).

46. Cann, D.B.; Lajoie, P. *Études Des Sols Des Comtés de Stanstead, Richmond, Sherbrooke et Compton Dans la Province de Québec*; Ministère de l'Agriculture: Ottawa, ON, Canada, 1943.

47. Fortier, J.; Truax, B.; Gagnon, D.; Lambert, F. Root biomass and soil carbon distribution in hybrid poplar riparian buffers, herbaceous riparian buffers and natural riparian woodlots on farmland. *SpringerPlus* **2013**, *2*. [CrossRef] [PubMed]

48. Simavi, M.A. Effet de Plantations de Bandes Riveraines d'arbres sur l'abondance et la Répartition de la Faune Aquatique dans des Ruisseaux Dégradés de Milieux Agricoles dans Les Cantons-de-l'Est. M.Sc. Thesis, Université du Québec à Montréal, Montréal, QC, Canada, 2012.

49. Lyons, J.; Thimble, S.W.; Paine, L.K. Grass versus trees: Managing riparian areas to benefit streams of central North America. *JAWRA* **2000**, *36*, 919–930. [CrossRef]

50. Wallace, J.B.; Eggert, S.L.; Meyer, J.L.; Webster, J.R. Multiple trophic levels of a forest stream linked to terrestrial litter inputs. *Science* **1997**, *277*, 102–104. [CrossRef]

51. Farrar, J.L. *Les Arbres du Canada*; Fides et le Service Canadien des Forêts, Ressources naturelles Canada: St-Laurent, QC, Canada, 2006.

52. Williams, R.D. Black walnut. In *Silvics of North America: 2. Hardwoods. Agriculture Handbook 654*; Burns, R.M., Honkala, B.H., Eds.; U.S. Department of Agriculture, Forest Service: Washington, DC, USA, 1990.

53. Kabrick, J.M.; Dey, D.C.; Van Sambeek, J.W.; Coggeshall, M.V.; Jacobs, D.F. Quantifying flooding effects on hardwood seedling survival and growth for bottomland restoration. *New For.* **2012**, *43*, 695–710. [CrossRef]

54. Johnson, P.S. Bur oak. In *Silvics of North America: 2. Hardwoods. Agriculture Handbook 654*; Burns, R.M., Honkala, B.H., Eds.; U.S. Department of Agriculture, Forest Service: Washington, DC, USA, 1990.

55. Qian, P.; Schoenau, J.J.; Huang, W.Z. Use of ion exchange membranes in routine soil testing. *Commun. Soil Sci. Plant Anal.* **1992**, *23*, 1791–1804. [CrossRef]

56. Fortier, J.; Truax, B.; Gagnon, D.; Lambert, F. Mature hybrid poplar riparian buffers along farm streams produce high yields in response to soil fertility assessed using three methods. *Sustainability* **2013**, *5*, 1893–1916. [CrossRef]

57. West, P. *Tree and Forest Measurement*; Springer: Berlin/Heidelberg, Germany, 2009.

58. Perron, J.-Y. Inventaire forestier. In *Manuel de Foresterie*; Les Presses de l'Université Laval: Ste-Foy, QC, Canada, 1996; pp. 390–473.

59. Petersen, R.G. *Design and Analysis of Experiments*; Marcel-Dekker: New York, NY, USA, 1985.

60. Warton, D.I.; Hui, F.K.C. The arcsine is asinine: The analysis of proportions in ecology. *Ecology* **2011**, *92*, 3–10. [CrossRef] [PubMed]

61. Day, R.W.; Quinn, G.P. Comparisons of treatments after an analysis of variance in ecology. *Ecol. Monogr.* **1989**, *59*, 433–463. [CrossRef]

62. Gotelli, N.J.; Ellison, A.M. *A Primer of Ecological Statistics*; Sinauer Associated, Inc.: Sunderland, MA, USA, 2004.

63. Bentrup, G. *Conservation Buffers: Design Guidelines for Buffers, Corridors, and Greenways*; U.S. Department of Agriculture, Forest Service, Southern Research Station: Asheville, NC, USA, 2008.

64. Périé, C.; Munson, A.D. Ten-year responses of soil quality and conifer growth to silvicultural treatments. *Soil Sci. Soc. Am. J.* **2000**, *64*, 1815–1826. [CrossRef]

65. Vitousek, P.M.; Porder, S.; Houlton, B.Z.; Chadwick, O.A. Terrestrial phosphorus limitation: Mechanisms, implications, and nitrogen-phosphorus interactions. *Ecol. Appl.* **2010**, *20*, 5–15. [CrossRef] [PubMed]

66. Johnson, D.W. Sulfur cycling in forests. *Biogeochemie* **1984**, *1*, 29–43. [CrossRef]

67. Sæbø, A.; Fløistad, I.S.; Netland, J.; Skúlason, B.; Edvardsen, Ø.M. Weed control measures in Christmas tree plantations of *Abies nordmanniana* and *Abies lasiocarpa* on agricultural land. *New For.* **2009**, *38*, 143–156. [CrossRef]

68. Laungani, R.; Knops, J.M.H. Species-driven changes in nitrogen cycling can provide a mechanism for plant invasions. *Proc. Natl. Acad. Sci. USA* **2009**, *106*, 12400–12405. [CrossRef] [PubMed]

69. Paris, P.; Cannata, F.; Olimpieri, G. Influence of alfalfa (*Medicago sativa* L.) intercropping and polyethylene mulching on early growth of walnut (*Juglans* spp.) in central Italy. *Agrofor. Syst.* **1995**, *31*, 169–180. [CrossRef]

70. Von Althen, F.W. Revitalizing a black walnut plantation through weed control and fertilization. *For. Chron.* **1985**, *61*, 71–74. [CrossRef]

71. Garrett, H.E.; Jones, J.E.; Kurtz, W.B.; Slusher, J.P. Black walnut (*Juglans nigra* L.) agroforestry—Its design and potential as a land-use alternative. *For. Chron.* **1991**, *67*, 213–218. [CrossRef]

72. Von Althen, F.W. Afforestation of former farmland with high-value hardwoods. *For. Chron.* **1991**, *67*, 209–212. [CrossRef]

73. Aber, J.D.; Magill, A.; Boone, R.; Melillo, J.M.; Steudler, P. Plant and soil responses to chronic nitrogen additions at the Harvard forest, Massachusetts. *Ecol. Appl.* **1993**, *3*, 156–166. [CrossRef] [PubMed]

74. Reich, P.B.; Schoettle, A.W. Role of phosphorus and nitrogen in photosynthetic and whole plant carbon gain and nutrient use efficiency in eastern white pine. *Oecologia* **1988**, *77*, 25–33. [CrossRef] [PubMed]

75. Steele, K.L.; Kabrick, J.M.; Dey, D.C.; Jensen, R.G. Restoring riparian forests in the Missouri Ozarks. *North. J. Appl. For.* **2013**, *30*, 109–117. [CrossRef]

76. Truax, B.; Gagnon, D.; Lambert, F.; Fortier, J. Multiple-use zoning model for private forest owners in agricultural landscapes: A case study. *Forests* **2015**, *6*, 3614–3664. [CrossRef]

77. Danner, B.T.; Knapp, A.K. Growth dynamics of oak seedlings (*Quercus macrocarpa* Michx. and *Quercus muhlenbergii* Engelm.) from gallery forests: Implications for forest expansion into grasslands. *Trees* **2001**, *15*, 271–277. [CrossRef]

78. Beketov, M.A.; Kefford, B.J.; Schäfer, R.B.; Liess, M. Pesticides reduce regional biodiversity of stream invertebrates. *Proc. Natl. Acad. Sci. USA* **2013**, *110*, 11039–11043. [CrossRef] [PubMed]

79. Stone, W.W.; Gilliom, R.J.; Ryberg, K.R. Pesticides in U.S. streams and rivers: Occurrence and trends during 1992–2011. *Environ. Sci. Technol.* **2014**, *48*, 11025–11030. [CrossRef] [PubMed]

80. Myers, J.P.; Antoniou, M.N.; Blumberg, B.; Carroll, L.; Colborn, T.; Everett, L.G.; Hansen, M.; Landrigan, P.J.; Lanphear, B.P.; Mesnage, R.; et al. Concerns over use of glyphosate-based herbicides and risks associated with exposures: A consensus statement. *Environ. Health* **2016**, *15*. [CrossRef] [PubMed]

81. Fortier, J.; Messier, C. Are chemical or mechanical treatments more sustainable for forest vegetation management in the context of the TRIAD? *For. Chron.* **2006**, *82*, 806–818. [CrossRef]

82. Steinmetz, Z.; Wollmann, C.; Schaefer, M.; Buchmann, C.; David, J.; Tröger, J.; Muñoz, K.; Frör, O.; Schaumann, G.E. Plastic mulching in agriculture. Trading short-term agronomic benefits for long-term soil degradation? *Sci. Total Environ.* **2016**, *550*, 690–705. [CrossRef] [PubMed]

83. Jiang, X.J.; Liu, W.; Wang, E.; Zhou, T.; Xin, P. Residual plastic mulch fragments effects on soil physical properties and water flow behavior in the Minqin Oasis, Northwestern China. *Soil Tillage Res.* **2017**, *166*, 100–107. [CrossRef]

84. Yan, C.; Liu, E.; Shu, F.; Liu, Q.; Liu, S.; He, W. Review of agricultural plastic mulching and its residual pollution and prevention measures in China. *J. Agric. Res. Environ.* **2014**, *31*, 95–102.

85. Knight, K.W.; Schultz, R.C.; Mabry, C.M.; Isenhart, T.M. Ability of remnant riparian forests, with and without grass filters, to buffer concentrated surface runoff. *JAWRA* **2010**, *46*, 311–322.

86. Hood, W.G.; Naiman, R.J. Vulnerability of riparian zones to invasion by exotic vascular plants. *Plant Ecol.* **2000**, *148*, 105–114. [CrossRef]

87. Pysek, P.; Prach, K. Plant invasions and the role of riparian habitats: A comparison of four species alien to central Europe. *J. Biogeogr.* **1993**, *20*, 413–420. [CrossRef]

88. Maillard, É.; Paré, D.; Munson, A.D. Soil carbon stocks and carbon stability in a twenty-year-old temperate plantation. *Soil Sci. Soc. Am. J.* **2010**, *74*, 1775–1785. [CrossRef]

89. Rolando, A.C.; Baillie, R.B.; Thompson, G.D.; Little, M.K. The risks associated with glyphosate-based herbicide use in planted forests. *Forests* **2017**, *8*. [CrossRef]

90. Environmental Protection Authority (EPA) of South Australia. *Safe and Effective Herbicide Use—A Handbook for Near-Water Applications*; EPA: Adelaide (SA), Australia, 2017.

91. Sweeney, B.W.; Czapka, S.J.; Petrow, L.C.A. How planting method, weed abatement, and herbivory affect afforestation success. *South J. Appl. Ecol.* **2007**, *31*, 85–92.

forests

MDPI

Article

Ungulate Browsing Limits Bird Diversity of the Central European Hardwood Floodplain Forests

Ivo Machar [1],*, Petr Cermak [2] and Vilem Pechanec [3]

[1] Department of Development and Environmental Studies, Faculty of Science, Palacky University Olomouc, 77147 Olomouc, Czech Republic
[2] Department of Forest Protection and Wildlife Management, Faculty of Forestry and Wood Technology, Mendel University in Brno, Zemedelska 3, 61300 Brno, Czech Republic; cermacek@mendelu.cz
[3] Department of Geoinformatics, Faculty of Science, Palacky University Olomouc, 77147 Olomouc, Czech Republic; vilem.pechanec@upol.cz
* Correspondence: ivo.machar@upol.cz; Tel.: +420-724-502-474

Received: 20 April 2018; Accepted: 13 June 2018; Published: 21 June 2018

Abstract: Temperate hardwood floodplain forests along lowland rivers are considered important forest biodiversity refugia in the European cultural landscape. The absence of apex predators combined with an artificial feeding of herbivore populations in winter seasons has caused an increase in browsing pressure on hardwood trees, nearly preventing their regeneration in some localities. There are still important knowledge gaps in understanding the relationships between deer abundance (and browsing pressure) and the abundance (and diversity) of forest bird species in unmanaged hardwood forests. We have studied the red deer and fallow deer browsing pressure in Central European unmanaged hardwood floodplain forests using a novel method based on monitoring browsing pressure along transects combined with bird census data in the Litovelské Pomoraví Protected Landscape Area (Czech Republic). The monitoring data suggested a very high browsing pressure on hardwood trees, causing a strong reduction of the shrub layer and young tree layer (30–210 cm above ground surface). The bird census data from the study area were collected using the territory mapping method. Our results revealed a bird diversity decline in all study plots and the bush nesters guild was found to be completely absent. As bird species from the bush nesters guild are generally common (usually dominant) in hardwood floodplain forest ecosystems with a rich shrub and young tree layer and low browsing pressure, we conclude that intense browsing by large herbivores represents a limiting factor to the bird diversity (especially bush nesters) of hardwood floodplain forests.

Keywords: deer abundance; forest diversity; avian guilds; protected landscape area; understorey; unmanaged forest

1. Introduction

Forest vegetation structure influences the diversity of forest avian communities [1–5]. The habitat characteristics of floodplain forests, modified by human activities, have influenced the bird diversity [6]. This is especially important at the local scale of riparian forest stands, because bird density has a direct relationship with site-scale resources, as pointed out by Zenzal et al. [7]. Bird species richness in hardwood floodplain forests differs among the habitat types, with mature forests supporting the largest number of species because of high stand heterogeneity based on diverse understorey tall shrubs and young trees [8] Natural forest regeneration, the presence or absence of understorey bush and herb cover, and even the general structure of forest ecosystems in the European hardwood temperate forests are affected by browsing pressure of large herbivores, typically deer [9]. The absence of large predators in hardwood forests of the European cultural landscape aggravates the effects of browsing pressure on forest biodiversity [10]. Ungulate browsing can interact with the local flooding regime of rivers to delay the recruitment of some tree species, resulting in shifts in successional trajectories, and leaving young forests vulnerable to invasion by exotic herbaceous species [11].

As shown in the study of Holt et al. [12], deer exclusion benefits birds which forage in the understorey layer. Several guilds or migrant species responded positively to deer exclusion and none responded negatively. The shrub-layer foraging guild was recorded less frequently in older and browsed vegetation, in both winter and spring. Exclusion of deer also increased the occurrence of ground-foraging species in both seasons, although these species showed no strong response to vegetation age. Newson et al. [13] have shown a strong association between deer densities and declines in understorey bird species. Their results indicate that deer-related habitat modification may be affecting some bird species on far larger scales than previously appreciated. Mainly through their effects on understorey vegetation, high deer populations are now likely to be affecting woodland biodiversity over large parts of lowland England and deer management plans, involving the integrated exclusion and culling of deer, need to be coordinated on large scales. It is suggested that such management plans could most usefully target areas that still support relatively high populations of species that are sensitive to deer. The density of understorey foliage is recognized as an important predictor of the distribution of forest birds [14]. Charchuk and Banes [15] have suggested that following understorey protection harvest, the retained forest regenerates quickly, rapidly providing a habitat to more mature forest species than natural disturbance harvest.

The understorey foliage can be seriously reduced by a high deer abundance, as Eichhorn et al. [16] revealed by a LiDAR survey. Their findings suggest that the reduction of deer populations is likely to have a strong impact on woodland structures and aid in restoring the complex understorey habitats required by many birds, whereas management interventions as currently practiced have limited and inconsistent effects. Thus, LiDAR seems to be a potentially important tool for forest bird conservation as it can help identify the full range of structural conditions associated with threshold responses [17].

Forest management decisions in hardwood forests should be made at a site level, and encompass factors such as browsing pressure and the dependence of species of conservation concern on particular habitats [18]. There are still important knowledge gaps in understanding relationships between deer abundance (and browsing pressure) and the abundance (and diversity) of forest bird species in unmanaged forests [19,20]. In order to address these knowledge gaps, we studied the impact of browsing pressure by red deer and fallow deer on avian communities in unmanaged hardwood floodplain forests with a high deer density in the Litovelské Pomoraví Protected Landscape Area, Czech Republic [21]. The main objective of this paper is to provide evidence of the relationship between deer browsing and bird diversity in hardwood floodplain forest ecosystems. We hypothesized that intensive ungulate browsing can be a significant limiting factor to bird diversity in hardwood floodplain forest in protected areas.

2. Materials and Methods

2.1. Study Area

The study area—Litovelske Pomoravi Protected Landscape Area (LPPLA)—is formed by a large segment of hardwood floodplain forests in the eastern part of the Czech Republic along the lowland (240–249 m a.s.l.) meandering Morava River [22]. According to the Czech national classification of forest habitats, the hardwood floodplain forests in the study area [23] are classified as *Ulmi-fraxineta carpini superior* [24]. According to the European classification of forest natural habitats under the Natura 2000 network [25], they are classified as riparian mixed forests along the great rivers (habitat code 91F0). The dominant species are *Quercus robur* L. and *Fraxinus excelsior* L., with admixture of *Tilia cordata* Mill., *Acer campestre* L., *Acer pseudoplatanus* L., *Acer platanoides* L., *Carpinus betulus* L., *Ulmus laevis* Pallas, and *Prunus padus* L.

In the study area, we established five study plots (100 × 100 m) to be used for field data collection (Figure 1). Such plots are considered suitable for characterizing local forest bird communities in ecological studies of Central European forest birds [26,27].

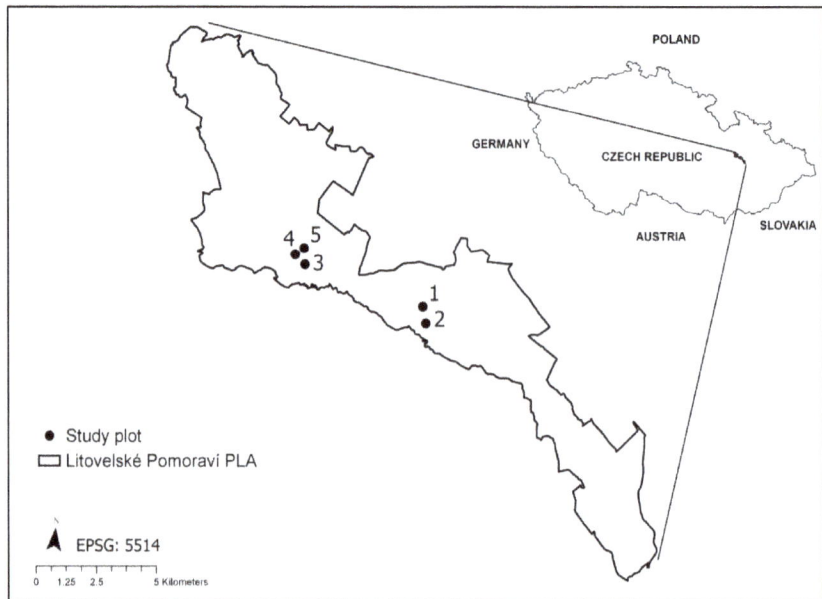

Figure 1. Study area and location of study plots 1–5.

2.2. Bird Census Data

In each study plot, birds were counted during nesting seasons (from the end of March to the end of June) in 2001, 2003, 2005, and 2010 using the 'territory mapping' method [28]. The field mapping involved seven to 10 repeated visits to each study plot. Only those birds were counted that were spotted no further than 50 m from the actual position of the surveying researcher to avoid mistakes based on the different detectability of birds in hardwood floodplain forests [29]. In order to obtain precise bird census results, we also searched all tree cavities and holes for bird nests in each of the study plots. To allow comparisons with other published studies, we used the field data to calculate the mean density (nesting pairs/10 ha). We excluded from our analyses bird species with obviously no relationship with the forest habitat, and which were observed only occasionally in study plots. Classification of bird species into four nesting guilds (ground nesters, bush nesters, canopy nesters, and tree hole nesters) was carried out a priori [30] based on field experience [31] and supported by relevant Czech ornithological literature [32]. Dominance values were calculated according to Aulak [33].

2.3. Ungulate Browsing Research

We established transects in all five study plots to assess herbivore browsing. All transects were 3 m wide and 30–70 m long (based on site conditions). The transect method is routinely used for unrepeated assessments of shrub layer density and browsing intensity [34,35].

The abundance of individual ungulates (IND) according to game management records (GMR 2003) was 63 IND/1000 ha for *Capreolus capreolus* and 11 IND/1000 ha for *Dama dama* in the study plots 1 and 2. In the study plots 3, 4, and 5, the abundance was higher: 98 IND/1000 ha for *Capreolus capreolus* and 63 IND/1000 ha for *Dama dama* [36].

We assessed all trees smaller than 2.1 m. Woody plants higher than 2.1 m were completely absent in the shrub layer in all study plots. The trees were divided into seven height classes (<30 cm, 31–60 cm, 61–90 cm, 91–120 cm, 121–150 cm, 151–180 cm, 181–210 cm). Browsing % is the percentage of individual woody plants that have been browsed. "Individual" was classified as the browsed individual if the terminal shoot was damaged or more than half of the lateral shoots were damaged. The field data collection was conducted in 2003 and it only focused on present browsing damage, i.e., browsing from the past winter (2002/2003) and from the present growing season (2003). In 2005 and 2010, the ungulate browsing in transects was visually verified and found to be without any visible changes.

2.4. Statistical Analysis and Control Plots

To calculate the Pearson's correlation between shrub layer parameters and browsing intensity and between deer density and bird density, we used the Statistica software (StatSoft s.r.o., Prague, Czech Republic) [37]. Statistically significant differences were compared using *t*-tests.

To analyse the similarity and diversity of nesting bird communities, we calculated the commonly used Jaccard similarity index [38]. We used rarefaction to standardize species richness of the bird communities in all five study plots and in all of the nesting seasons [39]. Based on rarefaction curves, this method makes it possible to compare the bird species diversity among different study plots and with different survey efforts. The rarefaction curve is constructed based on the expected number of species $E_{(Sn)}$ using the following equation:

$$E_{(Sn)} = \sum_{i=1}^{S} \left[\frac{\left(\frac{N=N_i}{n} \right)}{\frac{N}{n}} \right] \tag{1}$$

where S is the total number of (nesting bird) species found in the study plot, and N_i is the number of nesting pairs of a particular bird species i. We used the EstimateS 8.0.0. software (StatSoft s.r.o., Prague, Czech Republic) to perform the calculations [40].

As control plots for our original results from study plots, we used localities in hardwood floodplain forests in the Czech Republic, where the bird communities had recently been studied by the territory mapping method and from which data were available. Data related to deer densities in these control plots were collected from legal hunting statistics in the archives of Regional Offices in Olomouc, Hradec Kralove, Ostrava, and Plzen.

3. Results

3.1. Ungulate Browsing and Woody Plants in Herb and Shrub Layers

The density of woody plants was relatively high in the 0–30 cm height class. Woody plants higher than 120 cm were absent, except for very low densities in the height classes 150–180 cm (plot 2) and 180–210 cm (plot 4). Beyond height class 0–30 cm, the browsing percentage was greater than 50% (Figure 2). The percentage of damaged woody plants was negatively correlated with the density of woody plants (Table 1).

Young trees strongly outweighed shrubs in herb and shrub layers. The dominant species were *Fraxinus excelsior* (55.5%), *Acer campestre* (21.3%), and *Acer pseudoplatanus* (15.1%). The remaining tree species accounted for only 6.6% (*Carpinus betulus*, *Quercus petraea*, *Acer platanoides*, and *Tillia cordata*) and shrub species for 1.4% (*Swida sanguinea*, *Euonymus europaeus*, and *Crataegus laevigata*) of the total records.

plot	1		2		3		4		5	
	density	browsing	density	browsing	density	browsing	density	browsing	density	browsing
height	IND/m²	%	IND/m²	%	IND/m²	%	IND/m²	%	IND/m²	%
181–210 cm	0		0		0		0.01	100 %	0	
151–180 cm	0		0.01	50 %	0		0		0	
121–150 cm	0		0		0		0		0	
91–120 cm	0.02	100 %	0		0		0.04	100 %	0	
61–90 cm	0.12	100 %	0.02	50 %	0		0.13	100 %	0.02	75 %
31–60 cm	0.23	84 %	0.17	69 %	0.03	67 %	0.59	99 %	0.13	73 %
0–30 cm	1.38	20 %	0.86	38 %	2.6	41 %	0.8	58 %	0.21	46 %

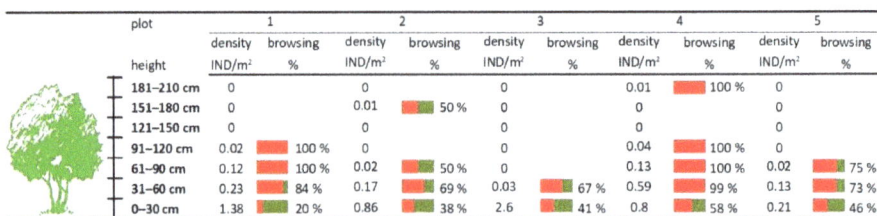

Figure 2. Ungulate density and browsing in study plots 1–5.

Table 1. Correlation matrix for woody plants height, density, and intensity of browsing damage (% of damaged woody plants).

	Height	Density	Browsing %
Height	-	−0.5653 [1]	0.4271
Density		-	−0.5318 [1]
Browsing %			-

[1] Significant values are shown in bold; significant level: $p < 0.05$.

3.2. Bird Community

A total of 21 bird species have been found nesting in the study plots (Table 2). The species composition of nesting birds in all study plots was very similar (Jaccard coefficient 78.9–95.0%).

Using the rarefaction method (Figure 3), we found out that for the minimum abundance of nesting pairs, the highest species richness was reached in study plot 1. However, the differences in bird species richness between individual study plots were not visually distinct.

The rarefaction revealed that the efforts in count surveys in all plots were sufficient. The number of species recorded in all study plots was therefore sufficiently large and representative with respect to the method used.

The highest density of nesting birds was recorded in study plot 2 (69.1 pairs/10 ha), and the lowest in study plot 4 (56.7 pairs/10 ha). The densities calculated for all study plots (Table 2) were thus atypically low for the floodplain forest ecosystem. The representation of dominant species in all study plots reflected the general model of species dominance in bird communities of floodplain forests, in which only a few species (*Sturnus vulgaris*, *Parus major*) dominate the community, and the remaining species are considered attendant or accessory species. However, the count survey surprisingly revealed low dominance values for a few species (*Phylloscopus collybyta*, *Erithacus rubecula*) that usually form the dominant component of the community.

A close examination of the avian guild structure revealed that in all study plots, the bush nesters guild was completely absent (Figure 4). None of the species of this guild typical for the floodplain forests (e.g., *Aegithalos caudatus*, *Coccothraustes*, *Hippolais icterina*, *Luscinia megarhynchos*, *Prunella modularis*, *Sylvia atricapilla*, *Troglodytes troglodytes*, *Turdus merula*) were found in any of the study plots.

The bird density values in all our study plots within the LPPLA were significantly lower than the values obtained by other authors using the same mapping method in hardwood floodplain forests in the Czech Republic (Table 3). High negative correlation rates were found between *Capreolus capreolus* density and bird density (correlation coefficient = −0.6515, *p*-value = 0.0008) and between both of the deer species together (*Capreolus capreolus* and *Dama dama*) and bird density (correlation coefficient = −0.5641, *p*-value = 0.0051). A moderately high negative correlation rate was detected between *Dama dama* density and bird density (correlation coefficient = −0.5201, *p*-value = 0.0269). It is possible that the absence of the bush nesters guild may be the reason for a lower diversity of birds by approximately 30% in our study plots compared to other floodplain forest sites in the Czech Republic (Table 3).

Table 2. Densities and dominance of bird nesting guilds in study plots.

Bird Species	Nesting Guilds [1]	Study Plot									
		1		2		3		4		5	
		DE [2]	DO [3]	DE	DO	DE	DO	DE	DO	DE	DO
Anthus trivialis	GN	0.5	0.8	0.4	0.6	0.7	1.2	0.3	0.5	0.5	0.8
Certhia brachydactyla	HN	2.1	3.4	2.7	3.9	0.4	0.6	2.9	5.1	0.0	0.0
Columba palumbus	CN	1.6	2.6	2.2	3.2	2.6	4.3	0.6	1.1	2.0	3.2
Cyanistes caeruleus	HN	3.2	4.9	0.9	1.3	2.7	4.5	3.9	6.9	3.0	4.7
Dendrocopos major	HN	2.8	4.3	3.9	5.6	1.5	2.5	2.8	4.9	3.1	4.9
Dendrocopos medius	HN	0.0	0.0	0.7	1.0	0.0	0.0	0.5	0.9	0.6	0.9
Erithacus rubecula	GN	2.4	3.8	2.7	3.9	2.1	3.5	1.3	2.3	2.9	4.6
Ficedula albicollis	HN	7.0	10.8	4.3	6.2	7.6	12.7	7.9	13.9	3.8	6.0
Fringilla coelebs	CN	4.3	6.6	3.2	4.6	3.9	6.5	1.9	3.4	2.7	4.2
Garrulus glandarius	CN	0.6	0.9	0.9	1.3	0.5	0.8	1.2	2.0	0.7	1.1
Muscicapa striata	HN	0.6	0.9	1.1	1.7	1.3	2.3	0.9	1.6	0.5	0.8
Oriolus oriolus	CN	1.8	2.8	1.1	1.7	0.7	1.1	1.1	1.9	1.6	2.5
Parus major	HN	10.3	16.1	7.1	10.2	8.2	13.7	8.9	15.7	9.6	15.1
Phylloscopus collybita	GN	2.2	3.4	2.9	4.2	2.2	3.7	3.7	6.5	1.8	2.8
Picus viridis	HN	0.3	0.4	0.9	1.3	0.0	0.0	1.0	1.8	0.0	0.0
Poecile palustris	HN	0.9	1.4	0.2	0.3	0.7	1.2	0.8	1.4	0.6	0.9
Sitta europaea	HN	5.3	8.2	6.7	9.7	5.1	8.5	6.9	12.2	5.0	7.9
Steptopelia turtur	CN	0.8	1.2	1.5	2.2	0.6	1.0	0.6	1.1	0.9	1.4
Sturnus vulgaris	HN	17.2	26.6	19.0	27.5	15.4	25.7	9.1	16.1	18.3	28.8
Turdus philomelos	CN	0.6	0.9	0.5	0.7	0.6	0.6	0.4	0.7	0.7	1.1
Turdus pilaris	CN	0.0	0.0	6.2	8.9	3.1	5.2	0.0	0.0	5.3	8.3
Total of DE and DO		64.5	100	69.1	100	59.9	100	56.7	100	63.6	100
Total of bird species		19		21		19		20		19	

[1] Nesting guilds: GN—ground nesters, CN—canopy nesters, HN—hole nesters, BN—bush nesters; [2] DE = density [amount of nesting pairs/10 ha]; [3] DO = dominance [%].

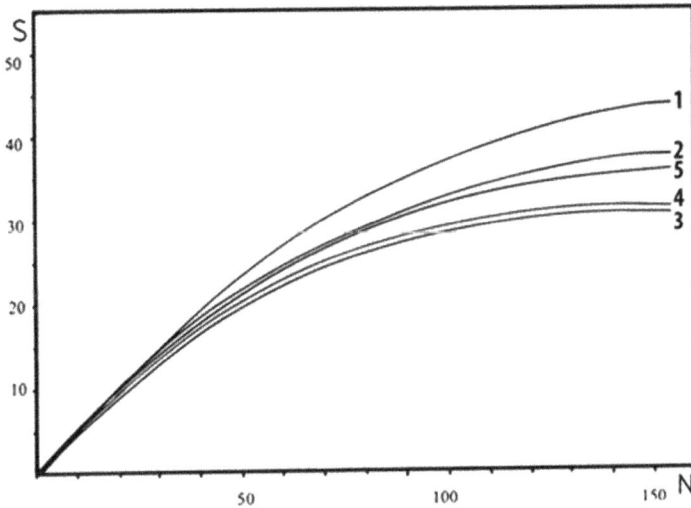

Figure 3. Standardized rarefaction curves for species richness in study plots (S = number of bird species, N = number of nesting pairs, 1–5 = rarefaction curves for study plots no. 1–5).

41

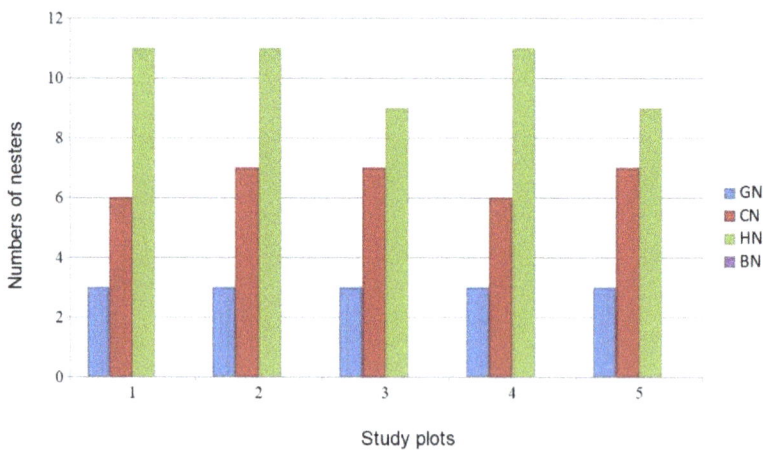

Figure 4. Structure of bird nesting guilds in study plots. (GN—ground nesters, CN—canopy nesters, HN—hole nesters, BN—bush nesters).

Table 3. Comparison of bird and deer density between LPPLA and other localities in hardwood floodplain forests in the Czech Republic.

Source [1]	Size of Study Plot (ha)	Deer Density (Individuals/1000 ha)		Species Richness of Nesting Birds in Study Plot	Density of Birds (Nesting Pairs/10 ha)
		Capreolus capreolus	*Dama dama*		
Results of this study in LPPLA (study plot 1)	1	63	11	19	64.5
Results of this study in LPPLA (study plot 2)	1	63	11	21	69.1
Results of this study in LPPLA (study plot 3)	1	98	63	19	59.9
Results of this study in LPPLA (study plot 4)	1	98	63	20	56.7
Results of this study in LPPLA (study plot 5)	1	98	63	19	63.6
[41]	15	21	5	39	161
[42]	5	21	5	48	79
[43]	9	39	-	44	39
[44]	10	18	15	37	177
[45]	12	44	-	28	93
[46] locality Bosin	32	32	3	44	103
[46] locality Dubno	51	29	3	40	101
[46] locality Choltice	52	37	3	42	109
[46] locality Zbytka	43	40	3	41	80.5
[47] locality Panensky les I	9	35	-	31	139
[47] locality Panensky les II	9	35	-	36	118
[48] locality Sargoun	12	21	5	33	135
[48] locality Vrapac	13	35	11	31	126
[49]	10	20	3	25	113
[50]	10	50	11	23	92
[51]	22	41	5	36	112
[52]	10	36	-	23	102
[53]	12	?	-	30	270
[54]	5	33	3	30	174

[1] Sources in brackets: see References. If there is more than one study site in cited literature, this is indicated.

4. Discussion

4.1. Impact of Ungulate Browsing on the Forest Understorey of Floodplain Forests

Browsing greater than 40% repeatedly led to important changes in woody plant composition in similar areas of the Czech Republic. More concretely, browsing led to a significant reduction of the abundance of the most affected species, especially if they had a low abundance [55]. Repeated browsing can significantly affect the forest dynamics [56]. Some authors [57] have observed that intensive ungulate browsing led to a significant prolongation of the regeneration period in mountain forests—new silver fir and Norway spruce were held to the height limit of 1–1.5 m in the long term (more than thirty years). Floodplain forests can generally better compensate for the lost woody-plant biomass than mountain forests, so the browsing has a lower impact on the growth of woody plants in the herb and shrub layers [58]. However, intensive browsing can promote an expansion of invasive herbaceous species. A strong herb competition can then complicate forest restoration in floodplain forests [59].

The shrub layer is usually well developed in the studied type of floodplain forest. In our study plots, however, the shrub layer (woody plants higher than 1 m) was poorly developed (plots 1 to 3 with a maximum of one woody plant individual per 20 m^2) or completely absent (plots 4 and 5) (Figure 2). Our opinion is that browsing is the main factor for the low abundance of shrubs. The situation of our study area may be compared with the floodplain forest near the Dyje river (locality Lednice hardwood floodplain forest) with the same type of hardwood forest, but with lower browsing pressure [60]. In Lednice, the percentage of browsing damaged woody plants (up to 150 cm of height) was 32% (more than 50% in our plots) and the relative abundance of shrub species was 9% (1.4% in our plots). The ungulates usually prefer some shrub woody plants (e.g., *Swida sanguinea*, especially during winter, or *Eonymus europaeus*) over others. *Crataegus laevigata* is less attractive, but it is strongly browsed, especially in locations where the woody-plant biomass of herb and shrub layers is limited. Most of the tree species present are also attractive [61].

The negative effect of browsing on shrub species can be intensified by their occurrence in herb layers, as browsers often prefer plant species occurring close to the forest floor [62,63]. The observed extreme browsing pressure was most likely the main factor leading to the absence or reduction of the shrub layer.

4.2. Bird Communities in Floodplain Forests

The avifauna of the LPPLA study area mainly consists of common bird species typical for the European hardwood floodplaorests [64]. Surprisingly, a few species otherwise commonly nesting in the understorey of this type of forest (*Sylvia atricapilla*, *Turdus merula*) [65] were found to be absent in all of the five study plots.

The rarefaction curves for all study plots approached their asymptotes and further field surveys would likely not reveal any new species [66]. The species composition of nesting birds was very similar in all study plots (measured by the Jaccard coefficient). These results are in contrast with the findings of Kolecek et al. [67], who have found relatively large differences in species composition between two floodplain-forest sites along the Morava River—the 'Království' site had about 30% more species than the 'Žebračka' site. As the authors themselves point out, these differences could have been caused by the different sizes of the studied sites, as larger sites can host more area-sensitive species [68]. Alternatively, the differences in species composition could have been caused by differences in environmental heterogeneity between the sites [69]. As a result of an intensive forest management, the 'Království' site (unlike the 'Žebračka' site) represents a diverse mosaic of various types of floodplain forest (clear-cut areas, young-growth and old-growth stands, forest margins). These consequences of forest management could explain the higher diversity of birds, because the edge effect on the diversity of nesting birds in the floodplain forests is very pronounced [70]. As all our study plots are located within the LPPLA conservation zone, no forest management has been applied in these sites since the establishment of the LPPLA in 1991. Implementation of multiple-use zoning [71] is generally considered as an important support tool for the conservation of forest diversity.

It is commonly known that past forest management of the European floodplain forests is very clearly reflected in the current structure of forest stands [72]. In our study plots, however, the historical development of forest stands was practically identical and therefore does not play an important role [73].

The structure of Central European hardwood forests dominated by oak (*Quercus* sp.) strictly influences the structure of breeding bird communities, especially in young forest stands [74]. Bird species typical for these stands are bush nesters (*Phylloscopus trochilus*, *Sylvia atricapilla*, *Sylvia curruca*, *Prunella modularis*, *Aegithalos caudatus*). In the LPPLA study plots, these bird species were absent. The comparison of guild structures in the LPPLA and in the hardwood floodplain forests in the Czech Republic (Figure 5) reveals a clear absence of common bird species from the bush nesters guild in the LPPLA study plots. Similarly, many studies have shown that the shrub layer is one of the key characteristics determining the alpha diversity and density of bird communities in the European oak-dominated hardwood forests [75–79].

Reductions in understorey foliage strongly indicate a browsing effect in forest areas of a high deer density. Terrestrial laser scanning applied to oak-dominated woods in the UK by Eichhorn et al. [16] revealed a reduction in understorey foliage of 68% at high deer density sites. In the context of the above-mentioned studies, the results from our study plots suggest that intensive ungulate browsing pressure can be considered a limiting factor to forest bird diversity. In our plots, the diversity of nesting birds was approximately 30% lower than the diversity found in a comparable forest type without an intense browsing pressure (Table 3).

Due to the absence of shrubs, our study plots had lower numbers of bird species otherwise commonly inhabiting the European temperate hardwood forests [80]. Although these species are considered neither rare/endangered in the Czech Republic [81] or habitat specialists, the results are significant [82]. As population trends for common forest bird species in the Czech Republic are otherwise remarkably stable in the long term (in contrast to the common bird species of the agricultural landscapes) [83], the intense ungulate browsing pressure in the LPPLA hardwood floodplain forests, reducing the diversity and density of common bird species, represents an extraordinary situation in terms of nature conservation. Here, we would like to point out that all five study plots are located within a protected area specifically aimed at the conservation of hardwood floodplain forest biodiversity [84].

To pursue the conservation targets for protected areas in the ecological conditions of Central European forests with extraordinary high deer densities, in which large predators are missing [85,86], it is necessary to take a step towards an intense targeted reduction of large herbivores by hunting. Understanding the types of relationships between deer abundance and habitat quality for birds, and other biodiversity, is an important knowledge gap that needs to be addressed if sound, collaborative deer management plans are to be developed.

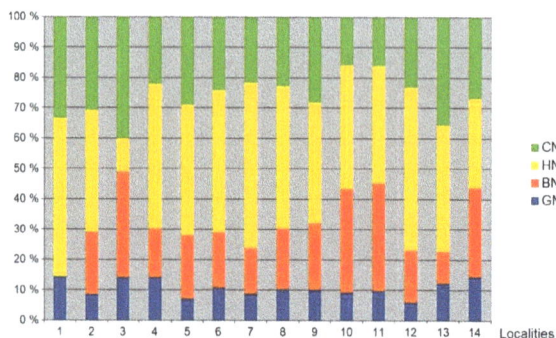

Figure 5. Comparison of guild structures in bird communities in the hardwood floodplain forests in the Czech Republic. (GN—ground nesters, CN—canopy nesters, HN—hole nesters, BN—bush nesters). Localities: 1 = average data from LPPLA, 2 = [41], 3 = [42], 4 = [44], 5 = [45], 6 = [46] Bosin, 7 = [46] Dubno, 8 = [46] Choltice, 9 = [46] Zbytka, 10 = [48] Sargoun, 11 = [48] Vrapac, 12 = [49], 13 = [49], 14 = [54]).

The authors of this paper suggest concentrating all forest and wildlife management practice on deer exclusion in regions with a high browsing pressure. This is especially important in protected areas focused on conserving the biodiversity of European hardwood forests, in which the absence of large carnivores limits the spontaneous succession of the forest ecosystem. A survey [87] on the near-natural hardwood floodplain forest of Cahnov-Soutok in the Czech Republic, which has been left to spontaneous development since the beginning of the 1930s, showed that (1) the most significant trend is a decreased representation of *Quercus robur* in all monitored indicators and conversely an expanding representation of *Acer campestre*, *Carpinus betulus*, and *Tilia cordata*; and (2) that the floodplain forest ecosystem demonstrates a high-level stability in the total volume of tree biomass with an essential change in the tree species composition, spatial structure, and average stem volume of individual trees. However, if the spontaneous gap dynamics and natural regeneration in the forest are limited by ungulate browsing, we cannot protect the biodiversity of hardwood floodplain forest, as pointed out in this paper, which seems to be a general problem in forest biodiversity conservation in the Czech Republic [88]. This problem is increased by the predictions of climate change impacts to European hardwood forests [89], because of an increase in the effect of factors limiting natural forest development and forest natural regeneration. The authors of this paper suggest that deer exclusion in European hardwood forests—especially in protected areas focusing on forest biodiversity conservation [90]—is necessary for putting the basic principles of sustainable forest management [91] into practice.

5. Conclusions

The European hardwood floodplain forests are characterized by a high avian diversity and density. In the unmanaged hardwood floodplain forest of the Litovelské Pomoraví PLA, however, an atypically low bird density was identified in five study plots with a very high degree of red deer and fallow deer browsing pressure on hardwood trees. The heavy browsing pressure has caused a strong reduction of the shrub and young tree layer in the understorey (30–210 cm above ground surface). In all five study plots, bird species from the bush nesters guild—usually dominant in hardwood floodplain forest ecosystems with a rich shrub and young tree layer and low browsing pressure—were found to be absent. As the decline in bird diversity is clearly a consequence of the browsing pressure, the intense ungulate browsing can be considered an important limiting factor to bird diversity (especially to bush nesters) in hardwood floodplain forests. Based on these results, we suggest that deer exclusion become an important component of forest management practice, in order to conserve hardwood floodplain forest biodiversity.

Author Contributions: I.M. and P.C. conceived and designed the sampling protocol; V.P. analyzed the data; I.M. wrote the paper.

Funding: This study was funded by the research grant "Significant Trees—Living Symbols of National and Cultural Identity" (no. DG18P02OVV027), within the NAKI II Program founded by the Czech Ministry of Culture.

Acknowledgments: Authors are grateful to Otto Cizek for providing data related to deer density in control plots.

Conflicts of Interest: The authors declare no conflict of interest.

References

1. Doyon, F.; Giroux, J.F.; Gagnon, D. Effects of different types of management on the diversity of birds and small mammals in a hardwood forest. In *Landscape Ecology in Land use Planning Methods and Practice, Proceedings of the 4th Workshop of the Canadian-Society-for-Landscape-Ecology-and-Management, Laval, QC, Canada, 5–7 June 1995*; Université Laval: Sanite Foy, QC, Canada; pp. 211–219.
2. Laiolo, P. Effects of habitat structure, floral composition and diverzity on a forest bird community in north-western Italy. *Folia Zool.* **2002**, *51*, 121–128.

3. Lindbladh, M.; Lindstrom, A.; Hedwall, P.O.; Felton, A. Avian diversity in Norway spruce production forests—How variation in structure and composition reveals pathways for improving habitat quality. *For. Ecol. Manag.* **2017**, *397*, 48–56. [CrossRef]

4. Perry, R.W.; Jenkins, J.M.A.; Thill, R.E.; Thompson, F.R. Long-term effects of different forest regeneration methods on mature forest birds. *For. Ecol. Manag.* **2018**, *408*, 183–194. [CrossRef]

5. Sallabanks, R.; Haufler, J.B.; Mehl, C.A. Influence of forest vegetation structure on avian community composition in west-central Idaho. *Wildl. Soc. B* **2006**, *34*, 1079–1093. [CrossRef]

6. Pierce, A.R.; King, S.L. A comparison of avian communities and habitat characteristics in floodplain forests associated with valley plugs and unchannelized streams. *River Res. Appl.* **2011**, *27*, 1315–1324. [CrossRef]

7. Zenzal, T.J.; Smith, R.J.; Ewert, D.N.; Diehl, R.H.; Buler, J.J. Fine-scale heterogeneity drives forest use by spring migrant landbirds across a broad, contiguous forest matrix. *Condor* **2018**, *120*, 166–184. [CrossRef]

8. Knutson, M.G.; McColl, L.E.; Suarez, S.A. Breeding bird assemblages associated with stages of forest succession in large river floodplains. *Nat. Areas J.* **2005**, *25*, 55–70.

9. Vacek, Z.; Vacek, S.; Bilek, L.; Kral, J.; Remes, J.; Bulusek, D.; Kralicek, I. Ungulate impact on natural regeneration in spruce-beech-fir stands in Cerny Dul nature reserve in the Orlicke Hory mountains, Case study from Central Sudetes. *Forests* **2014**, *5*, 2929–2946. [CrossRef]

10. Chapron, G.; Kaczensky, P.; Linnell, J.D.C.; von Arx, M.; Huber, D.; Andren, H.; Lopez-Bao, J.V.; Adamec, M.; Alvares, F.; Anders, O.; et al. Recovery of large carnivores in Europe's modern human-dominated landscapes. *Science* **2014**, *346*, 1517–1519. [CrossRef] [PubMed]

11. De Jager, N.R.; Cogger, B.J.; Thomsen, M.A. Interactive effects of flooding and deer (*Odocoileus virginianus*) browsing on floodplain forest recruitment. *For. Ecol. Manag.* **2013**, *303*, 11–19. [CrossRef]

12. Holt, C.A.; Fuller, R.J.; Dolman, P.M. Exclusion of deer affects responses of birds to woodland regeneration in winter and summer. *IBIS* **2014**, *156*, 116–131. [CrossRef]

13. Newson, S.E.; Johnston, A.; Renwick, A.R.; Baillie, S.R.; Fuller, R.J. Modelling large-scale relationship between changes in woodland deer and bird populations. *J. Anim. Ecol.* **2012**, *49*, 278–286. [CrossRef]

14. Hinsley, S.A.; Hill, R.A.; Fuller, R.J.; Bellamy, P.E.; Rothery, P. Bird species distributions across woodland canopy structure gradients. *Community Ecol.* **2009**, *10*, 99–110. [CrossRef]

15. Charchuk, C.; Bayne, E.M. Avian community response to understory protection harvesting in the boreal forest of Alberta, Canada. *For. Ecol. Manag.* **2018**, *407*, 9–15. [CrossRef]

16. Eichhorn, M.P.; Ryding, J.; Smith, M.J.; Gill, R.M.A.; Siriwardena, G.M.; Fuller, R.J. Effects of deer on woodland structure revealed through terrestrial laser scanning. *J. Appl. Ecol.* **2017**. [CrossRef]

17. Garabedian, J.E.; Moorman, C.E.; Peterson, M.N.; Kilgo, J.C. Use of LiDAR to define habitat thresholds for forest bird conservation. *For. Ecol. Manag.* **2017**, *399*, 24–36. [CrossRef]

18. Fuller, R.J. Searching for biodiversity gains through woodfuel and forest management. *J. Appl. Ecol.* **2013**, *50*, 1295–1300. [CrossRef]

19. Petty, S.J.; Avery, M.I. *Forest Bird Communities. A Review of the Ecology and Management of Forest Bird Communities in Relation to Silvicultural Practices in the British Uplands*; Forestry Commission Paper; Forestry Commission: Edinburgh, UK, 1990; Volume 26, pp. 1–41.

20. Fuller, R.J.; Smith, K.W.; Hinsley, S.A. Temperate western European woodland as a dynamic environments for birds: A resource-based review. In *Birds and Habitat: Relationship in Changing Landscapes*; Fuller, R.J., Ed.; Cambridge University Press: Cambridge, UK, 2012; pp. 352–380.

21. Machar, I. Conservation and Management of Floodplain Forests in the Protected Landscape Area Litovelske Pomoravi (Czech Republic) Introduction. In *Conservation and Management of Floodplain Forests in the Protected Landscape Area Litovelske Pomoravi (Czech Republic)*; Accession Number: WOS:000331015800001; Machar, I., Ed.; Palacky University: Olomouc, Czech Republic, 2009; pp. 7–108, ISBN 978-80-244-2355-5.

22. Kilianova, H.; Pechanec, V.; Svobodova, J.; Machar, I. Analysis of the evolution of the floodplain forests in the aluvium of the Morava river. In Proceedings of the 12th International Multidisciplinary Scientific Geoconference (SGEM 2012), Albena, Bulgaria, 17–23 June 2012; SGEM: Albena, Bulgaria48535300001, 2012; Accession Number: WOS:000348535300001; Volume IV, pp. 1–8, ISSN 1314-2704.

23. Simon, J.; Machar, I.; Bucek, A. Linking the historical research with the growth simulation model of hardwood floodplain forests. *Pol. J. Ecol.* **2014**, *62*, 273–288. [CrossRef]

24. Kusbach, A.; Friedl, M.; Zouhar, V.; Mikita, T.; Šebesta, J. Assessing Forest Classification in a Landscape-Level Framework: An Example from Central European Forests. *Forests* **2017**, *8*, 461. [CrossRef]

25. Miko, L. Nature and landscape protection in the European context. In *Ochrana prirody a krajiny v Ceske Republice, Vols I and II*; Accession Number: WOS:000334387900004; Machar, I., Drobilova, L., Eds.; Palacky University: Olomouc, Czech Republic, 2012; pp. 43–49, ISBN 978-80-244-3041-6.

26. Hanzelka, J.; Reif, J. Responses to the black locust (*Robinia pseudoacacia*) invasion differ between habitat specialist and generalist in central European forest birds. *J. Ornithol.* **2015**, *156*, 1015–1024. [CrossRef]

27. Kroftova, M.; Reif, J. Management implications of bird responses to variation in non-native/native tree ratios within central European forest stands. *For. Ecol. Manag.* **2017**, *391*, 330–337. [CrossRef]

28. Bibby, C.J.; Burges, N.D.; Hill, D.A.; Mustoe, S. *Bird Census Techniques*; Academic Press: London, UK, 2007; pp. 42–64, ISBN 978-0-12-095831-3.

29. Poprach, K.; Vrbkova, J.; Machar, I. Detectability as an important factor influencing the knowledge of bird diversity in a floodplain forest ecosystem. *J. For. Sci.* **2015**, *61*, 89–97. [CrossRef]

30. Wiens, J.A. *The Ecology of Bird Communities, Vol. 1, Foundation and Patterns*; Cambridge University Press: Cambridge, UK, 1989; pp. 1–539, ISBN 0-521-26030.

31. Machar, I. Changes in ecological stability and biodiversity in a floodplain landscape. In *Applying Landscape Ecology in Conservation and Management of the Floodplain Forest (Czech Republic)*; Accession Number: WOS:000325436900004; Machar, I., Ed.; Palacky University: Olomouc, Czech Republic, 2012; pp. 73–87, ISBN 978-80-244-2997-7.

32. Stastny, K.; Hudec, K. *Fauna of the Czech Republic. Birds 2 and 3*, 2nd ed.; Academia: Prague, Czech Republic, 2011; pp. 1–1178, ISBN 80-200-1113-7.

33. Aulak, W. Small mammal communities of the Białowieża National Park. *Acta Theriol.* **1970**, *15*, 465–515. [CrossRef]

34. Cermak, P.; Horsak, P.; Spirik, M.; Mrkva, R. Relationships between browsing damage and woody species dominance. *J. For. Sci. (Prague)* **2009**, *55*, 23–31.

35. Cermak, P.; Beranova, P.; Oralkova, J.; Horsak, P.; Plsek, J. Relationships between browsing damage and the species dominance by the highly food-attractive and less food-attractive trees. *Acta Univ. Agric. Mendel. Brun.* **2011**, *59*, 29–36. [CrossRef]

36. Cermak, P.; Mrkva, R. Browsing damage to broadleaves in some national nature reserves (Czech Republic) in 2000–2001. *Ekológia (Bratislava)* **2003**, *22*, 132–141.

37. StatSoft, s.r.o. *Statistica* [software, CD-ROM]. Ver. 12. Praha, 2013.

38. Jaccard, P. Étude comparative de la distribution florale dans une portion des Alpes et des Jura. *Bull. Soc. Vaud. Sci. Nat.* **1901**, *37*, 547–579.

39. James, F.C.; Rathbun, S. Rarefaction, relative abundance and diversity of avian communities. *Auk* **1981**, *98*, 785–800.

40. Author Colwell, R.K. EstimateS: Statistical Estimation of Species Richness and Shared Species from Samples. Version 8.0.0. 2005. Available online: http://viceroy.eeb.uconn.edu/estimates (accessed on 12 January 2018).

41. Bures, S.; Maton, K. Ptaci slozka segmentu skupiny typu geobiocenu Ulmi-fraxineta populi v navrhovane CHKO Pomoravi [Birds of hardwood floodplain forest in proclaimed PLA Pomoravi—In Czech]. *Sylvia* **1984**, *23–24*, 37–46.

42. Bures, S. Analyza ptaci slozky navrhovane SPR Sargoun [Birds of harwood floodplain forest in Sargoun locality—In Czech]. Unpublished Work. 1996.

43. Horak, Z. Ptactvo okolí Starého Labe u Cihelny u Pardubic [Birds of Stare Labe u Cihelny locality—In Czech]. *Panurus* **1988**, *9*, 53–61.

44. Chytil, J. Srovnání Produkce Savců a ptáků v Lužním lese [Comparison of the Production of Birds and Mammals in Floodplain Forest—In Czech]. Ph.D. Thesis, University JE Purkyne, Brno, Czech Republic, 1991.

45. Kubecka, D. Avifauna lužního lesa na lokalitě Horní Záseky v CHKO Litovelské Pomoraví [Birds of floodplain forest in Horni Zaseky locality—In Czech]. Ph.D. Thesis, Palacky University, Olomouc, Czech Republic, 2003.

46. Lemberk, V. Srovnání ornitocenóz čtyř lužních lesů ve východních Čechách [Comparison of avian communities in floodplain forests—In Czech]. *Panurus* **2001**, *11*, 69–79.

47. Machar, I. The effect of floodplain forest fragmentation on bird community. *J. For. Sci.* **2012**, *58*, 213–224. [CrossRef]

48. Machar, I. The impact of floodplain forest habitat conservation on the structure of bird breeding communities. *Ekológia (Bratislava)* **2011**, *30*, 36–50. [CrossRef]

49. Pavelka, J. Hnízdní ornitocenóza v lužním lese u řeky Odry [Birds in floodplain forest along Odra river—In Czech]. *Zprávy MOS (Přerov)* **1987**, *46*, 115–118.

50. Polasek, V. Výzkum Ornitocenózy lužního lesa v Litovelském Pomoraví [Birds of Floodplain Forest in Litovelske Pomoravi—In Czech]. Ph.D. Thesis, Palacky University, Olomouc, Czech Republic, 2001.

51. Pykal, J. Ornitocenosy různých typů přirozených lesních společenstev v pahorkatině jihozápadních Čech [Avian community in hardwood forests in Czechia—In Czech]. *Panurus* **1991**, *3*, 67–76.

52. Ruzicka, I. Ornitologický výzkum lokality Chrbovský les u Záříčí [Birds of Floodplain Forest Near Zarici—In Czech]. Ph.D. Thesis, Palacky University, Olomouc, Czech Republic, 2005.

53. Storch, D. Densities and territory of birds in two different lowland communities in eastern Bohemia. *Folia Zool.* **1998**, *47*, 181–188.

54. Toman, A. Avifauna SPR Zastudanci [Avian Community in Floodplain Forest Zastudanci Locality]. Ph.D. Thesis, Palacky University, Olomouc, Czech Republic, 2004.

55. Nascher, F.A. Zur Waldbaulichen Bedeutung des Rothirschverbisses in der Waldgesellschaft des subalpinen Fichtenwalds in der Umgebung des Schweizerischen Nationalparks. Ph.D. Thesis, ETH Zürich, Rämistr, Zürich Switzerland, 1979. Diss. Nr. 6373.

56. Heuze, P.; Schnitzler, A.; Klein, F. Is browsing the major factor of silver fir decline in the Vosges Mountains of France? *For. Ecol. Manag.* **2005**, *217*, 219–228. [CrossRef]

57. Barancekova, M.; Krojerova-Prokesova, J.; Homolka, M. Impact of deer browsing on natural and artificial regeneration in floodplain forest. *Folia Zool.* **2007**, *56*, 354–364.

58. Kubicek, F.; Simonovic, V.; Kollar, J.; Kanka, R. Herb layer biomass of the Morava river floodplain forests. *Ekológia (Bratislava)* **2008**, *27*, 23–30.

59. Cogger, B.J.; De Jager, N.R.; Thomsen, M.; Adams, C.R. Winter Browse Selection by White-Tailed Deer and Implications for Bottomland Forest Restoration in the Upper Mississippi River Valley, USA. *Nat. Areas J.* **2014**, *34*, 144–153. [CrossRef]

60. Cermak, P.; Mrkva, R.; Horsak, P.; Spirik, M.; Beranova, P.; Oralkova, J.; Plsek, J.; Kadlec, M.; Zarybnicky, O.; Svatos, M. *Impact of Ungulate Browsing on Forest Dynamics*, 1st ed.; Lesnická Práce, Folia Forestalia Bohemica: Kostelec nad Černými lesy, Czech Republic, 2011; pp. 61–66, ISBN 978-80-87154-94-6.

61. Frerker, K.; Sonnier, G.; Waller, D.M. Browsing rates and ratios provide reliable indices of ungulate impacts on forest plant communities. *For. Ecol. Manag.* **2013**, *291*, 55–64. [CrossRef]

62. Ammer, C. Impact of ungulates on structure and dynamics of natural regeneration of mixed mountain forests in the Bavarian Alps. *For. Ecol. Manag.* **1996**, *88*, 45–53. [CrossRef]

63. Stergar, M. Objedenost Mladja Drevesnih Vrstv Odvisnosti od Zgradbe Sestoja. Ph.D. Thesis, Biotehniška fakulteta, Univerza v Ljubljani, Ljubljana, Slovenia, 2005; p. 70.

64. Hubalek, Z. Seasonal variation of forest habitat preferences by birds in a lowland riverine ecosystem. *Folia Zool.* **2001**, *50*, 281–289.

65. Schlaghamersky, J.; Hudec, K. The fauna of temperate European floodplain forests. In *Floodplain Forests of the Temperate Zone of Europe*; Klimo, E., Hager, H., Matic, S., Anic, I., Kulhavy, J., Eds.; Lesnicka Prace: Kostelec, Czech Republic, 2008; pp. 160–230, ISBN 978-80-87154-16-8.

66. Walther, B.A.; Martin, J.L. Species richness estimation of bird communities: How to control for sampling effort? *IBIS* **2001**, *143*, 413–419. [CrossRef]

67. Kolecek, J.; Paclik, M.; Weidinger, K.; Reif, J. Abundance and species richness of birds in two lowland riverine forests in Central Moravia—Possibilities for analyses of point-count data. *Sylvia* **2010**, *46*, 71–85.

68. Freemark, K.E.; Collins, B. Landscape ecology of birds breeding in temperate forest fragments. In *Ecology ans Conservation of Neotropical Migrant Landbird*; Hagan, J.M., Johnston, D.W., Eds.; Smithsomian Institution Press: Washington, DC, USA, 1992; pp. 443–454.

69. Pechanec, V.; Brus, J.; Kilianova, H.; Machar, I. Decision support tool for the evaluation of landscapes. *Ecol. Inform.* **2015**, *30*, 305–308. [CrossRef]

70. Kornan, M. Comparison of bird assemblage structure between forest ecotone and interior of an alder swamp. *Sylvia* **2009**, *45*, 151–176.

71. Truax, B.; Gagnon, D.; Lambert, F.; Fortier, J. Multiple-Use Zoning Model for Private Forest Owners in Agricultural Landscapes: A Case Study. *Forests* **2015**, *6*, 3614–3664. [CrossRef]

72. Anic, I.; Mestrovic, S.; Matic, S. Important events in the history of forestry in Croatia. *Sumar. List* **2012**, *136*, 169–177.

73. Kilianova, H.; Pechanec, V.; Brus, J.; Kirchner, K.; Machar, I. Analysis of the development of land use in the Morava River floodplain, with special emphasis on the landscape matrix. *Morav. Geogr. Rep.* **2017**, *25*. [CrossRef]

74. Leso, P. Breeding bird communities of two succession stages of young oak forests. *Sylvia* **2003**, *39*, 67–78.

75. Fulller, R.J. *Bird Life of Woodland and Forest*, 1st ed.; Cambridge University Press: Cambridge, UK, 2003; pp. 126–127.

76. Glowacinski, Z. Succession of bird communities in the Niepolomice Forest (Southern Poland). *Ekol. Polska* **1975**, *23*, 231–263.

77. Kropil, R. Struktura a Produkcia Ornitocenoz Vybranych Prirodnych Lesov Slovenska [Structure and Production of Bird Communities in Natural Forests in Slovakia]. Ph.D. Thesis, Technical University, Zvolen, Zvolen, Slovakia, 1993.

78. Waliczky, Z. Bird community changes in different-aged oak forest stands in the Buda-hills. *Ornis Hung.* **1991**, *1*, 1–9.

79. Wesolowski, T.; Rowinski, P.; Mitrus, C.; Czeszczewik, D. Breeding bird community of a primeval temperate forest (Bialowieza National Park, Poland) at the beginning of the 21st century. *Acta Ornithol.* **2006**, *41*, 55–70. [CrossRef]

80. Tucker, G.M.; Evans, M.I. *Habitats for Birds in Europe*; Birdlife International: Cambridge, UK, 1997; pp. 1–464, ISBN 978-0-94688-8320.

81. Stastny, K.; Bejcek, V.; Hudec, K. *Atlas of Breeding Bird Distribution in the Czech Republic*; Academia: Prague, Czech Republic, 2006; pp. 1–463, ISBN 80-86858-19-7.

82. Reif, J.; Jiguet, F.; Šťastný, K. Habitat specialization of birds in the Czech Republic: Comparison of objective measures with expert opinion. *Bird Study* **2010**, *57*, 197–212. [CrossRef]

83. Reif, J.; Storch, D.; Šímová, I. The effect of scale-dependent habitat gradients on the structure of bird assemblages in the Czech Republic. *Acta Ornithol.* **2008**, *43*, 197–206. [CrossRef]

84. Machar, I. Attempt to summarize the problems: Is a sustainable management of floodplain forest geobiocenoses possible? In *Biodiversity and Target Management of Floodplain Forests in the Morava River Basin (Czech Republic)*; Accession Number: WOS:000328003200016; Machar, I., Ed.; Palacky University: Olomouc, Czech Republic, 2010; pp. 189–226, ISBN 978-80-244-2530-6.

85. Kovarik, P.; Kutal, M.; Machar, I. Sheep and wolves: Is the occurrence of large predators a limiting factor for sheep grazing in the Czech Carpathians? *J. Nat. Conserv.* **2014**, *22*, 5. [CrossRef]

86. Machar, I.; Harmacek, J.; Vrublova, K.; Filippovova, J.; Brus, J. Biocontrol of common vole populations by avian predators versus rodenticide application. *Pol. J. Ecol.* **2017**, *65*, 434–444. [CrossRef]

87. Janik, D.; Adam, D.; Vrska, T.; Hort, L.; Unar, P.; Kral, K.; Samonil, P.; Horal, D. Tree layer dynamics of the Cahnov-Soutok near-natural floodplain forest after 33 years (1973–2006). *Eur. J. For. Res.* **2008**, *127*, 337–345. [CrossRef]

88. Machar, I. Protection of nature and landscapes in the Czech Republic Selected current issues and possibilities of their solution. In *Ochrana Prirody a Krajiny v Ceske Republice, Vols I and II*; Accession Number: WOS:000334387900001; Machar, I., Drobilova, L., Eds.; Palacky University: Olomouc, Czech Republic, 2012; p. 9-+, ISBN 978-80-244-3041-6.

89. Machar, I.; Vlckova, V.; Bucek, A.; Vozenilek, V.; Salek, L.; Jerabkova, L. Modelling of Climate Conditions in Forest Vegetation Zones as a Support Tool for Forest Management Strategy in European Beech Dominated Forests. *Forests* **2017**, *8*, 82. [CrossRef]

90. Machar, I.; Simon, J.; Rejsek, K.; Pechanec, V.; Brus, J.; Kilianova, H. Assessment of Forest Management in Protected Areas Based on Multidisciplinary Research. *Forests* **2016**, *7*, 285. [CrossRef]

91. Spathelf, P. Sustainable Forest Management as a Model for Sustainbale Development: Conclusions Toward a Concrete Vision. *Sustain. For. Manag. Chang. World Manag. For. Ecosyst.* **2009**, *19*, 237–240. [CrossRef]

forests

MDPI

Article

Ecological Factors Affecting White Pine, Red Oak, Bitternut Hickory and Black Walnut Underplanting Success in a Northern Temperate Post-Agricultural Forest

Benoit Truax [1],*, Daniel Gagnon [1,2], Julien Fortier [1], France Lambert [1] and Marc-Antoine Pétrin [1]

[1]   Fiducie de Recherche sur la Forêt des Cantons-de-l'Est/Eastern Townships Forest Research Trust, 1 rue Principale, Saint-Benoît-du-Lac, QC J0B 2M0, Canada; daniel.gagnon@uregina.ca (D.G.); fortier.ju@gmail.com (J.F.); france.lambert@frfce.qc.ca (F.L.); marcantoinepetrin@gmail.com (M.-A.P.)

[2]   Department of Biology, University of Regina, 3737 Wascana Parkway, Regina, SK S4S 0A2, Canada

*   Correspondence: btruax@frfce.qc.ca; Tel.: +1-819-821-8377

Received: 9 July 2018; Accepted: 10 August 2018; Published: 16 August 2018

Abstract: This study took place in southern Québec (Canada) where young stands of white ash and grey birch have been underplanted with white pine, red oak, bitternut hickory and black walnut. The establishment success of white pine and red oak was measured with and without tree shelters (to protect from deer). Ecological factors affecting the height growth of the four species were also measured for protected trees. After 6 years, the survival and total height of unprotected oak was 29% and 44.3 cm vs. 80.5% and 138.5 cm for protected oak. White pine was less affected by browsing (survival of 79.5 and 93.5%; height of 138.5 and 217.9 cm for unprotected vs. protected pine). Height of white pine was higher in the grey birch stands, while height of all hardwoods was higher in the white ash stands, which had better soil drainage, higher fertility, and an understory dominated by *Rubus* species. Total height of all hardwoods was significantly ($p < 0.05$) correlated with *Rubus* cover and with soil fertility. Pine and walnut height were strongly correlated ($p < 0.001$) to shelterwood structure (canopy openness or total basal area). Pine was less sensitive to variations in shelterwood characteristics, while black walnut showed high sensitivity. This study provides evidence that underplanting is suitable for black walnut assisted migration northward and for bitternut hickory restoration, despite soil conditions that were less favorable than in bottomland habitats mainly supporting these species in eastern Canada. Tree shelters offering protection from deer browsing and species-specific site selection are recommended for underplanting in the southern Québec region.

Keywords: tree shelter; deer browsing; hardwood restoration; assisted migration; enrichment planting; shelterwood; *Pinus strobus* L.; *Quercus rubra* L.; *Carya cordiformis* (Wangenh.) K. Koch; *Juglans nigra* L.

---

1. Introduction

In eastern North America, nut producing hardwoods (*Quercus*, *Juglans* and *Carya* spp.) and eastern white pine (*Pinus strobus* L.) are major components of temperate hardwood forest ecosystems for biodiversity, but also for the production of high-value timber [1–3]. In the southern Québec region (southeastern Canada), multiple ecological and human factors have contributed to the decline of these important species. Historically, white pine was among the first species to be overexploited following settlement, and therefore this species is now much less abundant than it used to be [4,5]. Early settlers also clearcut several butternut (*Juglans cinerea* L.) stands because this species was associated with high quality soils for agricultural use [4]. Besides, butternut is now threatened by the butternut canker

(*Sirococcus clavigignenti-juglandacearum* Nair et al.), a virulent and deadly Asian fungal pathogen, which affects the species in all of its habitats [6].

The human control of forest fires is another factor that could have contributed to a reduction in the abundance of species that typically regenerate after fires including hickories, oaks and white pine [7–9]. Furthermore, in many regions of northeastern North America, the natural regeneration of several hardwood species and white pine is threatened by the overabundance of white-tailed deer (*Odocoileus virginianus* Zimm.), and changes in forest composition over the long-term are documented in areas supporting large deer populations [3,10–13]. Climate change is also expected to increase summer temperatures and lower soil water content in the southern Québec region, which could be detrimental to drought sensitive species, but potentially beneficial to drought tolerant species [14], including pines, hickories and oaks [15]. However, the migration of forest species into more suitable habitats is expected to occur at a slower rate than the rate of modification of regional climates [14], and the migration of bottomland species, such as hickories (*Carya cordiformis* (Wangenh.) K. Koch and *Carya ovata* (Mill.) K. Koch), may also be constrained by soil fertility factors in southern Québec [16].

Given the multiple past, present and future factors that will affect the regeneration and distribution of nut producing hardwoods and white pine, there is an urgent need to test restoration and/or migration strategies in the particular context of southern Québec. Field plantations have been often proposed for the restoration of white pine and nut producing hardwoods [17–19]. However, field plantations are often costly as they need site preparation (soil cultivation), intensive vegetation management (herbicide or mulch treatments) and tree pruning to achieve wood production objectives. In rural areas, the social acceptability of such plantations is often low because tree planting on cultivated land and old-fields competes with agricultural land use [20]. Furthermore, many stressors prevailing in open-field environments (high light, wind exposure, lack of mycorrhizal partners, herbaceous competitors, vole predation, etc.) can be detrimental to the planted species [18,21–23]. Although white pine has the ecophysiological capacity to become established in grasslands and old-fields [24], such environments increase its susceptibility to the pine weevil (*Pissodes strobi* Peck) and to the blister rust (*Cronartium ribicola* J.C. Fisch. ex Rabenh.) [25].

Underplanting (*i.e.*, enrichment or gap planting) in forest stands and tree plantations is a promising alternative to field plantations of oaks and white pine [3,25–32]. Such sylvicultural systems are often characterized by reduced abundance of grass species (Gramineae), which are strong competitors for nutrients and water [33]. Shelterwood environments can also contribute to hide seedlings from large herbivores such as deer [27], while being characterized by reduced populations of meadow voles, which are key consumers of tree seedlings in old-field habitats [22]. However, when deer populations are high in an area, tree shelters are generally required for successful underplanting [3,34]. Shelterwood environments can also increase the wood quality of more shade-tolerant species having high crown plasticity, by increasing height growth of the stem at the expense of lateral branch growth [35].

Past studies have identified several factors responsible for the success or failure of underplanting. For red oak (*Quercus rubra* L.) and white pine, two intermediate shade-tolerant species at the seedling stage [36,37], sufficient light availability in the understory is a critical factor to achieve optimal seedling development [25,30,31,34,38]. For oaks, the presence of shrubs (*i.e.*, *Rubus* spp.) in the understory is believed to have an indirect facilitation effect on seedling growth by protecting trees from large herbivores and by eliminating other competing plants [34,39,40]. In terms of site selection, young early-successional stands of *Populus tremuloides* Michx. located on mesic fertile soils were found to be optimal for red oak [41]. Also, red oak tends to become established well in all topographic locations (*i.e.*, ridge, middle slope and valley) when underplanted [42]. Grey birch stands have equally been used for red oak underplanting, but such shelterwoods often have imperfect drainage conditions and poorly drained microsites [40,43], which are inadequate for red oak [44]. However, grey birch is often an associated forest cover species of white pine, which grows well on imperfectly drained sites [37]. In the understory, competition from shrubs and hardwoods is also an issue with white pine, given its slow growth rate at the seedling stage [37]. Initial competition from aspen suckers and later competition from shrubs and red maple (*Acer rubrum* L.) led to very high pine mortality in young thinned and unthinned aspen stands [45]. Conversely, mesic sites having a balsam fir mid-story prior to the shelterwood treatment were found to be adequate for white pine because they have reduced hardwood competition in the understory [46].

For species of the Juglandaceae very few studies have evaluated their potential in underplanting systems [28,47]. Moreover, limited information exists about the regeneration ecology of hickory species, especially at the northern limit of their range [9]. In southern Québec, only two hickory species are found (shagbark hickory, *C. ovata*, and bitternut hickory, *C. cordiformis*), but mostly in the bottomlands and on the moraine ridges of the St. Lawrence Valley where soil fertility is high [43,48,49]. Surprisingly, in both southern Québec and southern Ontario (Canada), poor growth and survival have been observed with bitternut and shagbark hickory in old-field plantations [17,50]. Yet, theory suggests that shelterwood cuts can be used to create advance hickory regeneration, but experimental evidence is lacking [51]. Observations made before 1935 further suggest that bitternut hickory had a wider distribution as it was frequently found in hilly landscapes of southern Québec where soil fertility is lower than in the St-Lawrence Valley [52]. Bitternut hickory individuals have been observed in different areas of the Precambrian Shield foothills (Outaouais and Laurentides regions in Quebec), where seepage increases soil moisture and nutrient availability on lower slopes [53,54]. This suggests that bitternut hickory could be suited for underplanting in upland habitats of southern Québec, providing soil richness and soil moisture are adequate.

Butternut is the sole species from the *Juglans* genus native to Québec, and it has been designated an endangered species following high mortality caused by the butternut canker [55]. Black walnut (*Juglans nigra* L.), which is native to nearby southern Ontario (Canada), has been suggested as a replacement species for butternut pending the development of resistant butternut hybrids or the identification of resistant individuals [41]. Black walnut is known for its high sensitivity to soil conditions, as it generally grows on deep, well-drained, nearly neutral pH, fertile and mesic soils [17,56,57]. Besides, black walnut generally requires very intensive and long-term weed control in old-field environments, otherwise growth stagnation may occur due to nitrogen limitation [58,59]. Compared to most hardwoods, black walnut flushes later in the spring and drops it leaves earlier in the fall, which allow herbaceous competitors to thrive for many years in the plantation understory [58,59]. This is a potential indication that black walnut may be more suitable in gap plantations where herbaceous competition is reduced. Yet, black walnut is relatively shade-intolerant [56], so competition for light by overstory trees may be an important growth-limiting factor in underplanting systems.

This study took place in southern Québec on a privately owned property where young stands of white ash and grey birch, originating from agricultural abandonment, have been underplanted with white pine, red oak, bitternut hickory and black walnut. In 1991, red oak had been successfully underplanted without protection from deer in such shelterwoods [27,41]. However, two decades ago, deer was less abundant than it had become when the present study was initiated in 2012 [60]. The first objective of this study was to evaluate if tree shelters are needed for the successful underplanting of white pine and red oak, two regionally important species. The second objective was to evaluate ecological factors, other than deer, affecting the height growth of underplanted white pine, red oak, bitternut hickory and black walnut after 6 years. Since red oak and white pine seedlings are heavily browsed in habitats supporting high deer populations [12,61], we hypothesized that both species will be responsive to the tree shelter treatment in terms of height growth and survival. We also hypothesized that black walnut will be the most responsive species to variations in shelterwood characteristics.

2. Materials and Methods

2.1. Study Site Description

The study site is located on the land of a Benedictine monastery at St-Benoît-du-Lac, in the Estrie administrative region of southern Québec, Canada (45°10′ N; 72°16′ W), a few km north of the Vermont (United States) border in the Appalachian geographic region (Figure A1). This 216 ha privately owned property has a 150 ha forested area composed of a complex mosaic of young and older successional stands, and some old growth stands, with most young stands originating from agricultural abandonment [41]. The study site is located on the western shore of Lake Memphremagog, a large lake (95.3 km^2) [62] within a wide north-south valley flanked by hills. Thick till generally underlays glacio-lacustrine deposits on these lakeside hills [63]. In the study area, forest ecosystems are dominated by hardwoods on mesic sites and by conifers on xeric and hydric sites [63]. The study area belongs to the sugar maple-basswood ecoregion of Québec [63], and more generally to the northern hardwoods forest ecosystem [64,65]. A continental subhumid moderate climate [63], with mean annual precipitation of 1260 mm and mean annual temperature of 5.3 °C, characterizes the study site [66].

In 2010, a vegetation analysis of the forested area was undertaken to identify the forest communities. Digital topographical and ecoforest maps, and aerial photos (orthophotos) were used. Using these sources of information and ArcGIS (Esri, Redlands, CA, United States), a set of parallel transects were used to determine the location of 71 permanent plots (20 m × 20 m), where vegetation, soil and site characteristics where measured. A Detrended Correspondence Analysis of the 71 forest vegetation plots was done and two community types presenting a high potential for white pine and hardwood underplanting where identified: (1) the White ash community type, which was located on mesic sites dominated by young forests regenerated on old-fields (average largest tree age = 47) and (2) the very young Grey birch-balsam poplar-elm community type that also has regenerated on old fields (average largest tree age = 40) (Figure 1). Mean total basal area (trees + saplings) was 30 m^2/ha for the White ash community type and 18 m^2/ha for the Grey birch-balsam poplar-elm community type. Additional details related to site description can be found in Truax et al. [41].

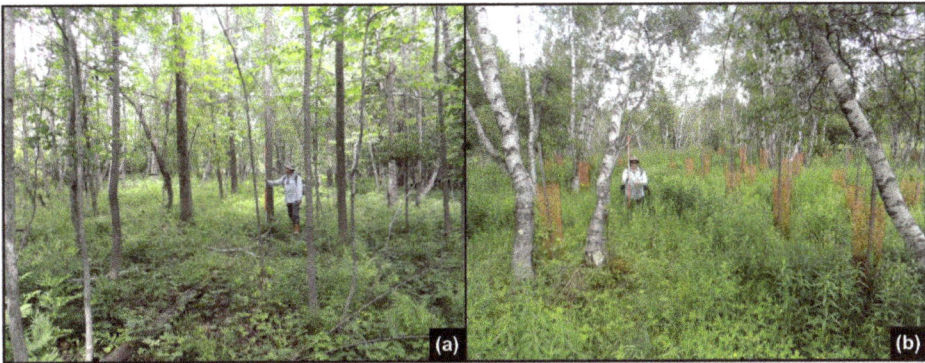

Figure 1. The two distinct forest community types used for underplanting: (**a**) the White ash community type and (**b**) the Grey birch-balsam poplar-elm community type.

2.2. Experimental Design

A complete randomized block design with 25 blocks and two factors (Tree species and Deer protection treatment) was established within the two selected forest community types. Among the 25 blocks, 8 were located in the White ash community type and 17 were located in the Grey birch-balsam poplar-elm community type. Each block measured 9 × 12 m and contained two species (red oak and white pine) and two deer protection treatments (a tree shelter treatment and a control treatment with no protection) for a total of 100 experimental plots (2 Tree species × 2 Deer protection treatments × 25 blocks). Each plot measured 4.5 × 6 m and contained 8 trees of a single species/treatment combination. In the middle of each plot, one bitternut hickory or one black walnut seedling was also planted, and these additional species were always protected with a tree shelter. Tree spacing was 1.5 × 2 m between all trees. Overall, the initial experimental design contained 400 red oak seedlings and 400 white pine seedlings, half of which were protected by tree shelters, but also 50 bitternut hickory seedlings and 50 black walnut seedlings, all protected by shelters. Figure 2 gives an overview of the experimental design.

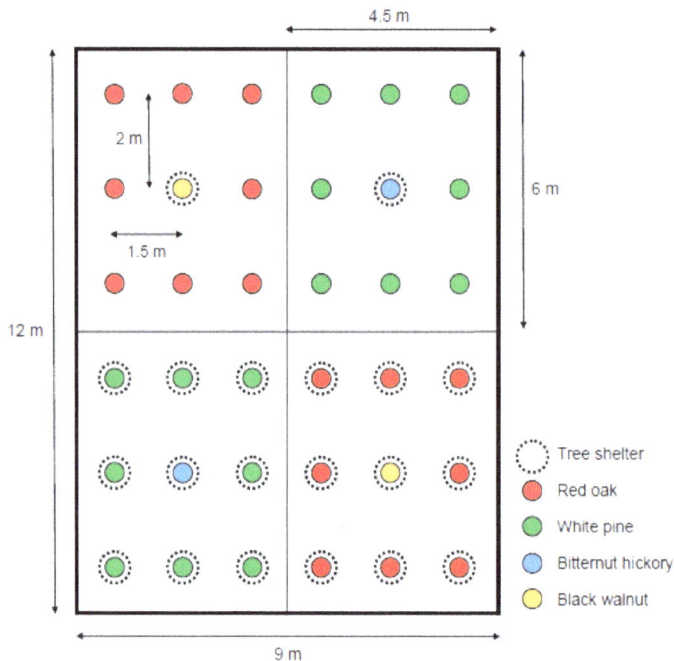

Figure 2. Schematic representation of a block with 4 experimental plots, one for each tree species (red oak or white pine)/deer protection treatments (tree shelter or control) combination. The position of each species/treatment combination was randomly assigned within each block. In the middle of each experimental plot, one bitternut hickory or one black walnut was planted and was protected with a tree shelter. The position of the two bitternut hickories and the two black walnuts within a block was always crisscrossed. The position of the first black walnut or bitternut hickory position was randomly assigned in the top left plot.

During the year preceding tree planting, a shelterwood cut was done in all blocks to increase light availability in the understory and remove overtopping trees located in the middle of the blocks. Large woody debris were also removed to facilitate the establishment of the experimental design. In early May 2012, all trees were planted manually and no vegetation treatment was done. Two-year-old bare root seedlings were used for red oak (2-0) and two-year-old container seedlings were used for white pine (2-0). One-year-old bare-root seedlings were used for bitternut hickory and black walnut (1-0). Height and basal diameter of seedlings at planting were respectively: 21 cm and 5.5 mm for white pine; 70 cm and 11.2 mm for red oak, 38 cm and 8.3 mm for bitternut hickory, and 48 cm and 9.8 mm for black walnut. Seedlings were provided by the Berthier nursery (Sainte-Geneviève-de-Berthier, QC) of the Ministère des Forêts, de la Faune et des Parcs (MFFP) of Québec.

During the week following tree planting, tree shelters were installed. Homemade tree shelters were conceived based on the model developed by Peter Kilburn (Model-K shelter), a private landowner who has successfully planted more than 5000 hardwoods on his property in southern Québec. The Model-K shelter was built using Vexar® construction plastic fence (MasterNet Ltd., Mississauga, ON, Canada), which has a mesh size of 5 × 5 cm. Fence sections approximately 1 m wide were cut, rolled and attached (using tie-wraps) to form a cylinder, which was then slid through a hardwood stake (2 m long above ground). Dimensions of the shelter were 120 cm in height by approximately 25–30 cm in diameter. The model-K shelter is ideal for underplanting because the large mesh size of the fencing material casts very little shade on the seedling compared to solid or small mesh-size commercial tree shelters developed for afforestation. As shown by Bardon et al. [67] the reduction in light availability caused by some commercial tree shelters can reduce the growth and survival of underplanted red oaks. The Model-K shelter can also be slid up on the wooden stake as tree height increases, which eliminates deer browsing of the main stem.

2.3. Measurements of Survival, Tree Growth and Deer Browsing

At the end of each growing season (except year 5), survival of each tree was recorded. In September of year 1 and 6 deer browsing on the main stem was evaluated for each living tree. Deer browsing data are expressed in percent data (number of browsed trees/number of trees alive × 100). In September of year 6, total tree height and basal diameter were measured on each living tree and DBH (1.3 m from ground-level) was measured when possible. For trees with no DBH, the simple cone volume formula was used to calculate stem volume [68]:

$$V = \pi \, D_B{}^2 H/12 \qquad\qquad (1)$$

where V is the stem volume (cm^3), D_B is the basal diameter (cm) and H is the tree height (cm). For trees with a DBH value, the stem volume was measured by summing the volume of two stem sections (1) from basal diameter to DBH and (2) from DBH to tree tip. For stem Section 2, Equation (1) was used, but D_B was replaced by a DBH value and H was replaced by the height of the stem section from DBH to the tree tip. For stem Section 1, the following formula was used [69]:

$$V = \pi/12(D_1{}^2 + D_2{}^2 + D_1 D_2) \, L \qquad\qquad (2)$$

where, V is the volume (cm^3) of a stem section, D_1 is the base diameter (cm) of the stem section, D_2 is the diameter (cm) at the top of the stem section, and L is the length (or height) of the stem section. Thus, the volume of stem Section 1 was measured by replacing D_1 by a basal diameter value, D_2 by a DBH value and L by 130 cm in Equation (2).

2.4. Measurements of Shelterwood Characteristics

Over the years it became obvious that the main factor affecting white pine and red oak growth and survival in the control treatment (no shelter) was deer browsing. Consequently, some ecological variables were only measured in the tree shelter treatment in order to evaluate which factors, other than deer, affect height growth. Descriptive statistics of ecological variables measured across the 25 blocks are presented in Table A1.

2.4.1. Overstory Structure and Composition

During summer 2012 (first growing season), residual basal area of all trees and saplings located within block boundaries was calculated using diameter at breast height (DBH, 1.3 m from ground-level) measurements. Since saplings (i.e., tree stems with DBH ranging 1.0–9.9 cm) were a minor component of the shelterwoods, their basal area was combined with the basal area of trees to form a single basal area index (i.e., total basal area) at the block level. Basal area was also calculated for dominant tree species (*Betula populifolia* Marsh., *Ulmus americana* L. and *Fraxinus americana* L.). In the center of each plot, hemispherical photographs were taken at the end of spring 2012. The camera was placed 90 cm above the ground level with its back always facing north. The same procedure was repeated once in the middle of the block position. Canopy openness data were obtained from hemispherical photographs using the software Gap Light Analyzer V 2.0 (Simon Fraser University, Vancouver, BC, Canada).

2.4.2. Relative Cover of Understory Vegetation

During August of the 6th growing season (2017), the relative vegetation cover in the understory was determined visually at the plot-level for the most abundant species, genera or functional group (i.e., *Rubus* spp., Gramineae spp., *Solidago* spp., *Carex* spp., *Fragaria virginiana* Duch., *Phalaris arundinacea* L. and *Onoclea sensibilis* L.). This sampling procedure was done only in plots of the tree shelter treatment.

2.4.3. Soil Characteristics

During summer 2012 (first growing season), a composite soil sample was collected in each plot (0–20 cm of depth). Soil samples were air dried and sieved (2 mm). Soil pH, clay, silt and sand content, percent organic matter, cation exchange capacity (CEC) and base saturation were determined by the Agridirect Inc. soil analysis lab in Longueuil (Québec). Methods used are those recommended by the Conseil des productions végétales du Québec [70]. The determination of soil pH was made using a 1:1 ratio of distilled water to soil. For particle size analyses, the Bouyoucos [71] method was used. Percent organic matter was determined by weight loss after ignition at 550 °C for 4 h. Cation exchange capacity and base saturation were calculated following the recommendations of the Centre de référence en agriculture et agroalimentaire du Québec [72], after Ca, K and Mg extraction with the Mehlich III method [73] and concentration determination using ICP emission spectroscopy [74]. Total soil C and N concentrations were determined by the combustion method at high temperature (960 °C) followed by thermal conductivity detection. These analyses were done by the CEF lab (Dr. R. Bradley and Dr. W. Parsons) at the University of Sherbrooke.

Soil macronutrient supply rates were determined using Plant Root Simulator (PRS^{TM}-Probes) technology from Western Ag Innovations Inc. (Saskatoon, SK, Canada). The PRS-probes consist of an ion exchange membrane encapsulated in a thin plastic probe, which is inserted into the ground with little disturbance of soil structure. Nutrient supplies observed with this method are strongly correlated with nutrients concentrations or stocks obtained with conventional soil extraction methods over a wide range of soil types [75]. On 22 June 2017 (6th growing season), four pairs of probes (an anion and a cation probe in each pair) were buried in the A horizon of each plot for a 41-day period. After probes were removed from the soil (2 August 2017), they were washed with distilled water, and returned to Western Ag Labs for analysis (NO_3, NH_4, P, K, Ca, Mg, S). Composite samples were made in each plot by combining the four pairs of probes. This sampling procedure was only done in red oak and white pine plots of the tree shelter treatment.

2.5. Statistical Analyses

2.5.1. Red Oak and White Pine Data

Main effects (Tree species and Deer protection treatment) and interaction effects (Tree species × Deer protection treatment) on measured variables were analyzed using two-way ANOVA in a fixed factorial design [76]. For survival data, the ANOVA was done with data from the 100 experimental plots (2 species × 2 treatment × 25 blocks = 100 experimental plots). However, for tree growth and main stem browsing data collected after 6 years, the ANOVA was done with data from only 21 blocks (84 experimental plots) given that no living red oak was found in the control treatment of 4 blocks. Following ANOVA, the normality of residuals was verified using the Shapiro-Wilk W-test. Survival data for year 4 were logit transformed to satisfy the ANOVA assumption of normality in residuals distribution [77]. However, all survival data are reported in percent values. Main effects or interaction effects were declared statistically significant for three levels of significance ($p < 0.05$, $p < 0.01$ and $p < 0.001$).

In this study, 8 blocks were located in the White ash community type and 17 blocks were located in the Grey birch-balsam poplar-elm community type (referred to as the Grey birch community in Tables). Thus, we evaluated if survival and height growth of white pine and red oak, in both treatments (deer protection and control), significantly differed between community types using Student's *t*-test. An individual *t*-test was done for each tree species/deer protection treatment combinations. Using the mean value at the block level for the different ecological variables (basal area, canopy openness, understory plant cover, soil characteristics), a *t*-test was also used to determine if ecological variables significantly varied between the two forest community types. All *t*-tests were run at an alpha level of 0.05.

To evaluate which ecological factors (measured as continuous variables) were significantly correlated with red oak or white pine total height after 6 years in the tree shelter treatment, a correlation matrix, with Pearson correlation coefficients (*r*), was used (Table A2). Linear and non-linear regressions between ecological factors and red oak or white pine height growth were then developed. Bivariate regression models were selected based on the normality of residuals distribution, which was evaluated using the Shapiro-Wilk W-test. Plot-level data were used for soil variables, understory plant cover variables and canopy openness, while block-level data where used for total basal area or species-specific basal area.

2.5.2. Bitternut Hickory and Black Walnut Data

The effect of tree species on growth and survival was analyzed using a one-way ANOVA in a fixed factorial design [76]. In each block, the two trees of a single species (bitternut hickory or black walnut) were considered as a plot in the ANOVA (see Figure 2). For survival data, the ANOVA was done with data from the 50 experimental plots (2 species × 25 blocks = 50 experimental plots). However, for tree growth data collected after 6 years, the ANOVA was done using data from only 23 blocks (46 experimental plots) given that no living black walnut was found in 2 blocks. We also evaluated if survival and height growth of bitternut hickory and black walnut significantly varied between forest community types using the Student *t*-test procedure for means separation. Using mean value at the block level for the different ecological variables, a correlation matrix, with Pearson correlation coefficients (*r*), was used to identify significant factors affecting height growth (Table A2). Linear and non-linear regressions between ecological factors and bitternut hickory or black walnut total height after 6 years were then developed. Models were selected based on the normality of residuals distribution, which was evaluated using the Shapiro-Wilk W-test. All statistical analyses were done using JMP 11 from SAS Institute (Cary, NC, United States).

3. Results

3.1. Effect of Tree Species and Deer Protection Treatments on Survival and Growth

For survival data of red oak and white pine, the two-way ANOVA showed significant ($p < 0.001$) Tree species × Deer protection treatment interaction effects for all years except year 1, where tree survival ranged 99%–100% across all species/treatment combinations (Figure 3a). After 6 years, the highest survival rate was observed for white pine in the shelter treatment (93.5%), while the lowest survival rate was observed for red oak in the control treatment (29%). Similar survival rates were observed for unprotected white pine (79.5%) and sheltered red oak (80.5%). Results from Figure 3a also show that tree mortality occurred gradually over the years for unprotected white pine and red oak. A significant Species effect on deer browsing of the main stem was observed for unprotected trees after 1 year, with 75% of red oak trees being browsed vs. only 0.5% for white pine (Figure 3b). However, during the 6th growing season, deer browsing was recorded on about half of living trees for both species.

For growth data of red oak and white pine after 6 years, the two-way ANOVA showed significant Tree species and Deer protection treatment effects on total height, basal diameter and stem volume, but non-significant interaction effects (Figure 4). White pine growth was significantly higher than red oak growth across treatments, while the growth of sheltered trees was significantly higher than the growth of unprotected trees across species. After 6 years, the mean height of sheltered white pine and red oak was 217.9 cm and 138.5 cm, respectively, while the mean height of unprotected white pine and red oak was 146.3 cm and 44.3 cm, respectively. Stem volume of white pine was 55% higher in the shelter treatment compared to the control, while stem volume of red oak was 610% higher in the shelter treatment compared to the control.

For growth and survival data related to bitternut hickory and black walnut growing in tree shelters, the one-way ANOVA showed significant Tree species effects on survival rate, basal diameter and stem volume after 6 growing seasons (Table 1). While survival of black walnut was inferior to that of bitternut hickory (64% vs. 90%, respectively), stem volume of walnut was 5.5 times higher than stem volume of hickory. However, height growth of both species was similar after 6 years.

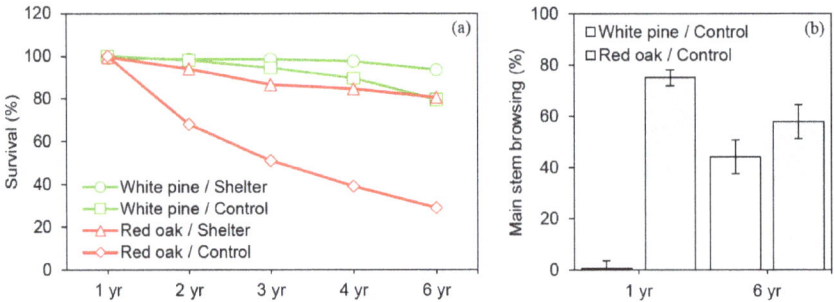

Figure 3. (a) Tree species × Deer protection treatment interaction effect on survival rate of underplanted red oak and white pine measured after 1, 2, 3, 4 and 6 years of growth in young forest stands. The interaction effect is significant ($p < 0.001$) for all years, except for year 1 ($p = 0.51$) and the standard error of the mean (SE) is 3.5% for year 6. (b) Tree species effect on deer browsing of the main stem of underplanted trees in the control treatment. The species effect on stem browsing is only significant for year 1 ($p < 0.001$).

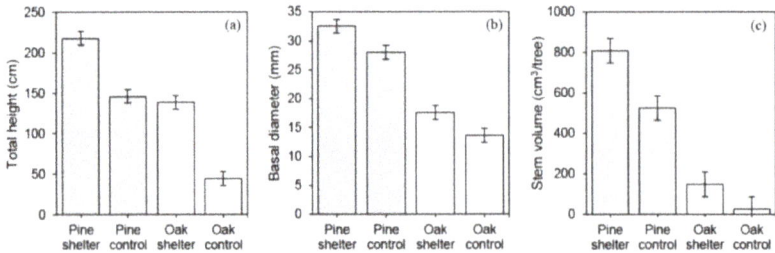

Figure 4. (a) Total height (b), basal diameter, and (c) stem volume of underplanted red oak and white pine in the tree shelter and control treatments (no deer protection) after 6 growing seasons in young forest stands. The interaction effect is not significant for total height ($p = 0.19$), basal diameter ($p = 0.81$) and stem volume ($p = 0.19$). The Species effect is significant at $p < 0.001$ for all variables. The Deer protection treatment effect is significant at $p < 0.001$ for total height and at $p < 0.01$ for basal diameter and stem volume. Pine = white pine and Oak = red oak.

Table 1. Tree species effect on survival, height, basal diameter, stem volume of underplanted bitternut hickory and black walnut after 6 growing seasons in young forest stands (all trees protected with shelters). SE = standard error of the mean.

Species	Survival (%)	Total Height (cm)	Basal Diameter (mm)	Stem Volume (cm³/tree)
Bitternut hickory	90	162	16.4	158
Black walnut	64	171	27.6	862
SE	3.6	17	2.1	316
p-value	<0.001	0.74	<0.001	<0.05

3.2. Effect of the Forest Community Type on Ecological Variables, and on Tree Survival and Growth

Results from Table 2 show that the mean value of several ecological variables was statistically different between the White ash and Grey birch-balsam poplar-elm community types. Compared to the Grey birch-balsam poplar-elm community type, the White ash community type was characterized by a lower total basal area and canopy openness, but higher soil fertility in terms of soil NO_3 supply rate and CEC. The understory of the White ash community type was also characterized by a lower cover of Gramineae and *Solidago* species, but a higher cover of *Rubus* and *Carex* species.

After 6 years, the total height of the four tree species protected with tree shelters was statistically different between the two forest community types (Table 3). The three hardwood species were taller in the White ash community type, while white pine was taller in the Grey birch-balsam poplar-elm community type. In terms of magnitude, the total height of white pine was the least affected by the forest community type, while the total height of black walnut was the most affected (Table 3). After 6 years, the survival of sheltered white pine was also higher in the Grey birch-balsam poplar-elm community type (97.1% ± 2.4%) vs. the White ash community type (85.9% ± 3.4%), while the survival rate of sheltered hardwood species was not statistically different between forest community types.

For unprotected white pine, total height, survival and main stem browsing were also statistically different between the two forest community types after 6 years (Table 4). Unprotected white pines were taller, less browsed and had higher survival rate in the Grey birch-balsam poplar-elm community type, compared to the White ash community type. For unprotected red oak, total height, survival and stem browsing were not statistically different between the forest community types.

Table 2. Mean value (± standard error) of selected environmental variables in the two forest community types (N = 8 blocks in the White ash community type and N – 17 blocks in the Grey birch-balsam poplar-elm community type). All means are statistically different between community types at the $\alpha = 0.05$ level following Student's *t*-test. d = days.

Community Type	Stand Structure		Soil		Understory Plants			
	Total Basal Area (m²/ha)	Canopy Openness (%)	CEC (meq/100 g)	NO_3 Supply (μg/10 cm²/41d)	Gramineae spp. (%)	*Rubus* spp. (%)	*Solidago* spp. (%)	*Carex* spp. (%)
White ash	10.8 ± 1.7	25.9 ± 1.8	18.5 ± 0.5	89.8 ± 13.6	5 ± 6	39 ± 4	25 ± 7	11 ± 2
Grey birch	16.3 ± 1.1	35.9 ± 1.2	16.4 ± 0.3	14.9 ± 9.3	26 ± 4	3 ± 3	45 ± 5	2 ± 1

Table 3. Mean total height growth (± standard error) after 6 years for underplanted tree species protected with tree shelters in the two forest community types (N = 8 plots in the White ash community type and N = 17 plots in the Grey birch-balsam poplar-elm community type for red oak, white pine and bitternut hickory. For black walnut, N = 8 plots in the White ash community type and N = 15 plots in the Grey birch-balsam poplar-elm community type. All means are statistically different between community types at the α = 0.05 level following Student's *t*-test.

Community Type	Total Height of Sheltered Trees (cm)			
	Red Oak	White Pine	Bitternut Hickory	Black Walnut
White ash	200.5 ± 23.1	192.8 ± 8.7	214.0 ± 20.0	287.9 ± 34.8
Grey birch	131.1 ± 15.9	230.4 ± 6.0	126.3 ± 13.7	108.0 ± 25.4

Table 4. Mean total height growth, survival rate and main stem browsing (± standard error) after 6 years for unprotected white pine in the two forest community types (N = 8 plots in the white ash community type and N = 17 plots in the Grey birch-balsam poplar-elm community type). All means are statistically different between community types at the α = 0.05 level following Student's *t*-test.

Community Type	Unprotected White Pine		
	Total Height (cm)	Survival (%)	Stem Browsing (%)
White ash	87.5 ± 14.1	59.4 ± 7.0	81.5 ± 6.7
Grey birch	163.5 ± 9.6	89.0 ± 4.8	34.5 ± 4.6

3.3. Relationships between Shelterwood Characteristics and Total Height after 6 Years for Red Oak, White Pine, Bitternut Hickory and Black Walnut Protected with Tree Shelters

For white pine, significant positive relationships were observed between canopy openness or the basal area of grey birch and total height, while soil CEC was found to be a significant negative predictor of total height (Figure 5). For red oak, significant positive relationships were observed between the cover of *Rubus* species or soil CEC and total height (Figure 6). For bitternut hickory, significant positive relationships were observed between the cover of *Rubus* species or soil NO_3 supply rate and total height, while a significant negative relationship was observed between the basal area of grey birch and total height (Figure 7). For black walnut, six ecological variables were found to be significantly correlated with total height (Figure 8). Soil NO_3 supply rate, soil CEC and the cover of *Rubus* species in the understory were positive predictors of walnut total height, while total tree basal area, the basal area of grey birch and the cover of Gramineae species in the understory were negative predictors of walnut total height (Figure 8).

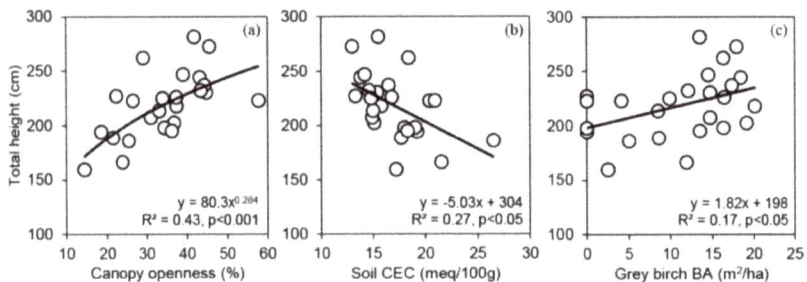

Figure 5. White pine (*Pinus strobus*) total height after 6 years in the tree shelter treatment as a function of (**a**) canopy openness, (**b**) soil cation exchange capacity (CEC) and (**c**) basal area (BA) of grey birch (*Betula populifolia*) in the overstory. N = 25 plots for each relationship.

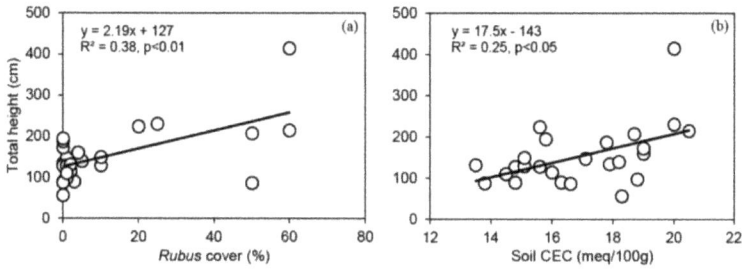

Figure 6. Red oak (*Quercus rubra*) total height after 6 years in the tree shelter treatment as a function of (**a**) cover of *Rubus* species in the understory and (**b**) soil cation exchange capacity (CEC). N = 25 plots for each relationship.

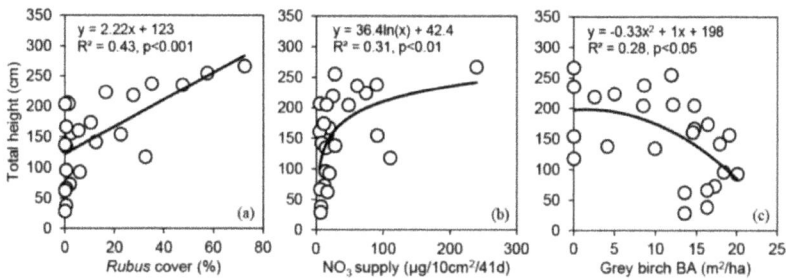

Figure 7. Bitternut hickory (*Carya cordiformis*) total height after 6 years as a function of (**a**) cover of *Rubus* species in the understory, (**b**) soil NO$_3$ supply rate and (**c**) the basal area (BA) of grey birch (*Betula populifolia*) in the overstory. N = 25 plots for each relationship.

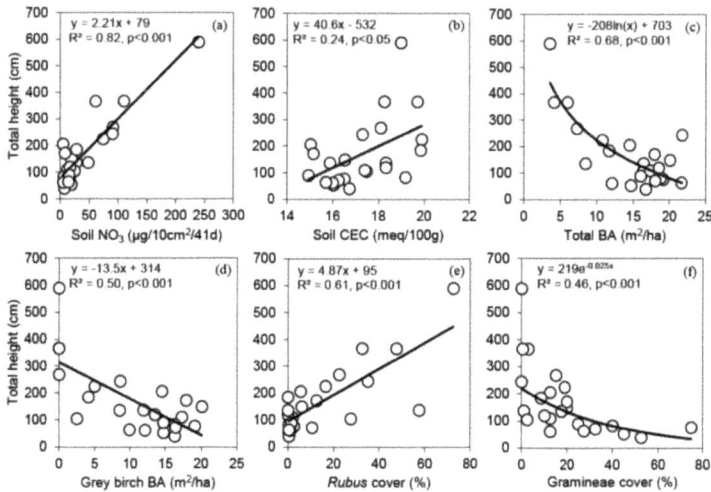

Figure 8. Black walnut (*Juglans nigra*) total height after 6 years as a function of (**a**) soil NO$_3$ supply rate, (**b**) soil cation exchange capacity (CEC), (**c**) total basal area (BA) of trees in the shelterwood, (**d**) basal area (BA) of grey birch (*Betula populifolia*) in the overstory, (**e**) cover of *Rubus* species in the understory and (**f**) total cover of grass (Gramineae) species in the understory. N = 23 plots for each relationship.

4. Discussion

4.1. Deer Impact on Red Oak and White Pine Survival and Growth

In northeastern North America, the overabundance of deer is a growing problem for the regeneration of several hardwood and coniferous species [3,10–12]. As hypothesized, the survival and growth of red oak and white pine were significantly decreased when these species were underplanted without tree shelters (Figures 3 and 4). Only 29% of unprotected red oaks survived after 6 growing seasons, and the average total height of survivors was lower (44.3 cm) than average seedling height at planting (70 cm). This contrasts with the high survival (80.5%) and total height (138.5 cm) achieved by red oak in the tree shelter treatment (Figures 3 and 4). Two decades ago, it was possible to achieve successful underplanting of red oak without tree shelters at the study site [27], however, this is no longer possible because of increased deer density.

For white pine, the impact of deer was less striking than for red oak (Figures 3 and 4). By the end of the first growing season, 75% of red oaks had their main stem browsed, while almost no deer browsing was observed on white pine seedlings (Figure 3b). Such a browsing pattern reflects the tendency of deer to browse heavily on deciduous species during the growing season, while conifers are more heavily browsed during the dormant season when other food sources are scarce [10]. Also, at planting, white pine seedlings were relatively small (21 cm of height) and less conspicuous in the understory vegetation, compared to taller red oak seedlings (70 cm). Small white pine seedlings were also protected by snow during the first winters (B. Truax, field observations). Thus, even though a similar proportion of red oaks and white pines had their main stem browsed during the 6th growing season (Figure 3b), the vulnerability of red oak to browsing was greater. However, browsing of white pine lateral branches in the control treatment remained severe despite that many 6 year-old trees had their terminal shoot above the browsing line (Figure 9a). Thus, even though pines had their lateral branches constrained in the tree shelters, this silvicultural treatment was highly efficient at increasing growth and survival (Figures 3, 4 and 9b). Furthermore, in the control treatment, no red oak was observed above the browsing line in June 2018 (7th growing season), which suggests that mortality induced by over-browsing will likely increase in the subsequent years.

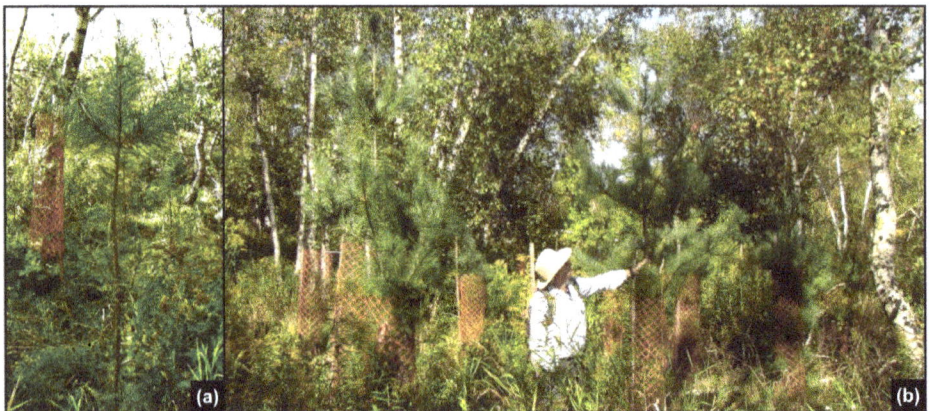

Figure 9. (**a**) Heavy browsing on white pine lateral branches in the control treatment. (**b**) White pines in the tree shelter treatment growing in the Grey birch-balsam poplar-elm community type (6th growing season).

4.2. The Effect of Shelterwood Characteristics on Underplanted Tree Growth

After 6 years, the forest community type had an important effect on the total height of all hardwoods and white pine protected with tree shelters (Table 3). White pine was the only species with a higher total height and survival in the Grey birch-balsam poplar-elm community type (Table 3, Section 3.2, Figure 9b). White pine was also the species with the smallest total height difference between the two community types (Table 3), an indication of its wider ecological amplitude compared to the studied hardwoods.

Previous observations have shown that white pine grows well on imperfectly drained and lower fertility soils when no hardwood competition is present, with grey birch being an associated forest cover of this species [37,46]. White pine also has a very efficient nitrogen retention strategy, which allows its establishment in environments dominated by herbaceous vegetation, despite the strong competition for soil nitrogen [24,78]. Although white pine is intermediate in shade-tolerance [37], controlled and field experiments have shown strong positive relationships between canopy openness, gap size or light availability and seedling height or shoot biomass growth [79–81]. In this study the Grey birch-balsam poplar-elm community type had regenerated on an imperfectly drained old-field and was characterized by the lowest soil fertility, the highest canopy openness, an understory dominated by forbs and grasses, and the absence of hardwood competition in the understory and mid-story (Table 2). Thus, it is not surprising that white pine outcompeted the three hardwood species in the Grey birch-balsam poplar-elm community type since these hardwood species are very sensitive to competition from herbaceous species, and they generally require good drainage conditions and higher soil fertility to reach optimal growth [44,50,51,56,57,59,78]. Consistent with the community type effect observed on white pine growth, and previous knowledge about white pine ecology, the total height of this species in the shelter treatment was positively related to canopy openness and the basal area of grey birch, but negatively related to soil CEC (Figure 5). Also, because 17 of the 25 blocks of the experimental design were located in the Grey birch-balsam poplar-elm community type, it is not surprising that white pine had better overall growth and survival than red oak (Figures 3 and 4).

While canopy openness was strongly related to white pine total height in the shelter treatment ($R^2 = 0.43$, $p < 0.001$) (Figure 5a), no such significant relationship was observed between these two variables in the control treatment ($R^2 = 0.15$, $p = 0.06$). This suggests that deer browsing overrides the canopy openness effect on unprotected white pine growth, which contrasts with observations made in northern Wisconsin (United States) [81]. A possible explanation would be the potentially higher deer population density at our study site. Unprotected white pine also had significantly higher total height and survival in the Grey birch-balsam poplar-elm community type, with height and survival differences between the two community types being much larger for unprotected vs. sheltered pine (Tables 3 and 4). Because they had a slower height growth, white pines in the White ash community type were likely to be subjected to more intense and repeated browsing of their terminal shoots (Table 4).

In the northern hardwood forest ecosystem, red oak, bitternut hickory, black walnut and white ash are associated species in several habitats, while grey birch is generally not an associated forest cover of these species [44,49,51,53,54,56]. Accordingly, all hardwood species reached higher total height in the White ash community type (Table 3, Figure 10), where soil fertility was higher and the understory dominated by *Rubus* species, and not by grasses and forbs (Table 2). Consistent with this community type effect, the height of the three hardwood species was positively and significantly correlated with *Rubus* cover and with at least one soil fertility indicator (CEC or NO_3 supply rate), while bitternut hickory and black walnut total height was also negatively correlated with the basal area of grey birch in the overstory (Figures 6–8).

Figure 10. Underplanted hardwoods growing in the White ash community type at the beginning of the 7th growing season (early June 2018); (**a**) red oak, (**b**) bitternut hickory, and (**c**) black walnut.

Many authors have suggested that a moderate *Rubus* cover has a facilitating effect on underplanted oaks, because such a vegetation cover temporarily hides seedlings from large herbivores without being too competitive for light and nutrients [29,34,40]. Conversely, a dense *Rubus* cover under canopy gaps can interfere with red oak [81]. In this study, neither the height, the survival nor the main stem browsing of unprotected red oaks was statistically different between the two community types, which suggests little facilitating effect of *Rubus* in the White ash community type. At high deer densities, *Rubus* species are also severely browsed [13], potentially reducing their sheltering effects on hardwood seedlings. On the other hand, habitats supporting *Rubus* species regionally have fairly good soil drainage [82], and relatively high soil fertility in terms of NO_3 availability given that *Rubus* species are nitrophilous [83]. Such soil conditions are beneficial to the studied hardwoods [51,57], despite that red oak, black walnut and hickories have a slight preference for NH_4 uptake [84–86]. Being highly sensitive to grass competitors, the studied hardwoods [50,59,87,88] may also benefit from the reduction in herbaceous plant cover following understory colonization by *Rubus* species [81].

Among the studied species, black walnut is probably the most sensitive to competition for soil resources by herbaceous vegetation. A negative relationship ($R^2 = 0.46$, $p < 0.001$) between Gramineae cover in the understory and total height was observed for black walnut (Figure 8f), but not for the other hardwoods. The strong and steep negative relationship ($R^2 = 0.68$, $p < 0.001$) between total basal area and total height of black walnut (Figure 8c) is consistent with its shade-intolerance [56]. On the other hand, canopy openness or total basal area were not correlated with total height of red oak and bitternut hickory, suggesting that competition for light by shelterwood trees was not a strong factor affecting these moderately shade-tolerant species [9,36,51].

As hypothesized, black walnut was the most sensitive species to environmental conditions prevailing in the studied shelterwoods, with six ecological factors being significantly related to its total height after 6 years (Figure 8). Among trees growing in shelters, black walnut was also the species with widest total height difference between the two community types (Table 3), and the lowest survival rate (64%). Such results are consistent with the relatively narrow niche occupied by black walnut at the northern limit of its range, where it mostly regenerates in open forest habitats on rich soils of well-drained bottomlands [43].

A potential limitation of this study pertains to the physical constraint of the shelters on tree architecture and development. The use of fence enclosures around plots would have allowed unrestricted tree development, and a potentially more accurate quantification of ecological factors affecting growth.

4.3. Management Implications for Forest Restoration and Hardwood Species Migration

This study is the first to document the major impact of deer on underplanted red oak and white pine in the southern Québec region. Moreover, to our knowledge, few studies have documented the establishment success of underplanted black walnut and bitternut hickory in the northern hardwood forest ecosystem. Despite the fact that conclusions from this study where only drawn from a single site, several management implications should be considered in the context of forest restoration and tree species migration into more northern forest ecosystems.

Considering that high deer densities are now common throughout the southern Québec region [89], it is clear from this study that without protection from deer, red oak restoration in young forest ecosystems will be challenging (Figures 3 and 4). The deer browsing situation in the study area is of lesser concern for white pine, although the use of tree shelters resulted in a 14% increase in survival and a 49% increase in total height after 6 years. While underplanted white pine grows much slower than it does on clearcut sites [3], open sites produce pines with a large branch biomass, which adversely affects wood quality [25]. Moreover, open sites increase vulnerability to damaging agents such as the white pine weevil, which affects the terminal shoot and leads to excessive branching [25]. Large openings or 50% canopy cover are thus recommended to maintain good growing conditions (Figure 5a), early branch self-pruning, and reduced damages by the blister rust and the pine weevil [25]. In southern Québec, many grey birch stands have regenerated on imperfectly drained pasture sites and along riparian corridors located on farmland (B. Truax and J. Fortier, personal observations). These grey birch stands could be targeted for white pine restoration.

For the studied hardwoods, we found that the mesic White ash forest community type was a more suitable shelterwood than the grey birch stands. Thus, white ash stands could be targeted for hardwood restoration, as white ash is an important forest associate of many northern hardwood species [90]. Moreover, considering that the emerald ash borer (*Agrilus planipennis* Farmaire) will very likely continue to spread in southern Québec, canopy gaps created following white ash mortality could provide suitable light conditions for hardwood underplanting.

This study provides rare evidence that underplanting can be suited for bitternut hickory and black walnut, two species often growing poorly in open environments where the soil has been strongly disturbed (e.g., old-fields and mined land) [17,50,58,59,91]. In the province of Québec (Canada), bitternut hickory stands are mostly found in the St-Lawrence Valley and on the Precambrian Shield foothills [43,48,49,53,54]. However, bitternut hickory is not found in the mountain forests of northern New England (United States), just south of the study site (Figure 1A) [51]. Yet, this study supports the notion that this species has the potential to grow on mesic sites of the Québec Appalachians, as depicted by the distribution map of *Carya* species made by Marie-Victorin in the early 1900s [52].

It was recently proposed that migration of bitternut and shagbark hickory will be constrained by soil fertility in the province of Québec [16]. However, in the eastern United States, the recent landscape and stand scale analysis of Lefland et al. [9] has shown that dry, acidic, and nutrient-poor sites favor the establishment of hickories. Moreover, at the northern limit of its range, on the Precambrian Shield foothills of the Outaouais region (Québec), bitternut hickory has been found in association with red oak and white oak (*Quercus alba* L.) on steep slopes having thin, nutrient-poor and dry soil, with pH ranging 3.9–4.7 [53]. Thus, there is no reason why bitternut hickory should not be underplanted in the Appalachian region of southern Québec, especially in a context where soil water content of forest ecosystems is expected to decline regionally with climate change [14]. Additional studies are also needed to evaluate the growth potential of shagbark hickory (*Carya ovata*) in the study area given that bitternut and shagbark hickories are associates on upland sites [9,51].

This study provides the first evidence of establishment success of black walnut in forest ecosystems located northward of its natural range. Black walnut growth was correlated to numerous ecological factors (Figure 8), an indication that proper site selection will lead to successful underplanting. Being shade-intolerant, black walnut may require larger canopy gaps than red oak or bitternut hickory to maintain good growth potential at the juvenile stage. In southern Illinois (United States), black walnut planted in 20 m^2 clearcut gaps achieved more than 90% survival after 7 years, but annual herbaceous vegetation control was necessary to achieve good growth (height > 5 m) [92].

In southern Québec, many post-agricultural forests have little hardwood regeneration because of over-browsing and/or of a lack of seed sources in the surrounding landscape [89,93,94]. While such a situation is worrisome for forest ecosystem integrity and resilience, it creates ideal conditions to underplant oak and white pine, which are adversely affected by tall hardwood competition (>1.5 m) in the understory [37,45,46,95]. Moreover, as shown by Lucas et al. [96], increased nutrient inputs through deer fecal and urine deposits along with competition reduction caused by deer browsing, had a positive effect on the growth rate of mature red oak. Thus, if trees are protected by shelters, they will likely have good growing conditions in the post-agricultural forests of southern Québec.

The type of tree shelter (Model-K) used was well-suited for underplanting because very little light is intercepted by the fencing material (Figures 9b and 10). Smaller mesh size or opaque commercial tree shelters have been found to have little positive effect on white pine in open environments, and to even have negative effects on underplanted red oak [67,97]. Yet, the large mesh size (5 cm) of the fencing material allowed some leaves and branches to grow outside the shelter and therefore be browsed by deer. Besides, during the 6th growing season, we had to replace many rotten stakes. If not replaced in a timely manner, rotten stakes break with snow packing and wind, and the shelter falls on the ground with the tree it contains. Such an occurrence was observed in the spring of the 7th growing season in blocks exposed to the dominant wind (B. Truax and J. Fortier, field observations). Lifting up the fencing material on the wooden stake is also recommended to maintain protection of the terminal shoot. Freeze and thaw cycles, and probably deer collisions, also reduced the stability of the wooden stakes, which had to be hammered down on a few occasions during the 6 years of the experiment. Ideally, to reduce shelter maintenance, we recommend the use of metal stakes.

5. Conclusions

This study showed that the use of tree shelters against deer browsing was essential to successfully establish underplanted red oak in southern Québec, where deer densities above 20 individuals/km^2 are now common in many areas [89]. Unprotected white pine achieved satisfactory growth, but the use of tree shelters was clearly beneficial.

Distinctive growth response for white pine and hardwoods (growing in tree shelters) was observed between the two forest community types underplanted. White pine achieved higher total height in the Grey birch-balsam poplar-elm community type, while hardwoods reached higher total height in the White ash community type, which was characterized by a better soil drainage, higher soil fertility, and low herbaceous competition. The height growth of all hardwoods was positively correlated with at least one soil fertility indicator, and with the cover of *Rubus* species, while the height growth of white pine was negatively correlated to soil fertility and positively correlated to canopy openness. Across the two community types, white pine had the smallest growth variation, while black walnut had the largest. Moreover, black walnut had the largest number of ecological factors significantly correlated with its total height, with the strongest relationships being observed with shelterwood total basal area and soil NO$_3$. This suggests that site selection for underplanting should be species-specific.

Lastly, underplanting was found to be a suitable method for starting black walnut migration northwards and for bitternut hickory restoration. Within the global change context, additional studies are required to determine the optimal shelterwood environments for black walnut and hickories at, or beyond, the northern limit of their actual range, given their high-value for biodiversity and timber production. White ash mortality induced by the emerald ash borer may provide suitable shelterwood conditions to restore or introduce nut producing hardwoods.

Author Contributions: B.T. and D.G. conceived and designed the experiment. B.T., F.L. and M.-A.P. were involved in sampling design and field sampling. J.F., B.T., F.L. and M.-A.P. analyzed the data. J.F. wrote the first draft of the manuscript. B.T., D.G., and F.L. critically revised the manuscript.

Funding: This research was funded by the Ministère des Forêts, de la Faune et des Parcs du Québec (Chantier sur la Forêt Feuillue) and Tree/Arbres Canada.

Acknowledgments: We gratefully acknowledge the Ministère des Forêts, de la Faune et des Parcs du Québec (Chantier sur la Forêt Feuillue) and Tree/Arbres Canada for the funding received. We wish to thank tree planters and field assistants for their help (J. Lemelin, A. Richard, L. Godbout, Y. Daigle, J.-D. Careau, S. Wood-Gagnon). We also thank La Communauté des Bénédictins de St-Benoît-du-Lac, and especially Brother Luc Lamontagne, who have kindly allowed the use of part of their property for the establishment of the experimental plantations. We acknowledge the Berthier nursery of the Ministère des Forêts, de la Faune et des Parcs of Québec for providing high quality planting stock. Dr. R. Bradley and Dr. W. Parsons (University of Sherbrooke) are acknowledged for soil C/N analysis.

Conflicts of Interest: The authors declare no conflict of interest.

Appendix A

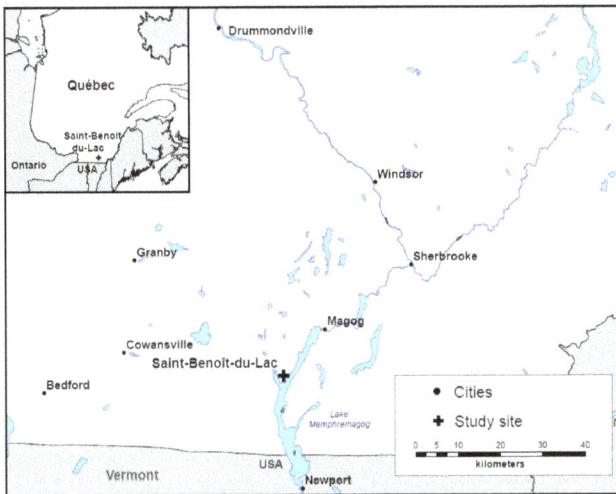

Figure A1. Study site location in southern Québec, Canada.

Table A1. Descriptive statistics for the ecological variables measured across all blocks. For each variable, descriptive statistics where obtained from block averages (N = 25 blocks).

Component	Variable [1]	Min.	Max.	Median	Mean	Std. dev.
Soil	pH	4.7	5.6	5.3	5.3	0.2
	CEC (meq/100 g)	14.3	19.9	16.7	17.1	1.7
	Base saturation (%)	31.0	58.7	45.4	45.3	8.1
	NO_3 (μg/10 cm^2/41d)	4.9	239.6	16.7	38.8	51.8
	NH_4 (μg/10 cm^2/41d)	3.0	39.0	4.4	5.9	7.0
	P (μg/10 cm^2/41d)	1.1	8.7	2.8	3.1	1.7
	K (μg/10 cm^2/41d)	7.4	92.4	26.4	34.1	22.3
	Ca (μg/10 cm^2/41d)	1561	2384	2095	2040	228
	Mg (μg/10 cm^2/41d)	286	567	432	430	64
	S (μg/10 cm^2/41d)	17	128	55	58	27
	Organic matter (%)	6.9	13.4	10.6	10.3	1.6
	C/N	9.4	12.0	10.8	10.8	0.7
	Clay (%)	0	62	26	30	13
	Silt (%)	18	83	44	44	11
	Sand (%)	0	58	29	26	11
Stand	Canopy openness (%)	20.8	46.2	34.0	32.7	6.9
structure and	Total BA (m^2/ha)	3.5	21.8	16.3	14.5	5.3
composition	*Betula pop.* BA (m^2/ha)	0	20.1	13.5	11.0	6.7
	Ulmus am. BA (m^2/ha)	0	11.8	0.0	1.2	3.1
	Fraxinus am. BA (m^2/ha)	0	10.8	0.0	1.0	2.6
Understory	*Rubus* spp. (% cover)	0	73	3	14	20
plant cover	Gramineae spp. (% cover)	0	75	15	19	18
	Solidago spp. (% cover)	8	80	38	38	21
	Fragaria virginiana (% cover)	1	45	18	20	13
	Carex spp. (% cover)	0	28	2	5	7
	Phalaris arundinacea (% cover)	0	33	5	9	10
	Onoclea sensibilis (% cover)	0	16	0	3	4

1. Abbreviations used in Table A1: CEC (cation exchange capacity), BA (basal area), pop. (*populifolia*), am. (*americana*), min. (minimum value), max. (maximum value), std. dev. (standard deviation), d (days).

Table A2. Correlation matrix between total height after 6 years for the four studied tree species and selected ecological variables.

Red oak	Total height	*Rubus* cover	Soil CEC
Total height	1.00	0.61	0.50
Rubus cover	0.61	1.00	0.49
Soil CEC	0.50	0.49	1.00

White pine	Total height	Canopy openness	Soil CEC	Grey birch BA
Total height	1.00	0.60	−0.52	0.41
Canopy openness	0.60	1.00	−0.29	0.49
Soil CEC	−0.52	−0.29	1.00	−0.40
Grey birch BA	0.41	0.49	−0.40	1.00

Bitternut hickory	Total height	*Rubus* cover	Soil NO_3	Grey birch BA
Total height	1.00	0.65	0.48	−0.50
Rubus cover	0.65	1.00	0.76	−0.62
Soil NO_3	0.48	0.76	1.00	−0.70
Grey birch BA	−0.50	−0.62	−0.70	1.00

Black walnut	Total height	Soil NO_3	Soil CEC	Total BA	Grey birch BA	*Rubus* cover	Gramineae cover
Total height	1.00	0.91	0.49	−0.72	−0.71	0.78	−0.57
Soil NO_3	0.91	1.00	0.45	−0.67	−0.69	0.75	−0.43
Soil CEC	0.49	0.45	1.00	−0.42	−0.59	0.46	−0.34
Total BA	−0.72	−0.67	−0.42	1.00	0.76	−0.48	0.40
Grey birch BA	−0.71	−0.69	−0.59	0.76	1.00	−0.60	0.61
Rubus cover	0.78	0.75	0.46	−0.48	−0.60	1.00	−0.57
Gramineae cover	−0.57	−0.43	−0.34	0.40	0.61	−0.57	1.00

Correlation coefficients (*r*) in bold are significant at *p* < 0.05.

References

1. Burns, R.M.; Honkala, B.H. *Silvics of North America: Vol. 1 Conifers*; Forest Service Agriculture (USDA): Washington, DC, USA, 1990; Agriculture Handbook; p. 654.
2. Fralish, J.S. *The Keystone Role of Oak and Hickory in the Central Hardwood Forest*; Gen. Tech. Rep. SRS-73; U.S. Department of Agriculture, Forest Service, Southern Research Station: Asheville, NC, USA, 2004; pp. 78–87.
3. Ward, J.S.; Mervosh, T.L. Strategies to reduce browse damage on eastern white pine (*Pinus strobus*) in southern New England, USA. *For. Ecol. Manag.* **2008**, *255*, 1559–1567. [CrossRef]
4. Booth, J.D. Timber utilization on the agricultural frontier in southern Québec. *J. East. Townsh. Stud.* **1994**, *4*, 15–30.
5. Simard, H.; Bouchard, A. The precolonial 19th century forest of the Upper St. Lawrence Region of Quebec; a record of its exploitation and transformation through notary deeds of wood sales. *Can. J. For. Res.* **1996**, *26*, 1670–1676. [CrossRef]
6. Tanguay, C. Distribution, Abondance et état de Santé du Noyer cEndré (*Juglans cinerea*) en Relation avec les Gradients écologiques dans les Cantons-de-l'Est. Master's Thesis, Université du Québec à Montréal, Montréal, QC, Canada, 2011.
7. Abrams, M.D. Fire and the development of oak forests. *BioSc.* **1992**, *42*, 346–353. [CrossRef]
8. Weyenberg, S.A.; Frelich, L.E.; Reich, P.B. Logging versus fire: How does disturbance type influence the abundance of *Pinus strobus* regeneration? *Silva Fenn.* **2004**, *38*, 179–194. [CrossRef]
9. Lefland, A.B.; Duguid, M.C.; Morin, R.S.; Ashton, M.S. The demographics and regeneration dynamic of hickory in second-growth temperate forest. *For. Ecol. Manag.* **2018**, *419–420*, 187–196. [CrossRef]
10. Côté, S.D.; Rooney, T.P.; Tremblay, J.-P.; Dussault, C.; Waller, D.M. Ecological impacts of deer overabundance. *Ann. Rev. Ecol. Evol. Syst.* **2004**, *35*, 113–147. [CrossRef]
11. Kittredge, D.B.; Ashton, P.M.S. Impact of deer browsing on regeneration in mixed stands in southern New England. *North. J. Appl. For.* **1995**, *12*, 115–120.
12. White, M.A. Long-term effects of deer browsing: Composition, structure and productivity in a northeastern Minnesota old-growth forest. *For. Ecol. Manag.* **2012**, *269*, 222–228. [CrossRef]
13. Horsley, S.B.; Stout, S.L.; deCalesta, D.S. White-tailed deer impact on the vegetation dynamics of a northern hardwood forest. *Ecol. Appl.* **2003**, *13*, 98–118. [CrossRef]
14. Ouranos. *Vers l'adaptation. Synthèse des connaissances sur les changements climatiques au Québec. Partie 1: Évolution climatique au Québec*; Ouranos: Montréal, QC, Canada, 2015.
15. Niinemets, Ü.; Valladares, F. Tolerance to shade, drought and waterlogging of temperate northern hemisphere trees and shrubs. *Ecol. Monogr.* **2006**, *76*, 521–547. [CrossRef]
16. Lafleur, B.; Pare, D.; Munson, A.D.; Bergeron, Y. Response of northeastern North American forests to climate change: Will soil conditions constrain tree species migration? *Environ. Rev.* **2010**, *18*, 279–289. [CrossRef]
17. Cogliastro, A.; Gagnon, D.; Bouchard, A. Experimental determination of soil characteristics optimal for the growth of ten hardwoods planted on abandoned farmland. *For. Ecol. Manag.* **1997**, *96*, 49–63. [CrossRef]
18. von Althen, F.W. Afforestation of former farmland with high-value hardwoods. *For. Chron.* **1991**, *67*, 209–212. [CrossRef]
19. Peichl, M.; Arain, M.A. Above- and belowground ecosystem biomass and carbon pools in an age-sequence of temperate pine plantation forests. *Agric. For. Meteorol.* **2006**, *140*, 51–63. [CrossRef]
20. Neumann, P.D.; Krahn, H.J.; Krogman, N.T.; Thomas, B.R. 'My grandfather would roll over in his grave': Family farming and tree plantation on farmland. *Rural Sociol.* **2007**, *72*, 111–135. [CrossRef]
21. Hatch, A.B. The role of mycorrhizae in afforestation. *J. For.* **1936**, *34*, 22–29.
22. Ostfeld, R.S.; Canham, C.D. Effects of meadow vole population density on tree seedling survival in old fields. *Ecol.* **1993**, *74*, 1792–1801. [CrossRef]
23. Margolis, H.A.; Brand, D.G. An ecophysiological basis for understanding plantation establishment. *Can. J. For. Res.* **1990**, *20*, 375–390. [CrossRef]
24. Laungani, R.; Knops, J.M.H. Species-driven changes in nitrogen cycling can provide a mechanism for plant invasions. *PNAS* **2009**, *106*, 12400–12405. [CrossRef] [PubMed]
25. Ostry, M.E.; Laflamme, G.; Katovich, S.A. Silvicultural approaches for management of eastern white pine to minimize impacts of damaging agents. *For. Pathol.* **2010**, *40*, 332–346. [CrossRef]

26. Gardiner, E.S.; Stanturf, J.A.; Schweitzer, C.J. An afforestation system for restoring bottomland hardwood forests: Biomass accumulation of nuttall oak seedlings interplanted beneath eastern cottonwood. *Rest. Ecol.* **2004**, *12*, 525–532. [CrossRef]

27. Truax, B.; Lambert, F.; Gagnon, D. Herbicide-free plantations of oaks and ashes along a gradient of open to forested mesic environments. *For. Ecol. Manag.* **2000**, *137*, 155–169. [CrossRef]

28. Williston, H.L.; Huckenpahler, B.J. Hardwood underplanting in North Mississippi. *J. For.* **1957**, *55*, 287–290.

29. Johnson, P.S. Responses of planted northern red oak to three overstory treatments. *Can. J. For. Res.* **1984**, *14*, 536–542. [CrossRef]

30. Tworkoski, T.J.; Smith, D.W.; Parrish, D.J. Regeneration of red oak, white oak, and white pine by underplanting prior to canopy removal in the Virginia Piedmont. *South. J. Appl. Ecol.* **1986**, *10*, 206–210.

31. Craig, J.M.; Lhotka, J.M.; Stringer, J.W. Evaluating initial responses of natural and underplanted oak reproduction and a shade-tolerant competitor to midstory removal. *For. Sci.* **2014**, *60*, 1164–1171. [CrossRef]

32. Truax, B.; Gagnon, D.; Chevrier, N. Nitrate reductase activity in relation to growth and soil N-forms in red oak and red ash planted in three different environments: Forest, clear-cut and field. *For. Ecol. Manag.* **1994**, *64*, 71–82. [CrossRef]

33. Balandier, P.; Collet, C.; Miller, J.H.; Reynolds, P.E.; Zedaker, S.M. Designing forest vegetation management strategies based on the mechanisms and dynamics of crop tree competition by neighbouring vegetation. *Forestry* **2006**, *79*, 3–27. [CrossRef]

34. Dey, D.C.; Gardiner, E.S.; Schweitzer, C.J.; Kabrick, J.M.; Jacobs, D.F. Underplanting to sustain future stocking of oak (*Quercus*) in temperate deciduous forests. *New For.* **2012**, *43*, 955–978. [CrossRef]

35. Pretzsch, H.; Rais, A. Wood quality in complex forests versus even-aged monocultures: Review and perspectives. *Wood Sci. Technol.* **2016**, *50*, 845–880. [CrossRef]

36. Crow, T.R. Reproductive mode and mechanisms for self-replacement of northern red oak (*Quercus rubra*)—A review. *For. Sci.* **1988**, *34*, 19–40.

37. Wendel, G.W.; Clay Smith, H. Eastern white pine. In *Silvics of North America: 1. Conifers. Agriculture Handbook 654*; Burns, R.M., Honkala, B.H., Eds.; U.S. Department of Agriculture, Forest Service: Washington, DC, USA, 1990; Volume 2, pp. 972–999.

38. Parker, W.C.; Dey, D.C.; Newmaster, S.G.; Elliott, K.A.; Boysen, E. Managing succession in conifer plantations: Converting young red pine (*Pinus resinosa* Ait.) plantations to native forest types by thinning and underplanting. *For. Chron.* **2001**, *77*, 721–734. [CrossRef]

39. Götmark, F.; Schott, K.M.; Jensen, A.M. Factors influencing presence–absence of oak (*Quercus* spp.) seedlings after conservation-oriented partial cutting of high forests in Sweden. *Scand. J. For. Res.* **2011**, *26*, 136–145. [CrossRef]

40. Paquette, A.; Bouchard, A.; Cogliastro, A. Successful under-planting of red oak and black cherry in early-successional deciduous shelterwoods of North America. *Ann. For. Sc.* **2006**, *673*, 823–831. [CrossRef]

41. Truax, B.; Gagnon, D.; Lambert, F.; Fortier, J. Multiple-use zoning model for private forest owners in agricultural landscapes: A case study. *Forests* **2015**, *6*, 3614–3664. [CrossRef]

42. Frey, B.R.; Ashton, M.S. Growth, survival and sunfleck response of underplanted red oaks (*Quercus* spp., section *Erythrobalanus*) along a topographic gradient in southern New England. *For. Ecol. Manag.* **2018**, *419–420*, 179–186. [CrossRef]

43. Farrar, J.L. *Les arbres du Canada*; Fides et le Service Canadien des Forêts, Ressources Naturelles Canada: St-Laurent, QC, Canada, 2006.

44. Sander, I.L. Northern red oak. In *Silvics of North America: 2. Hardwoods. Agriculture Handbook 654*; Burns, R.M., Honkala, B.H., Eds.; Department of Agriculture, Forest Service, U.S.: Washington, DC, USA, 1990; Volume 2, pp. 1401–1414.

45. Clements, J.R. Development of a white pine underplantation in thinned and unthinned aspen. *For. Chron.* **1966**, *42*, 244–250. [CrossRef]

46. Smidt, M.F.; Puettmann, K.J. Overstory and understory competition affect underplanted eastern white pine. *For. Ecol. Manag.* **1998**, *105*, 137–150. [CrossRef]

47. Oliver, L.B.; Jeremy, P.S.; Comer, C.E.; Williams, H.M.; Symmank, M.E. Weed control and overstory reduction improve survival and growth of under-planted oak and hickory seedlings. *Rest. Ecol.* **2018**, 1–12. [CrossRef]

48. Doyon, F.; Bouchard, A.; Gagnon, D. Tree productivity and successional status in Québec northern hardwoods. *Écoscience* **1998**, *5*, 222–231. [CrossRef]

49. St-Jacques, C.; Gagnon, D. La végétation forestière du secteur nord-ouest de la vallée du Saint-Laurent, Québec. *Can. J. Bot.* **1988**, *66*, 793–804. [CrossRef]

50. Von Althen, F.W. *Sowing and Planting Shagbark and Bitternut Hickories on Former Farmland in Southern Ontario*; Information Report O-X-403; Forestry Canada, Ontario Region: Sault Ste. Marie, ON, Canada, 1990; p. 11.

51. Smith, H.C. Bitternut hickory. In *Silvics of North America: 2. Hardwoods. Agriculture Handbook 654*; Burns, R.M., Honkala, B.H., Eds.; Department of Agriculture, Forest Service, U.S.: Washington, DC, USA, 1990; Volume 2, pp. 389–417.

52. Marie-Victorin, F.; Rouleau, E.; Brouillet, L.; Hay, S.G.; Goulet, I. *Flore Laurentienne—3ᵉ edition*; Gaëtan Morin éditeur ltée: Montréal, QC, Canada, 2002; p. 1093.

53. Gagnon, D.; Bouchard, A. La végétation de l'escarpement d'Eardley, parc de la Gatineau, Québec. *Can. J. Bot.* **1981**, *59*, 2667–2691. [CrossRef]

54. Gauthier, S.; Gagnon, D. La végétation des contreforts des Laurentides: Une analyse des gradients écologiques et du niveau successionnel des communautés. *Can. J. Bot.* **1990**, *68*, 391–401. [CrossRef]

55. The Committee on the Status of Endangered Wildlife in Canada (COSEWIC). Butternut *Juglans cinerea*. Available online: http://www.cosewic.gc.ca/eng/sct1/searchdetail_e.cfm?id=793&StartRow=1&boxStatus=All&boxTaxonomic=All&location=All&change=All&board=All&commonName=butternut&scienceName=&returnFlag=0&Page=1 (accessed on 22 May 2015).

56. Williams, R.D. Black walnut. In *Silvics of North America: 2. Hardwoods. Agriculture Handbook 654*; Burns, R.M., Honkala, B.H., Eds.; Department of Agriculture, Forest Service, U.S.: Washington, DC, USA, 1990; pp. 771–789.

57. Cogliastro, A.; Gagnon, D.; Daigle, S.; Bouchard, A. Improving hardwood afforestation success: An analysis of the effects of soil properties in southwestern Quebec. *For. Ecol. Manag.* **2003**, *177*, 347–359. [CrossRef]

58. Van Sambeek, J.W.; Schlesinger, R.C.; Roth, P.L.; Bocoum, I. Revitalizing slow-growth black walnut plantings. In *Proceedings of the Seventh Central Hardwood Forest Conference*; Rink, G., Budelsky, C.A., Eds.; USDA Forest Service; North Central Forest Experiment Station: Carbondale, IL, USA, 1989; pp. 108–114.

59. Von Althen, F.W. Revitalizing a black walnut plantation through weed control and fertilization. *For. Chron.* **1985**, *61*, 71–74. [CrossRef]

60. Huot, M.; Lebel, F. *Le plan de gestion du cerf de Virginie au Québec 2010–2017*; Direction de l'expertise sur la faune et ses habitats, Ministère des Ressources naturelles et de la Faune du Québec (MRNF): Québec, QC, Canada, 2010.

61. Rooney, T.P.; Waller, D.M. Direct and indirect effects of white-tailed deer in forest ecosystems. *For. Ecol. Manag.* **2003**, *181*, 165–176. [CrossRef]

62. Association Maritime du Québec. Lac Memphrémagog. Available online: http://www.navigationquebec.com/fiche_lac.php?l_id=46 (accessed on 14 June 2015).

63. Robitaille, A.; Saucier, J.-P. *Paysages régionaux du Québec méridional*; Les publications du Québec: Ste-Foy, QC, Canada, 1998; p. 213.

64. Westveld, M. Natural forest vegetation zones of New England. *J. For.* **1956**, *54*, 332–338.

65. Cogbill, C.V.; Burk, J.; Motzkin, G. The forests of presettlement New England, USA: Spatial and compositional patterns based on town proprietor surveys. *J. Bio geogr.* **2002**, *29*, 1279–1304. [CrossRef]

66. Government of Canada. Station results-1981–2010 climate normals and averages. Available online: http://climate.weather.gc.ca/climate_normals/station_select_1981_2010_e.html?searchType=stnProv&lstProvince=QC (accessed on 16 February 2017).

67. Bardon, R.E.; Countryman, D.W.; Hall, R.B. Tree shelters reduced growth and survival of underplanted red oak seedlings in southern Iowa. *North. J. Appl. For.* **1999**, *16*, 103–107.

68. West, P. *Tree and Forest Measurement*; Springer-Verlag: Berlin, Germany, 2009; p. 190.

69. Perron, J.-Y. Inventaire forestier. In *Manuel de foresterie*; Ordre des ingénieurs forestiers du Québec, Ed. Les Presses de l'Université Laval: Ste-Foy, QC, Canada, 1996; pp. 390–473.

70. Conseil des Productions Végétales du Québec. Méthodes D'analyse des Sols, des Fumiers et des Tissus végétaux. Available online: https://www.agrireseau.net/documents/96351/methode-d_analyse-des-sols-des-fumiers-et-des-tissus-vegetaux-agdex-533-mai-1988 (accessed on 15 August 2018).

71. Bouyoucos, G.J. Hydrometer method improved for making particle size analysis of soils. *Agron. J.* **1962**, *54*, 464–465. [CrossRef]

72. Centre de référence en agriculture et agroalimentaire du Québec (CRAAQ). *Guide de référence en fertilization, 1re ed.*; Centre de référence en agriculture et agroalimentaire du Québec: Ste-Foy, QC, Canada, 2003; p. 40.

73. Tran, T.S.; Simard, R.R. Mehlich III-Extractable elements. In *Soil Sampling and Methods of Analysis*; Carter, M.R., Ed.; Lewis Publishers and CRC Press: Boca Raton, FL, USA, 1993; pp. 43–49.

74. Association of Official Agricultural Chemists (AOAC). *Official Methods of Analysis. Method 984.27: Calcium, Copper, Iron, Magnesium, Manganese, Phosphorus, Potassium, Sodium and Zinc in Infant Formula—Inductively Coupled Plasma Emission Spectroscopic*, 16th ed.; AOAC International: Rockville, MD, USA, 1999; p. 1200.

75. Qian, P.; Schoenau, J.J.; Huang, W.Z. Use of ion exchange membranes in routine soil testing. *Comm. Soil Sci. Plant Anal.* **1992**, *23*, 1791–1804. [CrossRef]

76. Petersen, R.G. *Design and Analysis of Experiments*; Marcel-Dekker: New York, NY, USA, 1985; p. 429.

77. Warton, D.I.; Hui, F.K.C. The arcsine is asinine: the analysis of proportions in ecology. *Ecol.* **2011**, *92*, 3–10. [CrossRef]

78. Bowersox, T.W.; McCormick, L.W. Herbaceous communities reduce the juvenile growth of northern red oak, white ash, yellow poplar, but not white pine. In *Proceedings of the Central Hardwood Forest Conference VI*; Hoy, R.L., Woods, F.W., DeSelm, H., Eds.; University of Tennessee: Knoxville, TN, USA, 1987; pp. 39–43.

79. Saunders, M.R.; Puettmann, K.J. Effects of overstory and understory competition and simulated herbivory on growth and survival of white pine seedlings. *Can. J. For. Res.* **1999**, *29*, 536–546. [CrossRef]

80. Boucher, J.F.; Bernier, P.Y.; Munson, A.D. Radiation and soil temperature interactions on the growth and physiology of eastern white pine (*Pinus strobus* L.) seedlings. *Plant Soil* **2001**, *236*, 165–174. [CrossRef]

81. Kern, C.C.; Reich, P.B.; Montgomery, R.A.; Strong, T.F. Do deer and shrubs override canopy gap size effects on growth and survival of yellow birch, northern red oak, eastern white pine, and eastern hemlock seedlings? *For. Ecol. Manag.* **2012**, *267*, 134–143. [CrossRef]

82. Meilleur, A.; Véronneau, H.; Bouchard, A. Shrub communities as inhibitors of plant succession in southern Quebec. *Environ. Manag.* **1994**, *18*, 907–921. [CrossRef]

83. Truax, B.; Gagnon, D.; Lambert, F.; Chevrier, N. Nitrate assimilation of raspberry and pin cherry in a recent clearcut. *Can. J. Bot.* **1994**, *72*, 1343–1348. [CrossRef]

84. Kim, T.; Mills, H.A.; Wetzstein, H.Y. Studies on effects of nitrogen form on growth, development, and nutrient uptake in pecan. *J. Plant Nutr.* **2002**, *25*, 497–508. [CrossRef]

85. Nicodemus, M.; Salifu, K.; Jacobs, D. Nitrate reductase activity and nitrogen compounds in xylem exudate of *Juglans nigra* seedlings: Relation to nitrogen source and supply. *Tree Struct. Funct.* **2008**, *22*, 685–695. [CrossRef]

86. Truax, B.; Lambert, F.; Gagnon, D.; Chevrier, N. Nitrate reductase and glutamine synthetase activities in relation to growth and nitrogen assimilation in red oak and red ash seedlings: Effects of N-forms, N concentration and light intensity. *Tree Struct. Funct.* **1994**, *9*, 12–18. [CrossRef]

87. Truax, B.; Fortier, J.; Gagnon, D.; Lambert, F. Black plastic mulch or herbicide to accelerate bur oak, black walnut, and white pine growth in agricultural riparian buffers? *Forests* **2018**, *9*, 258. [CrossRef]

88. Truax, B.; Gagnon, D.; Lambert, F.; Fortier, J. Riparian buffer growth and soil nitrate supply are affected by tree species selection and black plastic mulching. *Ecol. Eng.* **2017**, *106*, 82–93. [CrossRef]

89. Boucher, S.; Crête, M.; Ouellet, J.-P.; Daigle, C.; Lesage, L. Large-scale trophic interactions: White-tailed deer growth and forest understory. *Écoscience* **2004**, *11*, 286–295. [CrossRef]

90. Burns, R.M.; Honkala, B.H. *Silvics of North America*; USDA, Forest Service: Washington, DC, USA, 1990; Volume 2, Harwoods, Agriculture Handbook 654.

91. Davis, V.; Burger, J.A.; Rathfon, R.; Zipper, C.E.; Miller, C.R. Chapter 7: Selecting tree species for reforestation of Appalachian mined lands. In *The Forestry Reclamation Approach: Guide to Successful Reforestation of Mined Lands*; Adams, M.B., Ed.; U.S. Department of Agriculture, Forest Service, Northern Research Station: Newtown Square, PA, USA, 2017; Gen. Tech. Rep. NRS-169; pp. 1–10.

92. Krajicek, J.E. *Planted Black Walnut Does Well on Cleared Forest Sites—If Competition Is Controlled*; USDA, Forest Service, North Central Forest Experiment Station: St. Paul, MN, USA, 1975; pp. 1–4.

93. Daigle, C.; Crête, M.; Lesage, L.; Ouellet, J.-P.; Huot, J. Summer diet of two white-tailed deer, *Odocoileus virginianus*, populations living at low and high density in southern Québec. *Can. Fld.-Nat.* **2004**, *118*, 360–367. [CrossRef]

94. D'Orangeville, L.; Bouchard, A.; Cogliastro, A. Post-agricultural forests: Landscape patterns add to stand-scale factors in causing insufficient hardwood regeneration. *For. Ecol. Manag.* **2008**, *255*, 1637–1646. [CrossRef]

95. Lorimer, C.G.; Chapman, J.W.; Lambert, W.D. Tall understorey vegetation as a factor in the poor development of oak seedlings beneath mature stands. *J. Ecol.* **1994**, *82*, 227–237. [CrossRef]

96. Lucas, R.W.; Roberto, S.-G.; Cobb, D.B.; Waring, B.G.; Anderson, F.; McShea, W.J.; Casper, B.B. White-tailed deer (*Odocoileus virginianus*) positively affect the growth of mature northern red oak (*Quercus rubra*) trees. *Ecosphere* **2013**, *4*, 1–15. [CrossRef]

97. Ward, J.S.; Gent, M.P.N.; Stephens, G.R. Effects of planting stock quality and browse protection-type on height growth of northern red oak and eastern white pine. *For. Ecol. Manag.* **2000**, *127*, 205–216. [CrossRef]

forests

Article

Ten-Year Responses of Underplanted Northern Red Oak to Silvicultural Treatments, Herbivore Exclusion, and Fertilization

Graham S. Frank [1,2,]*, **Ronald A. Rathfon [1]** and **Michael R. Saunders [1]**

[1] Department of Forestry and Natural Resources, Purdue University, 715 W. State Street, West Lafayette, IN 47907, USA; ronr@purdue.edu (R.A.R.); msaunder@purdue.edu (M.R.S.)

[2] Department of Forest Ecosystems and Society, Oregon State University, 252 Richardson Hall, Corvallis, OR 97331, USA

* Correspondence: graham.frank@oregonstate.edu or frankg@purdue.edu; Tel.: +1-971-533-4433

Received: 12 August 2018; Accepted: 11 September 2018; Published: 15 September 2018

Abstract: Establishing adequate advanced oak reproduction prior to final overstory removal is crucial for regenerating oak forests in the eastern U.S. Many management approaches exist to this end, but benefits associated with any individual technique can depend on the suite of techniques employed and the geographic location. At four mixed-hardwood upland forest sites in central and southern Indiana, we tested factorial combinations of deer fencing, controlled-release fertilization, and various silvicultural techniques (midstory removal, crown thinning, and a shelterwood establishment cut) for promoting the growth and survival of underplanted red oak seedlings. Crown thinning resulted in slow growth and low survival. Midstory removal and the shelterwood establishment cut were nearly equally effective for promoting seedling growth. Seedling survival was strongly influenced by fencing, and differences in survival between silvicultural treatments were minimal when fencing was employed. Fertilization had minimal effects overall, only increasing the probability that unfenced seedlings were in competitive positions relative to surrounding vegetation. We suggest that underplanting oak seedlings can augment natural reproduction, but the practice should be accompanied by a combination of midstory removal and fencing, at a minimum, for adequate growth and survival.

Keywords: *Quercus rubra*; oak regeneration; Central Hardwood Forest region; shelterwood; deer herbivory

---

1. Introduction

Oaks (*Quercus* spp.) have been a foundational species in forests of eastern North America, but have been failing to regenerate in recent decades [1,2]. Forests dominated by oaks are ecologically and economically valuable as a source of timber, as a food source and habitat for wildlife, and for relatively open stand characteristics that promote diverse understory flora [3]. These forests were historically maintained by a frequent fire regime, and current oak dominance often reflects legacies of heavy logging, burning, and grazing by early European settlers [2]. However, fire suppression, land use changes, and the expansion of deer populations have dramatically altered disturbance regimes, shifting forests towards less open stand conditions and resulting in a lack of large advance oak reproduction in the understory to take advantage of periodic canopy openings [4,5]. Oak declined in importance across 81% of forested area in the Central Hardwood Forest Region between 1980–2008 [6], and oak regeneration becomes increasingly challenging as forest compositions shift towards more shade-tolerant species, reducing light penetration to the forest floor, and further suppressing fire [7].

An array of silvicultural practices has been developed to counter the numerous obstacles to oak regeneration. Oak are only moderately shade tolerant, and seedlings beneath dense midstory canopies grow slowly and have high mortality rates [8]. Moreover, single-tree selection and other partial cutting practices favor the growth of shade-tolerant advance regeneration [9–11]. Conversely, oaks are slow-growing and can be outcompeted in high light environments on productive sites following more intensive harvests, such as commercial or silvicultural clearcuts [12,13]. Shelterwood techniques have been used successfully, providing adequate light for oak growth and survival, while reducing the advantage of shade-intolerant competitors [14–17].

Despite the relative effectiveness of shelterwood techniques, examples of regeneration failure in shelterwoods still exist [13,18], and can be influenced by deer herbivory, site quality, and competitive pressure [19]. The relative importance of these factors can vary geographically, causing uncertainty in the outcomes of prescriptions designed to promote oak regeneration [20]. For example, stronger understory competition on productive sites may make regeneration failure more likely in Indiana than in Missouri [21]. Deer herbivory pressure can vary according to landscape-level factors, including the amount of surrounding forest acreage [22].

Management context can be as important as geographic context for determining the effectiveness of any single management action for promoting oak regeneration. Deer herbivory pressure on hardwood seedlings can depend on the size of canopy gaps following different silvicultural treatments [23], the amount of early successional habitat created by harvest openings [24,25], or the application of understory competition control [18]. Fertilization improved height growth of planted red oak seedlings in one study where those seedlings were protected from herbivory, particularly in conjunction with understory competition control, but fertilization was confounded with lime application [26]. However, red oak plantings do not always respond to nitrogen fertilization [27,28], and fertilization can increase herbivory damage when seedlings are not protected, eliminating any potential benefits to growth [29]. Interactive effects of different treatments coupled with geographic variability in treatment effects make studies that examine multiple treatments in combination particularly important, and such studies should be implemented in a variety of locations and conditions [20].

Temporal variability in acorn production creates an additional source of uncertainty when developing prescriptions for oak regeneration [30], and asynchrony between management actions and acorn mast crop can result in limited oak reproduction present to take advantage of release [31]. Underplanting oak seedlings (i.e., artificial regeneration, enrichment planting) may be a viable approach to supplement poor natural reproduction and allow managers to precisely plan the timing of silvicultural practices [19], but the approach increases the costs of securing oak regeneration. This greater investment increases the economic consequences for managers and family forest landowners in the event of regeneration failure.

The economic efficiency of planting oak seedlings depends upon not only the cost of planting and other silvicultural treatments, but also the survival and competitive ability of those seedlings during early stages of stand development [32]. In this study, we aimed to improve oak regeneration prescriptions by directly testing the utility of commonly prescribed management actions for promoting competitive advance oak reproduction, in combination with one another and at an operational scale in central and southern Indiana. We expected that (1) a shelterwood establishment cut coupled with deer exclusion fencing would be the most effective approach for promoting growth and survival of natural and underplanted oak seedlings; (2) fertilization would increase growth and survival of underplanted seedlings; and (3) the strength of any individual treatment effect would depend upon the combination of treatments applied.

2. Materials and Methods

2.1. Site Descriptions

We conducted this study at four mature, mixed-oak forest sites in central and southern Indiana, US. Study sites had no evidence of recent (at least 40–50 years) major disturbance, based on observations of stand condition and ground evidence (i.e., lack of recent stumps, slash, etc.; RAR, pers. obs.).

All sites had a well-developed midstory canopy of tree species dominated by sugar maple (*Acer saccharum* Marshall), red maple (*A. rubrum* L.), and American beech (*Fagus grandifolia* Ehrh.). Three sites were selected from properties managed by The Nature Conservancy (TNC) in southern Indiana: (1) Wulfman Tract, a dry-mesic forest located in southeastern Harrison County; (2) Knapp Tract, a dry-mesic forest located in north central Washington County; and (3) McKinney Tract, a dry to dry-mesic forest located in eastern Brown County. The first two sites are located in the Mitchell Karst Plain Section of the Highland Rim Natural Region of Indiana. The McKinney Tract is located in the Brown County Hills Section of the same natural region [33]. These three sites are all located within the unglaciated region of the state. A fourth site was located on the Purdue University Nelson-Stokes property in northern Putnam County on the Tipton Till Plain Section of the Central Till Plain Natural Region. Soils at Nelson-Stokes are formed in deep glacial till supporting densely stocked mesic to dry-mesic forest.

Study sites represented a range of basal areas (BA), stocking levels, and overstory oak abundance. Pre-treatment BA ranged from 23.9 m^2/ha on the McKinney tract to 34.7 m^2/ha on the Nelson-Stokes tract (Table 1). Stocking levels at these sites ranged from 89% full stocking on the McKinney tract to 121% full stocking on the Nelson-Stokes tract. Oak abundance ranged from 21% of the total BA on the Wulfman tract to 49% on the Knapp tract. Beech and maple accounted for 27% of the BA across all tracts.

Table 1. Mean ± SE basal area (BA) estimates for silvicultural treatments in each sampling year. Initial BA change was calculated as the percent change between Y0 and Y2 sampling periods for each plot. Therefore, values displayed in the Initial BA Change column reflect means of the percent change in BA, rather than the percent change in the means.

Treatment	Basal Area				Initial BA Change
	2007 (Y0) (m^2 ha^{-1})	2009 (Y2)	2011 (Y4) [†]	2018 (Y10)	Percent
Control	26.7 ± 3.2	28.5 ± 3.0	28.4 ± 2.7	23.9 ± 2.0	+7.0 ± 5.3 [a]
Midstory Removal	27.4 ± 3.1	21.3 ± 2.8	21.2 ± 2.6	20.8 ± 1.9	−24.1 ± 5.0 [b]
Thinning	30.1 ± 3.4	20.4 ± 3.2	20.7 ± 2.9	19.3 ± 2.1	−29.2 ± 5.7 [b]
Shelterwood	30.0 ± 3.1	20.2 ± 2.8	19.9 ± 2.6	17.9 ± 1.8	−36.1 ± 5.0 [b]

Letters indicate significant differences at α = 0.05; only pretreatment BA and percent change tested; [†] Nelson-Stokes site re-measured 2010 (Y3).

2.2. Experimental Design

At each site, three to four plots, each ranging in size from 0.6 to 1.1 ha (1.5 to 2.7 acres) in size, were randomly assigned one of three silvicultural treatments or as a control. The first treatment, a light crown thinning with a residual stocking target of 70%–80% full stocking, was accomplished through a marked commercial timber harvest followed by timber stand improvement to remove non-commercial trees. Crown thinning removed lower-value timber and late successional species, and retained oak species. The second treatment, a midstory removal treatment, removed all trees from 5.1 cm dbh (diameter at breast height) (2.0 inches) to 20.3 to 30.5 cm (8.0 to 12.0 inches) dbh using chainsaw felling, girdling, and herbicide application. Oaks and other desirable species were coppiced. The third treatment was a shelterwood establishment cut that included light crown thinning to 70%–80% full stocking, followed by the removal of all non-merchantable-sized trees down to 5.1 cm (2.0 inches) dbh. Controls received no silvicultural treatment. All sites, other than the Knapp Tract, included each treatment; the Knapp Tract did not include the light crown thinning treatment because a midstory removal treatment had been previously implemented across the entire site, and addition of the light crown treatment would have been akin to a second replication of the shelterwood establishment cut at the site. In addition, control plots for the Knapp Tract were included on adjacent private land to avoid the previous midstory removal treatment, prohibiting the inclusion of a fenced control plot.

Half of each silvicultural treatment plot, including the control, was fenced to exclude deer using an eight-feet-high polypropylene woven fence, resulting in a split-plot design (Figure 1). Within each split plot, 100 1-0 northern red oak (*Quercus rubra* L.) seedlings were hand planted using KBC bars on a grid with approximately 9.1 m × 9.1 m (30 ft × 30 ft) spacing. Immediately after planting, every other seedling was fertilized with 60 g (2.1 oz) of Osmocote® Exact Lo-Start 15N-9P-10K plus minors, 16- to 18-month release fertilizer (ICL Specialty Fertilizers—North America: Dublin, OH, US), resulting in 50 fertilized and 50 unfertilized seedlings per split-plot. Fertilizer was applied directly adjacent to seedling roots using a Pottiputki seedling planter (BAP Equipment, Ltd.: Baileyville, ME, US). Fences were routinely maintained throughout the first 10 years of the study on the three TNC properties. The fence on the Nelson-Stokes property was maintained through the fifth year.

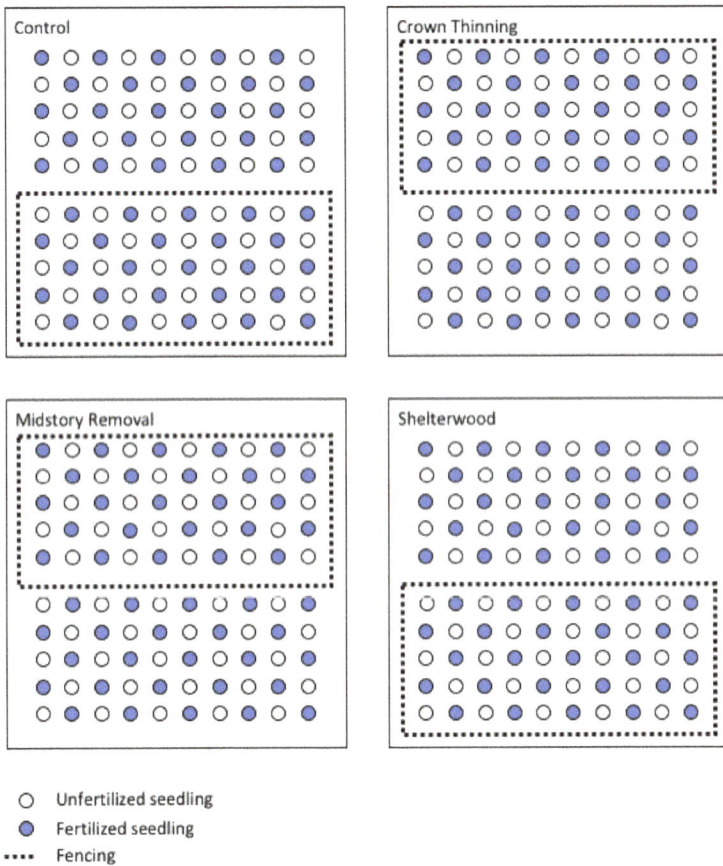

○   Unfertilized seedling
●   Fertilized seedling
····   Fencing

Figure 1. Schematic layout of a single study site, showing the split-plot design created by fencing half of the seedlings underplanted beneath each silvicultural treatment. We applied fertilizer to every other seedling at planting.

2.3. Data Collection

Overstory composition and BA were sampled using four 405-m^2 (0.1-acre) circular plots placed systematically within each silvicultural treatment plot, with two in each fenced or unfenced split-plot. Within 405-m^2 plots, we measured diameter and recorded the species of all trees ≥5.1 cm (2.0 inches) dbh. At the same plot center points, a 40.5-m^2 (0.01-acre) circular plot was used to tally all tree saplings

≥1 m (3.3 ft) tall and <5.1 cm (2.0 inches) dbh. Four 1-m^2 (10.8-ft^2) quadrats were permanently marked 3.6 m (11.8 ft) in cardinal directions from plot center to sample all tree seedlings <1 m (3.3 ft) tall. Sampling occurred prior to treatment and again at years 2, 4, and 10.

Data on planted red oak seedlings were collected at year 10. We measured height to the nearest 1 cm (0.4 inch) using a height pole and measured ground line diameter (GLD) to the nearest 1 mm (0.04 inch) using calipers for all surviving seedlings. We also assessed the competitive status of surviving seedlings by recording the crown class of each seedling, relative to the same cohort of woody regeneration, as well as its free-to-grow status. Free-to-grow status was assessed by projecting a 90-degree cone from the apical bud of each seedling 1.2 m (4.0 ft) upwards and counting the number of cone quadrants occupied by a woody competitor [34].

2.4. Statistical Analysis

We used generalized linear mixed-effects models (GLMMs) to analyze differences in height, GLD, survival, and competitiveness of planted seedlings, as well as density of naturally regenerated oak seedlings and saplings. Height and GLD were analyzed with a linear link (Gaussian distribution); competitiveness and survival were analyzed with a logistic link (binomial distribution); and counts of naturally regenerated seedlings and saplings were analyzed with a log link (Poisson distribution). Competitiveness was treated as binary, and planted seedlings were considered competitive if their crown class position was dominant or co-dominant, or if their crown class position was intermediate, but they were free-to-grow (i.e., zero competitors in cone) because of sparse local competition. GLMMs were initially fit with all possible two-way interactions, which were tested individually for inclusion in each model using likelihood ratio tests ($\alpha = 0.05$). GLMMs for the effects of treatments on planted seedlings included random site intercepts as well as random treatment effects for each site. To account for pretreatment differences between treatment plots, seedling densities were analyzed with a repeated-measures approach. Limited numbers of sapling-size oak in earlier sampling years restricted our analysis of naturally regenerated sapling densities to year 10. To test for significant differences between treatment effects, we used planned contrasts with adjusted *p*-values for multiple comparisons ($\alpha = 0.05$). All analyses were conducted with R programming software, version 3.4.4 [35], using the packages lme4 to build and run GLMMs [36] and multcomp to test planned contrasts [37]. R code and input data used in analyses can be found in Supplementary Materials.

3. Results

3.1. Overstory Conditions

Initial stand BA was slightly higher in plots randomly selected for crown thinning and shelterwood treatments, but these differences were not statistically significant (F3,51.8 = 0.53, *p* = 0.66; Table 1). Mean percent BA reductions between pretreatment values and the first post-treatment sampling year were greatest for the shelterwood treatment, intermediate for the thinning treatment, and lowest for the midstory removal treatment (Table 1). After initial treatments, stand BA declined further during the interval between 2011 and 2017 samples, primarily due to storm damage (RAR, personal observation). The greatest reductions in overstory BA during this interval occurred in control plots (Table 1). Despite the similarities in total BA between thinning and midstory removal treatments at all sampling periods following initial treatments, overstory size class distributions between these treatments were markedly different both before and after perceived storm damage (Figure 2).

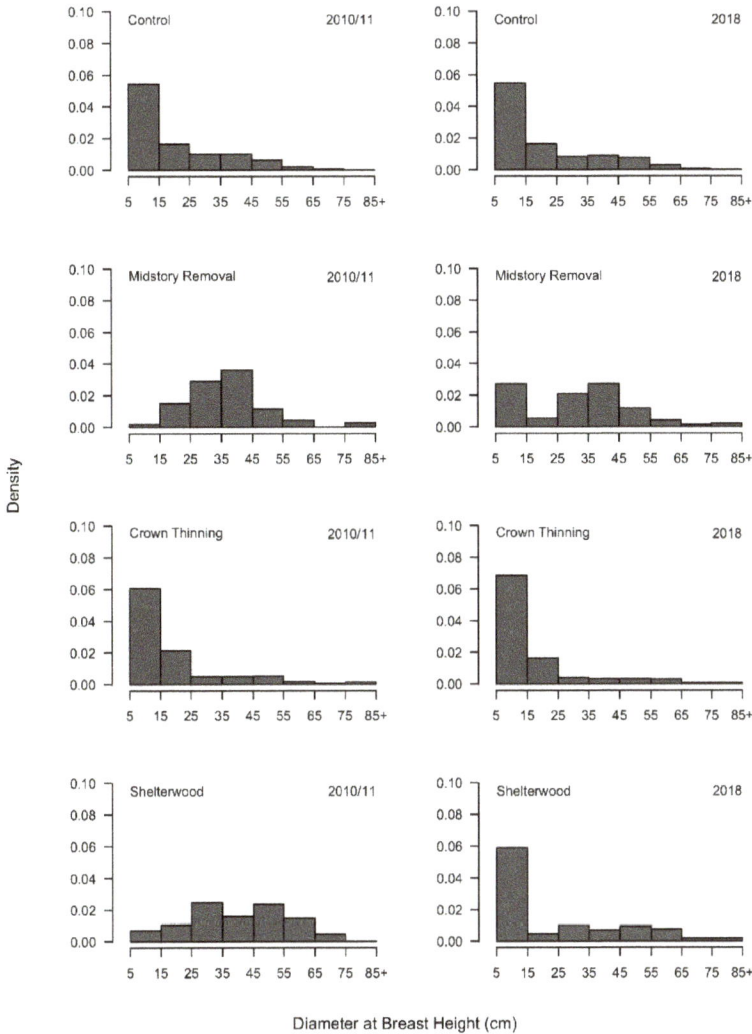

Figure 2. Size class distributions for all trees 5.1 cm dbh (diameter at breast height) and greater, three to four years and ten years after implementation of silvicultural treatments.

3.2. Artificial Regeneration

Of the 2900 seedlings planted at the commencement of this study in 2007, 866 survived to March 2018 across all study sites and treatments. All silvicultural treatments increased seedling survival relative to the control. With no additional treatments, midstory removal and the shelterwood increased the odds of survival 3.9 and 4.7 times more than the thinning treatment, respectively (Table 2). Compared to the crown thinning treatment alone, deer fencing had the strongest effect of any additional treatment on seedling survival, including the effect of the shelterwood cut (Figure 3a). However, a significant interaction between fencing and silvicultural treatment indicated that the strength of this fencing effect varied depending on the treatment combination (χ^2_3 = 27.03, p < 0.001). Fencing was less consequential in the shelterwood than in the thinning treatment and had intermediate effects when paired with midstory removal (Table 2). Fertilizing planted seedlings did not affect their odds of surviving to 10 years ($\chi^2_1 = 0.03$, $p = 0.86$). Overall, 10-year survival approached or exceeded 50 percent in all treatment combinations that included deer fencing and some silvicultural treatment (Table 3).

Table 2. Results of mixed-effects logistic regression analyzing the effects of silvicultural treatments, fencing, and fertilization on 10-year survival and competitiveness of underplanted northern red oak (*Quercus rubra* L.) seedlings. Coefficients are presented on the log-odds scale.

Response	Fixed Effects	Estimate	SE	z Value	Significance	95% Conf. Int.	
						Lower	Upper
Survival							
	Silvicultural Treatment						
	Midstory Removal	3.817 [a]	0.630	6.055	***	2.233	5.401
	Thinning	2.781 [b]	0.659	4.218	***	1.125	4.438
	Shelterwood	4.317 [a]	0.629	6.865	***	2.737	5.897
	Fencing	2.838	0.686	4.136	***	1.114	4.562
	Fertilization	0.015	0.092	0.160		−0.217	0.246
	Silvi. × Fence						
	MR × Fn	−1.328 [a]	0.641	−2.071		−2.939	0.283
	TH × Fn	−0.712 [a]	0.663	−1.075		−2.377	0.952
	SW × Fn	−1.968 [b]	0.639	−3.082	*	−3.572	−0.364
Competitiveness							
	Silvicultural Treatment						
	Midstory Removal [†]	-	-				
	Thinning	−2.000 [a]	0.820	−2.439	.	−4.181	0.181
	Shelterwood	0.341 [b]	0.437	0.780		−0.821	1.503
	Fencing	1.790	0.401	4.464	***	0.724	2.856
	Fertilization	0.974	0.313	3.106	*	0.140	1.807
	Silvi × Fence						
	TH × Fn	1.831 [a]	0.778	2.353		−0.238	3.899
	SW × Fn	−0.318 [b]	0.358	−0.890		−1.269	0.633
	Fert × Fence	−0.795	0.357	−2.227		−1.744	0.154

[†] Reference level; Superscripts indicate significant differences at $\alpha = 0.05$ between levels of a factor. Significance levels are displayed as $p < 0.10$, * $p < 0.05$, ** $p < 0.01$, *** $p < 0.001$.

Table 3. Predicted probabilities of seedling survival and competitiveness generated from the most parsimonious mixed-effects logistic regression models for each response after ten years. Probability of competitiveness is relative to the number of seedlings initially planted. Treatments are listed in descending order by predicted probability of competitiveness. *MR* midstory removal; *TH* thinning; *SW* shelterwood.

Silvicultural Treatment	Fence	Fertilized	Predicted Probability of Survival	Predicted Probability of Competitiveness
SW	Y	Y	0.531	0.161
MR	Y	Y	0.565	0.158
SW	Y	N	0.527	0.138
TH	Y	Y	0.461	0.137
MR	Y	N	0.562	0.135
TH	Y	N	0.457	0.117
SW	N	Y	0.321	0.089
MR	N	Y	0.223	0.065
SW	N	N	0.318	0.035
MR	N	N	0.220	0.025
TH	N	Y	0.092	0.009
TH	N	N	0.091	0.004
Control	Y	Y	0.097	–
Control	Y	N	0.096	–
Control	N	Y	0.006	–
Control	N	N	0.006	–

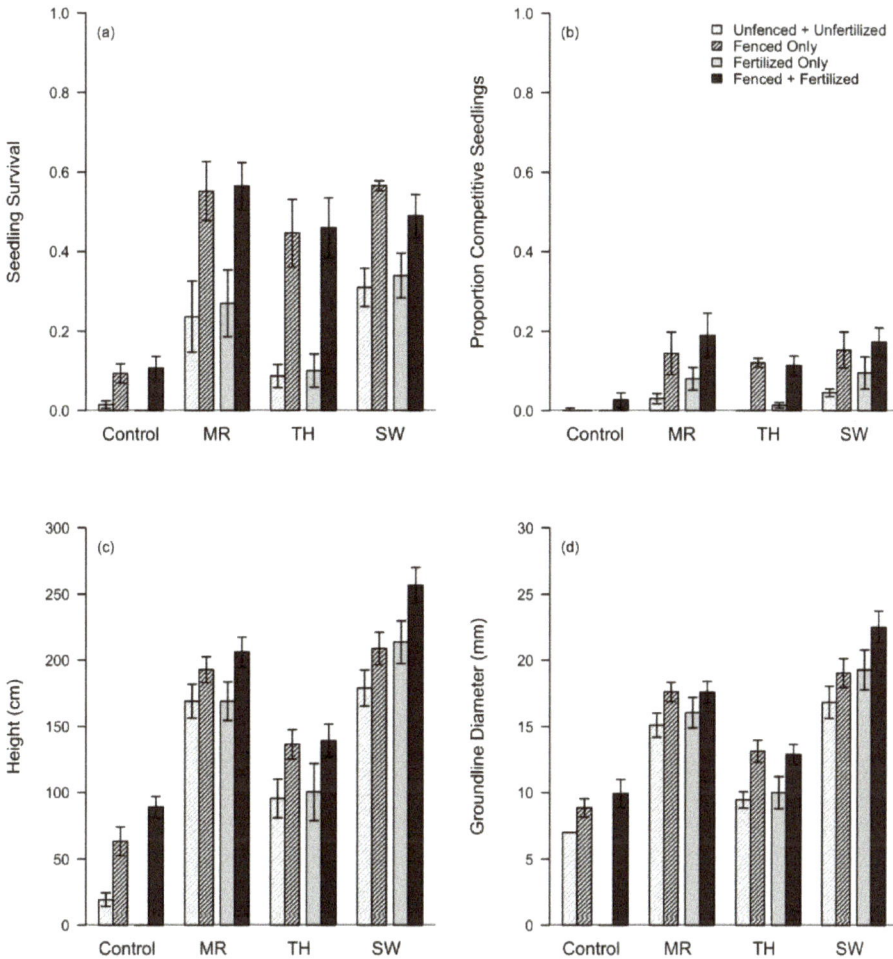

Figure 3. Mean (±SE) survival (**a**), competitiveness (**b**), height (**c**), and ground-line diameter (**d**) of underplanted northern red oak (*Quercus rubra* L.) seedlings ten growing seasons after planting with different silvicultural, fencing, and fertilization treatments. *MR* midstory removal; *TH* light crown thinning; *SW* shelterwood establishment cut.

In addition to increasing survivorship, deer fencing resulted in significantly greater height and diameter growth of surviving seedlings, and the absence of significant interactions between fencing and other treatments indicated that these effects were consistent across treatment combinations (Table 4). Mean increases in 10-year height and diameter growth attributable to fencing ranged from 24–43 cm (9.4–16.9 inches) and 1.5–3.6 mm (0.06–0.14 inches), respectively, in different treatment combinations. The effect of fencing on seedling height varied between study sites, but was consistently positive. Mean heights of all surviving seedlings planted in fenced areas relative to unfenced areas ranged from 13.5 cm (6.6%) greater at the Nelson-Stokes tract to 50.1 cm (81.8%) greater at the McKinney tract and 55.8 cm (31.3%) greater at the Knapp tract. The shelterwood resulted in the greatest height and diameter growth of any silvicultural treatment and had stronger effects on growth than fencing alone, relative to thinning (Figure 3c,d). Fertilized seedlings were larger than unfertilized seedlings in the control, but this may have been an artifact of low survival, as this effect was not evident in any of the treatment plots (Table 4). The combination of fencing and either midstory removal or the shelterwood treatment—which also included a midstory removal component—resulted in 10-year seedling heights of approximately 2 m (6.6 ft) or more (Figure 3c).

Table 4. Results of mixed-effects linear regression analyzing the effects of silvicultural treatments, fencing, and fertilization on 10-year height and ground line diameter (GLD) growth of underplanted northern red oak (*Quercus rubra* L.) seedlings. Responses were transformed with natural log prior to analysis.

Response	Fixed Effects	Estimate	SE	*t* Value	Signif.	95% Conf. Int.	
						Lower	Upper
Height (cm)							
	Silvicultural Treatment						
	Midstory Removal	1.135	0.218	5.205	***	0.560	1.710
	Thinning	0.877	0.226	3.881	***	0.282	1.472
	Shelterwood	1.288	0.217	5.929	***	0.715	1.860
	Fencing	0.299	0.068	4.379	***	0.119	0.479
	Fertilization	0.537	0.228	2.357		−0.063	1.138
	Silvi. × Fert.						
	MR × Fert.	−0.502	0.227	−2.215		−1.100	0.095
	TH × Fert.	−0.539	0.235	−2.289		−1.160	0.081
	SW × Fert.	−0.310	0.226	−1.370		−0.907	0.286
GLD (mm)							
	Silvicultural Treatment						
	Midstory Removal	0.516 [ab]	0.117	4.416	***	0.214	0.818
	Thinning	0.282 [a]	0.121	2.336		−0.030	0.595
	Shelterwood	0.664 [b]	0.117	5.692	***	0.363	0.965
	Fencing	0.184	0.045	4.054	***	0.067	0.301
	Fertilization	0.049	0.050	0.979		−0.081	0.179

Superscripts indicate significant differences at $\alpha = 0.05$ between levels of a factor. Significance levels are displayed as $p < 0.10$, * $p < 0.05$, ** $p < 0.01$, *** $p < 0.001$.

Of the 866 seedlings surviving after 10 years, 226 maintained potentially competitive positions, only one of which occurred in control plots and two of which occurred in unfenced plots beneath crown thinning (Figure 3b). Without fencing, the odds of a planted seedling being in a competitive position after 10 years were 8.0- and 10.4-fold higher in midstory removal and shelterwood treatments than in the thinning treatment ($p = 0.04$ and $p = 0.01$, respectively). Reflecting the interactive effects between fencing and silvicultural treatments on seedling survival, the odds of competitiveness did not differ between silvicultural treatments within deer fencing (Figure 3b). Among surviving seedlings only, the odds of being in a competitive position were no different between silvicultural treatments, but were increased 94 ± 24 percent (mean ± standard error) by fencing and 65 ± 18 percent by fertilization at planting (Table 3).

3.3. Natural Regeneration

Oak seedling densities were affected little by silvicultural treatments or fencing in this study (Figure 4). Overall, treatments explained less than 10% of the variation in oak seedling densities (marginal $R^2 = 0.062$), whereas location (site and plot) effects were much more important and explained well over half the variation (conditional $R^2 = 0.653$). As individual factors, neither fencing ($z = -0.08$, $p > 0.99$) nor the implementation of silvicultural treatments ($z = 1.62$, $p = 0.20$) had any detectable effect on the change in oak seedling density.

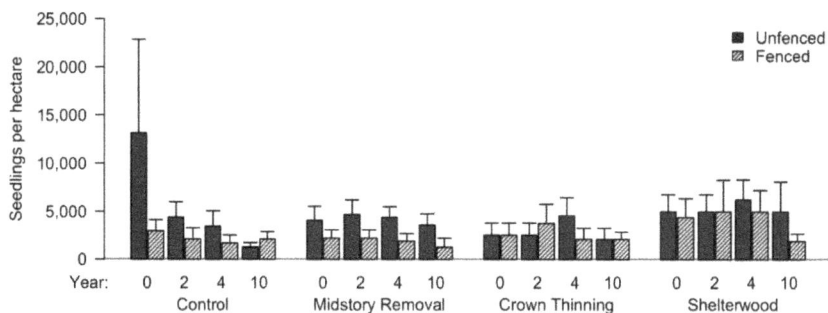

Figure 4. Mean (+SE) density of oak (*Quercus* spp.) seedlings per hectare in response to silvicultural treatments and deer fencing. Year 0 represents pretreatment data.

The density of naturally regenerated oak saplings greater than 1 m tall after 10 growing seasons was generally low across all treatments and was highly variable, with no clear effect of fencing (χ^2_1 = 1.71, p = 0.19; Figure 5). The shelterwood treatment had the most oak saplings, with an average of 41 ± 12 per ha across fenced and unfenced plots. There were more naturally regenerated oak saplings in the shelterwood and midstory removal treatments than the control ($p < 0.01$, $p = 0.03$). Considered in terms of stocking (i.e., whether a 40.5 m^2 plot contained at least one naturally regenerated sapling), both midstory removal and shelterwood treatments increased the odds of a stocked plot 2.5-fold over the control (both z = 2.91, $p < 0.05$), whereas thinning alone increased those same odds only 1.4-fold, which was not statistically different from the control (z = 1.62, p = 0.4).

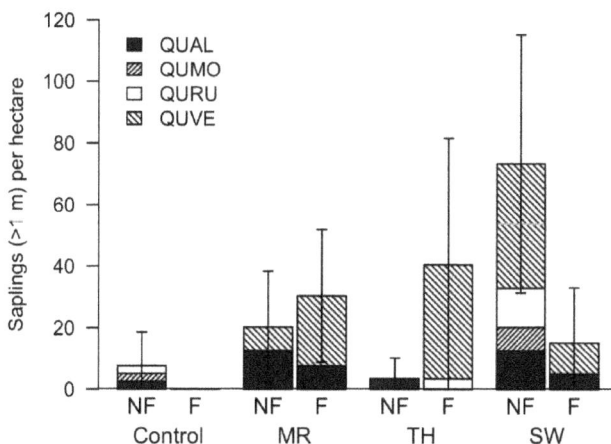

Figure 5. Mean (±SE) density of oak (*Quercus*) saplings (>1 m) per ha, ten growing seasons after midstory removal (MR), crown thinning (TH), and shelterwood (SW) treatments, with deer fencing (F) and without fencing (NF). Only year ten is displayed due to very few oak individuals tall enough to be counted in the sapling plots in previous sampling years.

4. Discussion

Our study sought to determine the best combination of common management approaches for promoting the growth, survival, and competitive status of advance oak reproduction. As we hypothesized, the combination of a shelterwood establishment cut and deer fencing maximized these measures of seedling success, though the effects of midstory removal were generally similar to those of the shelterwood. The benefits of additional treatments are not always additive, and interactive effects between many of the management prescriptions included in this study underscore the need to study them in combination. The numerous management actions commonly recommended to forest landowners for regenerating oak forests can be a deterrent to implementing these practices [38], particularly due to uncertainty surrounding the relative efficiency of different combinations of management actions.

Natural regeneration of oaks was sparse in this study. Sander et al. [39] recommended a minimum of 546 advance oak saplings 1.37 m (4.5 ft) or taller per ha (221 per acre) to obtain 30% oak stocking in the Missouri Ozarks. We only found 67 oak saplings greater than 1 m (3.3 ft) per ha (27 per acre) in our most effective treatment, the shelterwood. Low oak sapling densities may be explained by numerous factors limiting the recruitment of oak seedlings into the sapling layer, including ground-layer competition and browse pressure. Although we did not include understory competition control as a treatment in this study, fencing did not increase oak seedling densities and we found no increases in oak seedling densities even shortly after applying treatments. Acorn production varies dramatically between years and individuals [40], and timing silvicultural treatments to align with years of high acorn production can be critical [31]. Underplanting can help to ensure the presence of oak seedlings and circumvent the need to rigorously structure the timing of silvicultural treatments to coincide with mast years [19].

Underplanting oak seedlings, paired with midstory removal, has been recommended as an initial step in a shelterwood system [41]. Studies consistently show increased height and diameter growth of underplanted oak following midstory removal, and most show increases in survival and competitive ability [34,41], with exceptions likely related to insufficiently large planting stock [42]. Our results are consistent with these findings; mean 10-year heights in all treatment combinations that included midstory removal or shelterwood exceeded the 1.37 m (4.5 ft) minimum recommended by Sander et al. [39] for oaks to be competitive upon release. Notably, overstory BA reductions due to storm damage that affected control plots more than silvicultural treatment plots may have resulted in more conservative comparisons between treatments and the control than would have occurred otherwise. Nonetheless, we expect that comparisons between different silvicultural treatments were affected little due to the relatively small reductions in overstory BA observed between the 2011 and 2018 sampling years.

In addition to midstory removal, further reducing stand density by removing overstory trees is generally recommended as a necessary component of a shelterwood prescription to provide adequate light resources for growth of advance oak reproduction [19]. However, we found no significant differences between midstory removal alone and the shelterwood treatment for any of the metrics examined, suggesting that midstory removal alone may be adequate for underplanted oaks on productive sites. Miller et al. [43] found a shelterwood to be superior to midstory removal alone for survival, growth, and competitiveness of naturally reproduced northern red oak in northcentral Pennsylvania. Despite similar prescriptions for the midstory removal treatments in our study and that of Miller et al. [43], ours resulted in twice the relative BA reduction, 24% compared with 12%. This comparison suggests that complete removal of the midstory may create sufficient growing space for advance red oak reproduction where a dense subcanopy exists, but removal of some larger overstory trees may be necessary where midstory trees make up a smaller proportion of stand BA.

While midstory removal was the most important single treatment for promoting seedling growth, deer fencing was more important for seedling survival, but the strength of this effect varied between silvicultural treatments. Deer are the primary source of browse damage for planted seedlings in the central U.S. [44], and numerous studies have shown the deleterious effects of modern deer abundances on woody reproduction in Indiana [28,45,46]. Northern red oak is highly preferred as browse for deer, relative to co-occurring woody plants in Indiana [47], and seedlings likely lacked the resources necessary to recover from browse damage where the dense midstory layer was left intact. Another perspective on this interaction between treatments is that higher light availability may have reduced browse pressure on individual oak seedlings in the midstory removal and shelterwood treatments by increasing the abundance of alternative browse, thereby making deer fencing less beneficial [3,24,48,49]. Precise estimates of white-tailed deer populations across Indiana are not currently available, but our results show that while deer may impact seedling growth more strongly in some areas than others, browse pressure on unprotected seedlings was evident throughout the region encompassed by this study. We observed the weakest effect of deer fencing on seedling height growth at the Nelson-Stokes tract. Nelson-Stokes is located further north than other sites, but also had apparent damage to fencing that was not repaired between years five and ten of the study, preventing us from drawing conclusions about the effects of geography on browse pressure.

The presence of deer also mediated the effects of fertilization, which increased the odds that a seedling would be in a competitive position after 10 years, but only for seedlings not protected by fencing. In some cases, fertilizing seedlings at planting can protect against herbivory by allowing seedlings to grow above the browse line more rapidly, providing nutrients for the production of secondary defense compounds, and aiding in recovery from browse damage, but can make them more palatable to ungulates by increasing foliar nutrient content [50]. However, seedling responses to fertilization can also vary according to site-specific soil chemistry conditions [29], and red oak often shows no evidence of elevated tissue N or positive growth response to fertilization [27,28,51]. Consistent with these studies, we found no effect of fertilization on seedling growth. Therefore, higher nutrient supply did not help seedlings to more rapidly attain escape heights in this study, nor does existing evidence indicate that fertilization affects foliar tannin concentrations or browsing incidence for red oak [28]. Fertilization may have helped seedlings recover from minor or sporadic browse damage, but is unlikely to be a worthwhile investment if browsing is chronic or severe. Fencing is a better, and likely only option on high browse-pressure sites.

Though the shelterwood approach to oak regeneration is designed to reduce understory competition, relative to clearcutting, direct control of understory competition may have further improved outcomes for planted seedlings in some treatment combinations [43,52]. Competition control may be a more palatable approach for private landowners wary of the costs associated with deer fencing, but may have the unintended consequence of increasing browse pressure on planted seedlings [53], and may only be warranted where combined with fencing or reduced deer populations. The results of our study underscore the potential for variable effects of different treatments when applied in combination, and the interactions between deer browse and competition control warrant further investigation.

Increasing adoption of oak regeneration practices may be an even greater barrier than the silvicultural challenges to oak regeneration, in large part due to the economic costs and resulting risk associated with approaches such as fencing [38]. Continuing to develop geographically-specific prescriptions by testing a multitude of treatment combinations can reduce the risk associated with managing for oak regeneration. However, a rigorous economic analysis weighing the risks associated with each approach against their potential benefits is currently lacking in the Midwest and is the next necessary step in advancing the implementation of oak regeneration practices.

5. Conclusions

Regenerating oak forests in the eastern U.S. is a challenge with broad ecological implications and economic consequences. The many impediments to oak regeneration will accordingly require multifaceted solutions. Underplanted oak seedlings require careful management to ensure their success on many sites, but our results emphasize their usefulness for augmenting existing oak reproduction and potentially reducing instances of oak regeneration failure, even without re-entries to reduce understory competition via prescribed fire or herbicides. While natural regeneration will remain the primary means of regenerating oak forests in much of the eastern U.S. [19], artificial regeneration may be particularly suited to privately owned forest tracts, which tend to be relatively small and are more likely to be under-managed [54]. Interactive effects between treatments in our study show that the utility of various management actions to promote oak regeneration can depend particularly on the installation of deer fencing, or potentially on deer population densities. Fencing was far more effective than fertilization, and fertilization had little effect on seedlings protected by fencing. Following a selective crown thinning typical of private timber sales in the eastern U.S. [55], subsequent midstory removal to emulate a shelterwood establishment cut, coupled with deer fencing, can maximize growth and survival of underplanted oak seedlings.

Supplementary Materials: R code used in statistical analyses and the associated input data is available online at http://www.mdpi.com/1999-4907/9/9/571/s1..

Author Contributions: Conceptualization, R.A.R.; Methodology, G.S.F., R.A.R., and M.R.S.; Validation, G.S.F., R.A.R., and M.R.S.; Formal Analysis, G.S.F.; Investigation, G.S.F. and R.A.R.; Resources, R.A.R. and M.R.S.; Data Curation, G.S.F. and R.A.R.; Writing—Original Draft Preparation, G.S.F.; Writing—Review & Editing, R.A.R. and M.R.S.; Visualization, G.S.F.; Supervision, M.R.S.; Project Administration, G.S.F., R.A.R., and M.R.S.; Funding Acquisition, R.A.R.

Acknowledgments: The authors wish to acknowledge the support and cooperation of The Nature Conservancy, access to sites provided by participants in The Forest Bank program, and the time and resources of numerous individuals, including Brian Beheler, Chad Bladow, Don Carlson, Emily Frazier, Cassie Hauswald, Doug Jacobs, John McKim, Andy Meier, Chris Neggers, Charlotte Owings, Allen Pursell, and Dan Shaver.

Conflicts of Interest: Several employees of The Nature Conservancy, which provided funding for this project, assisted in data collection and implementation of the study. However, The Nature Conservancy agreed prior to the outset of the study that the results be published, and had no role in study design, data analysis and interpretation, or writing of the manuscript.

References

1. Hanberry, B.B.; Nowacki, G.J. Oaks were the historical foundation genus of the east-central United States. *Quat. Sci. Rev.* **2016**, *145*, 94–103. [CrossRef]
2. Clark, F.B. An historical perspective of oak regeneration. In *Oak Regeneration: Serious Problems Practical Recommendations*; Loftis, D.L., McGee, C.E., Eds.; USDA Forest Service Gen. Tech. Rep. SE-84; U.S. Department of Agriculture, Forest Service, Southeastern Forest Experiment Station: Asheville, NC, USA, 1993.
3. Lashley, M.A.; Harper, C.A.; Bates, G.E.; Keyser, P.D. Forage availability for white-tailed deer following silvicultural treatments in hardwood forests. *J. Wildl. Manag.* **2011**, *75*, 1467–1476. [CrossRef]
4. Lorimer, C.G. Causes of the oak regeneration problem. In *Oak Regeneration: Serious Problems, Practical Recommendations*; Loftis, D.L., McGee, C.E., Eds.; USDA Forest Service Gen. Tech. Rep. SE-84; U.S. Department of Agriculture, Forest Service, Southeastern Forest Experiment Station: Asheville, NC, USA, 1993; pp. 14–39.
5. Russell, F.L.; Zippin, D.B.; Fowler, N.L. Effects of White-Tailed Deer (Odocoileus Virginianus) on Plants, Plant Populations and Communities: A Review. *Am. Midl. Nat.* **2001**, *146*, 1–26. [CrossRef]
6. Fei, S.; Kong, N.; Steiner, K.C.; Moser, W.K.; Steiner, E.B. Change in oak abundance in the eastern United States from 1980 to 2008. *For. Ecol. Manag.* **2011**, *262*, 1370–1377. [CrossRef]
7. Nowacki, G.J.; Abrams, M.D. The Demise of Fire and "Mesophication" of Forests in the Eastern United States. *Bioscience* **2008**, *58*, 123–138. [CrossRef]
8. Lhotka, J.M.; Loewenstein, E.F. Influence of canopy structure on the survival and growth of underplanted seedlings. *New For.* **2008**, *35*, 89–104. [CrossRef]
9. Jenkins, M.A.; Parker, G.R. Composition and diversity of woody vegetation in silvicultural openings of southern Indiana forests. *For. Ecol. Manag.* **1998**, *109*, 57–74. [CrossRef]
10. Keyser, T.L.; Loftis, D.L. Long-term effects of single-tree selection cutting on structure and composition in upland mixed-hardwood forests of the southern Appalachian Mountains. *Forestry* **2013**, *86*, 255–265. [CrossRef]
11. Raymond, P.; Munson, A.D.; Ruel, J.C.; Nolet, P. Group and single-tree selection cutting in mixed tolerant hardwood-white pine stands: Early establishment dynamics of white pine and associated species. *For. Chron.* **2003**, *79*, 1093–1106. [CrossRef]
12. Morrissey, R.C.; Jacobs, D.F.; Seifert, J.R.; Kershaw, J.A. Overstory species composition of naturally regenerated clearcuts in an ecological classification framework. *Plant Ecol.* **2010**, *208*, 21–34. [CrossRef]

13. Schweitzer, C.J.; Dey, D.C. Forest structure, composition, and tree diversity response to a gradient of regeneration harvests in the mid-Cumberland Plateau escarpment region, USA. *For. Ecol. Manag.* **2011**, *262*, 1729–1741. [CrossRef]
14. Loftis, D.L. A shelterwood method for regenerating red oak in the Southern Appalachians. *For. Sci.* **1990**, *36*, 917–929.
15. Kellner, K.F.; Swihart, R.K. Timber harvest and drought interact to impact oak seedling growth and survival in the Central Hardwood Forest. *Ecosphere* **2016**, *7*, 1–15. [CrossRef]
16. Brose, P.H.; Van Lear, D.H. Responses of hardwood advance regeneration to seasonal prescribed fires in oak-dominated shelterwood stands. *Can. J. For. Res.* **1998**, *28*, 331–339. [CrossRef]
17. Brose, P.H.; Gottschalk, K.W.; Horsley, S.B.; Knopp, P.D.; Kochenderfer, J.N.; McGuiness, B.J.; Miller, G.W.; Ristau, T.E.; Stoleson, S.H.; Stout, S.L. *Prescribing Regeneration Treatments for Mixed-Oak Forests in the Mid-Atlantic Region*; USDA Forest Service Gen. Tech. Rep. NRS-33; U.S. Department of Agriculture, Forest Service, Northern Research Station: Newtown Square, PA, USA, 2008.
18. Gordon, A.M.; Simpson, J.A.; Williams, P.A. Six-year response of red oak seedlings planted under a shelterwood in central Ontario. *Can. J. For. Res.* **1995**, *25*, 603–613. [CrossRef]
19. Dey, D.C.; Gardiner, E.S.; Schweitzer, C.J.; Kabrick, J.M.; Jacobs, D.F. Underplanting to sustain future stocking of oak (*Quercus*) in temperate deciduous forests. *New For.* **2012**, *43*, 955–978. [CrossRef]
20. Dey, D.C.; Spetich, M.A.; Weigel, D.R.; Johnson, P.S.; Graney, D.L.; Kabrick, J.M. A suggested approach for design of oak (*Quercus* L.) regeneration research considering regional differences. *New For.* **2009**, *37*, 123–135. [CrossRef]
21. Sander, I.L.; Graney, D.L. Regenerating oaks in the central states. In *Oak Regeneration: Serious Problems, Practical Recommendations.*; Loftis, D.L., McGee, C.E., Eds.; USDA Forest Service Gen. Tech. Rep. SE-84; U.S. Department of Agriculture, Forest Service, Southeastern Forest Experiment Station: Asheville, NC, USA, 1993.
22. Jarnemo, A.; Minderman, J.; Bunnefeld, N.; Zidar, J.; Mansson, J. Managing landscapes for multiple objectives: Alternative forage can reduce the conflict between deer and forestry. *Ecosphere* **2014**, *5*. [CrossRef]
23. Murray, B.D.; Webster, C.R.; Jenkins, M.A.; Saunders, M.R.; Haulton, G.S. Ungulate impacts on herbaceous-layer plant communities in even-aged and uneven-aged managed forests. *Ecosphere* **2016**, *7*, 1–20. [CrossRef]
24. Miller, B.F.; Campbell, T.A.; Laseter, B.R.; Ford, W.M.; Miller, K.V. White-tailed deer herbivory and timber harvesting rates: Implications for regeneration success. *For. Ecol. Manag.* **2009**, *258*, 1067–1072. [CrossRef]
25. Royo, A.A.; Kramer, D.W.; Miller, K.V.; Nibbelink, N.P.; Stout, S.L. Spatio-temporal variation in foodscapes modifies deer browsing impact on vegetation. *Landsc. Ecol.* **2017**, *32*, 2281–2295. [CrossRef]
26. Schreffler, A.M.; Sharpe, W.E. Effects of lime, fertilizer, and herbicide on forest soil and soil solution chemistry, hardwood regeneration, and hardwood growth following shelterwood harvest. *For. Ecol. Manag.* **2003**, *177*, 471–484. [CrossRef]
27. Tripler, C.E.; Canham, C.D.; Inouye, R.S.; Schnurr, J.L. Soil nitrogen availability, plant luxury consumption, and herbivory by white-tailed deer. *Oecologia* **2002**, *133*, 517–524. [CrossRef] [PubMed]
28. Burney, O.T.; Jacobs, D.F. Species selection—A fundamental silvicultural tool to promote forest regeneration under high animal browsing pressure. *For. Ecol. Manag.* **2018**, *408*, 67–74. [CrossRef]
29. Cha, D.H.; Appel, H.M.; Frost, C.J.; Schultz, J.C.; Steiner, K.C. Red oak responses to nitrogen addition depend on herbivory type, tree family, and site. *For. Ecol. Manag.* **2010**, *259*, 1930–1937. [CrossRef]
30. Kellner, K.F.; Riegel, J.K.; Swihart, R.K. Effects of silvicultural disturbance on acorn infestation and removal. *New For.* **2014**, *45*, 265–281. [CrossRef]
31. Zaczek, J.J.; Lhotka, J.M. Seedling reproduction established with soil scarification within an oak overwood after overstory removal. *North J. Appl. For.* **2004**, *21*, 5–11.
32. Spetich, M.A.; Dey, D.C.; Johnson, P.S. Shelterwood-planted northern red oaks: Integrated costs and options. *J. Appl. For.* **2009**, *33*, 182–186.
33. Homoya, M.A.; Abrell, D.B.; Aldrich, J.R.; Post, T.W. The natural regions of Indiana. *Proc. Indiana Acad. Sci.* **1984**, *94*, 245–268. [CrossRef]
34. Parrott, D.L.; Lhotka, J.M.; Stringer, J.W.; Dillaway, D.N. Seven-Year Effects of Midstory Removal on Natural and Underplanted Oak Reproduction. *North J. Appl. For.* **2012**, *29*, 182–190. [CrossRef]

35. R Core Team. *R: A Language and Environment for Statistical Computing*; R Foundation for Statistical Computing: Vienna, Austria, 2018; Available online: https://www.R-project.org/ (accessed on 15 March 2018).
36. Bates, D.; Maechler, M.; Bolker, B.; Walker, S. Fitting linear mixed-effects models using lme4. *J. Stat. Softw.* **2015**, *67*, 1–48. [CrossRef]
37. Hothorn, T.; Bretz, F.; Westfall, P. Simultaneous inference in general parametric models. *Biometr. J.* **2008**, *50*, 346–363. [CrossRef] [PubMed]
38. Knoot, T.G.; Schulte, L.A.; Rickenbach, M. Oak conservation and restoration on private forestlands: Negotiating a social-ecological landscape. *Environ. Manag.* **2010**, *45*, 155–164. [CrossRef] [PubMed]
39. Sander, I.L.; Johnson, P.S.; Rogers, R. *Evaluating Oak Advance Reproduction in the Missouri Ozarks*; USDA Forest Service Gen. Tech Rep. NC-251; U.S. Department of Agriculture, Forest Service, North Central Forest Experiment Station: St. Paul, MN, USA, 1984; 16p.
40. Healy, W.M.; Lewis, A.M.; Boose, E.F. Variation of red oak acorn production. *For. Ecol. Manag.* **1999**, *116*, 1–11. [CrossRef]
41. Lhotka, J.M.; Loewenstein, E.F. Effect of midstory removal on understory light availability and the 2-year response of underplanted cherrybark oak seedlings. *South J. Appl. For.* **2009**, *33*, 171–177.
42. Craig, J.M.; Lhotka, J.M.; Stringer, J.W. Evaluating Initial Responses of Natural and Underplanted Oak Reproduction and a Shade-Tolerant Competitor to Midstory Removal. *For. Sci.* **2014**, *60*, 1164–1171. [CrossRef]
43. Miller, G.W.; Brose, P.H.; Gottschalk, K.W. Advanced Oak Seedling Development as Influenced by Shelterwood Treatments, Competition Control, Deer Fencing, and Prescribed Fire. *J. For.* **2017**, *115*, 179–189. [CrossRef]
44. Stange, E.E.; Shea, K.L. Effects of Deer Browsing, Fabric Mats, and Tree Shelters on Quercus rubra Seedlings. *Restor. Ecol.* **1998**, *6*, 29–34. [CrossRef]
45. Jenkins, L.H.; Jenkins, M.A.; Webster, C.R.; Zollner, P.A.; Shields, J.M. Herbaceous layer response to 17 years of controlled deer hunting in forested natural areas. *Biol. Conserv.* **2014**, *175*, 119–128. [CrossRef]
46. Jenkins, L.H.; Murray, B.D.; Jenkins, M.A.; Webster, C.R. Woody regeneration response to over a decade of deer population reductions in Indiana state parks 1. *J. Torrey Bot. Soc.* **2015**, *142*, 205–219. [CrossRef]
47. Wakeland, B.; Swihart, R.K. Ratings of white-tailed deer preferences for woody browse in Indiana. *Proc. Indiana Acad. Sci.* **2009**, *118*, 96–101.
48. Kellner, K.F.; Swihart, R.K. Simulation of oak early life history and interactions with disturbance via an individual-based model, SOEL. *PLoS ONE* **2017**, *12*, 1–23. [CrossRef] [PubMed]
49. Johnson, A.S.; Hale, P.E.; Ford, W.M.; Wentworth, J.M.; French, J.R.; Anderson, O.F.; Pullen, G.B. White-tailed deer foraging in relation to successional stage, overstory type and management of southern Appalachian forests. *Am. Midl. Nat.* **1995**, *133*, 18–35. [CrossRef]
50. Burney, O.T.; Jacobs, D.F. Ungulate herbivory of boreal and temperate forest regeneration in relation to seedling mineral nutrition and secondary metabolites. *New For.* **2013**, *44*, 753–768. [CrossRef]
51. Schuler, J.L.; Robison, D.J. Performance of northern red oak enrichment plantings in naturally regenerating Southern Appalachian hardwood stands. *New For.* **2010**, *40*, 119–130. [CrossRef]
52. Lorimer, C.G.; Chapman, J.W.; Lambert, W.D. Tall understorey vegetation as a factor in the poor development of oak seedlings beneath mature stands. *J. Ecol.* **1994**, *82*, 227–237. [CrossRef]
53. Dalgleish, H.J.; Lichti, N.I.; Schmedding, N.; Swihart, R.K. Exposure to herbivores increases seedling growth and survival of American chestnut (*Castanea dentata*) through decreased interspecific competition in canopy gaps. *Restor. Ecol.* **2015**, *23*, 655–661. [CrossRef]
54. Ross-Davis, A.; Broussard, S. A typology of family forest owners in North Central Indiana. *North J. Appl.* **2007**, *24*, 282–289.
55. Kittredge, D.B.; Finley, A.O.; Foster, D.R. Timber harvesting as ongoing disturbance in a landscape of diverse ownership. *For. Ecol. Manag.* **2003**, *180*, 425–442. [CrossRef]

Article

Relationships between Tree Vigor Indices and a Tree Classification System Based upon Apparent Stem Defects in Northern Hardwood Stands

Edouard Moreau [1], Steve Bédard [2], Guillaume Moreau [1] and David Pothier [1,*]

[1]   Centre d'étude de la forêt, Département des sciences du bois et de la forêt, Pavillon Abitibi-Price, 2405 rue de la Terrasse, Université Laval, Québec, QC G1V 0A6, Canada; edouard.moreau1@gmail.com (E.M.); guillaume.moreau.3@ulaval.ca (G.M.)

[2]   Direction de la recherche forestière, Ministère des Forêts, de la Faune et des Parcs du Québec, 2700 rue Einstein, Québec, QC G1P 3W8, Canada; steve.bedard@mffp.gouv.qc.ca

*   Correspondence: david.pothier@sbf.ulaval.ca; Tel.: +1-418-656-2131 (ext. 12908)

Received: 10 September 2018; Accepted: 18 September 2018; Published: 21 September 2018

Abstract: Many northern hardwood stands include several low-vigor trees as a result of past management. To restore these degraded stands, partial cuts are applied with partly validated tree classification systems that are based upon apparent stem defects. We sampled 214 sugar maple (*Acer saccharum* Marsh.) and 84 yellow birch (*Betula alleghaniensis* Britt.) trees from six sites covering the northern hardwood forest zone of the Province of Quebec, Canada. We evaluated their vigor with a four-class system, and quantified the growth efficiency index and several indices that were based solely upon radial growth. The growth efficiency index increased non-significantly with increasing tree vigor class. The five-year basal area increment (BAL_{1-5}) was significantly different between the lowest and highest tree vigor classes. Yet, temporal changes in BAL_{1-5} helped classify correctly only 16% of high-vigor trees that became poorly vigorous 8–10 years later. Overall, these results suggest that the tree classification system is weakly related to actual tree vigor and its application likely generates few significant gains in future stand vigor. Modifying and simplifying the tree vigor system must be considered to facilitate the tree marking process that is required to improve the vigor of degraded stands.

Keywords: sugar maple; yellow birch; tree vigor; growth efficiency index; tree selection

---

1. Introduction

Northern hardwood forests cover large areas of southeastern Canada and the northeastern USA, from the Great Lakes region to the Atlantic Ocean. These forest ecosystems have considerable socio-economic importance because they are located near densely populated areas and are composed of valuable tree species that are used by the appearance wood-products industry [1,2]. The traditionally high demand for these high-value products (veneers, flooring and furniture, among others) has long led to the application of diameter-limit cuts, which are aimed at mainly harvesting trees with high-quality saw logs [3]. Several decades of selection that is biased towards the most valuable trees within a stand, an unsustainable practice that is also known as 'high-grading', has resulted in large areas being covered by degraded, low-vigor forest stands [4–6].

In 2005, a new tree-vigor classification system was introduced in the Province of Quebec, Canada [1,2,7]. The system was implemented with the clear objective of restoring the overall vigor of degraded hardwood stands by prioritizing the harvesting of low-vigor trees in partial cut operations [1]. To achieve this objective, the classification system defined tree vigor as the risk of mortality during the upcoming cutting cycle (i.e., a 25-year period), based upon the presence or absence of precursors for wood decay and other tree defects [1,2]. Although this four-class tree-vigor system is based upon an extensive knowledge of forest pathology [7], its further application requires empirical validation. Among the studies that have attempted to realize such validations, Fortin et al. [8] and Guillemette et al. [9] observed that mortality risk of the lowest tree-vigor class was effectively higher than that of the other classes, while these other classes were only slightly different from one another. Furthermore, Hartmann et al. [10] observed that differences in radial growth occurred between the lowest and highest classes of vigor, without observing any differences between these extremes and the two intermediate classes. Overall, these results would suggest that this tree classification system might not achieve the intended purpose because its inherent complexity would not warrant its widespread application, given its lack of desired accuracy.

To obtain a better overview of the application potential of classification systems that are based upon apparent stem defects, validation using a proven quantitative tree vigor index would be appropriate. The growth efficiency index [11] appears to be suitable for this purpose since it corresponds to the ratio of annual stemwood production, which is one of the last priorities of tree resource allocation [12], to tree leaf area, which represents tree carbon acquisition. Therefore, a low value of this index indicates that most of the acquired carbon is used to sustain high priority allocations such as shoot and root growth, and little is left for stemwood production and defensive compounds [12]. Thus, trees associated with a low growth efficiency index should be poorly vigorous and more prone to mortality. Accordingly, the growth efficiency index is recognized as a reliable tree vigor index [13] and has been successfully used to study tree mortality [12] and stand growth dynamics [14]. In addition, this index has efficiently predicted the vigor of lodgepole pine (*Pinus contorta* Douglas) [15–17] and sugar maple (*Acer saccharum* Marshall) [18] trees, together with the probability of mortality of balsam fir (*Abies balsamea* (L.) Miller) in relation to insect defoliation [19], and the response of conifer stands to commercial thinning [20].

Because calculation of the growth efficiency index requires the estimation of tree leaf area, its use for retrospective tree vigor evaluations is difficult. Such retrospective evaluations are more easily achieved using indices that are based upon growth-ring analysis [10]. Accordingly, such growth-based indices are widely used and recognized as good indicators of tree vigor (e.g., [10,21–24]). Therefore, it would be relevant to establish the degree of agreement between the growth efficiency index, indices that are based solely on tree growth, and tree vigor classes that have been deduced from apparent stem defects.

The general objective of this study was to validate a tree classification system that was based upon apparent stem defects by using two types of tree vigor indices. In agreement with this objective, we stated three hypotheses: (1) growth efficiency index values increase with increasing tree vigor class; (2) temporal decreases in tree vigor class are paralleled by similar decreases in growth-base indices; and (3) tree and stand state variables are adequate predictors of the growth efficiency index. To test these hypotheses, we used a large sample of sugar maple and yellow birch (*Betula alleghaniensis* Britton) trees, within which vigor classes were evaluated twice on several individuals at intervals of 8 to 10 years. The results of this study could help improve tree selection for partial cutting with a goal of increasing the overall vigor of degraded northern hardwood stands.

2. Materials and Methods

2.1. Study Sites

Trees were sampled in six sites that were distributed across the northern hardwood zone of the Province of Quebec, Canada. Three sites were located on public lands (Mont-Laurier: 46°39′ N, 75°38′ W; Duchesnay: 46°39′ N, 75°38′ W; and Biencourt: 48°01′ N, 68°30′ W), while the three others were located on private woodlots (45°28′–46°28′ N, 70°20′–71°45′ W) that are owned by Domtar Corporation (Montreal, QC, Canada) (Figure 1).

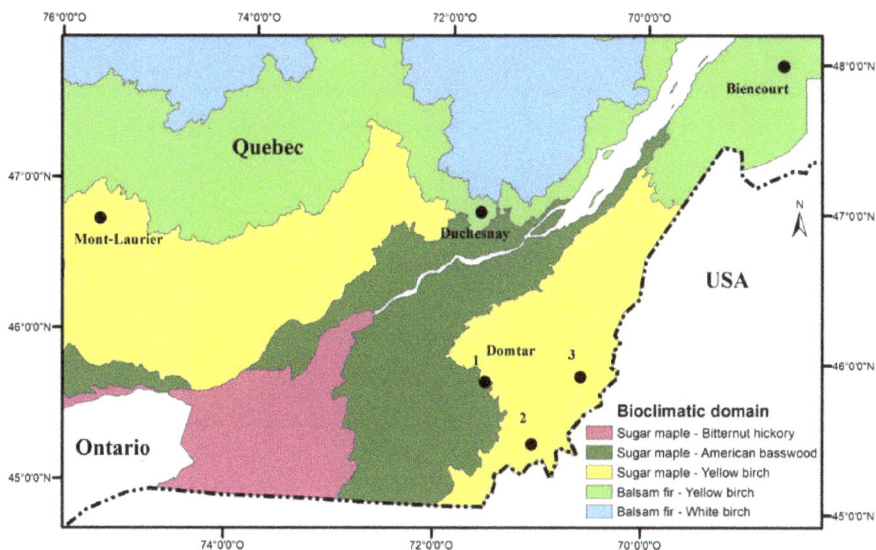

Figure 1. Locations of the six study sites in northern hardwood forests of Quebec.

The Mont-Laurier site lies within the sugar maple–yellow birch bioclimatic domain, whereas the Duchesnay and Biencourt sites are part of the balsam fir–yellow birch bioclimatic domain [25]. Two sites that were located on the private woodlots of Domtar (sites Domtar 2 and 3, Figure 1) are also part of the sugar maple–yellow birch bioclimatic domain, while the third privately owned site (site Domtar 1, Figure 1) is part of the sugar maple–American basswood (*Tilia americana* L.) bioclimatic domain. The latter is characterized by mean annual temperatures between 4 and 5 °C, and mean annual precipitation ranging between 1000 and 1150 mm; the length of the growing season ranges from 160 to 190 days. The sugar maple–yellow birch domain is characterized by mean annual temperatures between 2.5 and 5 °C, and mean annual precipitation between 950 and 1100 mm, with a growing season of 160–180 days. The balsam fir–yellow birch domain is characterized by mean annual temperatures between 1.5 and 2.5 °C, and mean annual precipitation between 900 and 1100 mm, with a growing season of 160–170 days [25]. The topography of all sites is characterized by hills and an average slope of 15°, with main surface deposits composed of shallow or deep tills [26]. The study sites were covered by naturally established, uneven-aged stands that were mainly composed of sugar maple, yellow birch and red maple (*Acer rubrum* L.), with minor components of American beech (*Fagus grandifolia* Ehrhart), balsam fir, and red spruce (*Picea rubens* Sargent).

2.2. Tree Sampling

The Mont-Laurier and Biencourt sites were sampled respectively in 2010 and 2011, while trees from the Duchesnay site were first sampled in 2007 and 2008, and a second time in 2016. In these three sites, trees were individually selected to cover a broad range of diameter and vigor classes. Trees from the three Domtar sites were located in permanent sample plots (400 m^2) that were first measured in 2006 and a second time in 2016. In total, we sampled 154 trees (91 sugar maple (sM) and 63 yellow birch (yB) trees) from Duchesnay, 31 (31 sM) from Mont-Laurier, 30 (30 sM) from Biencourt, 19 (11 sM, 8 yB) from Domtar 1, 33 (20 sM, 13 yB) from Domtar 2, and 31 (31 sM) from Domtar 3. The number of sampled trees decreased with increasing diameter class, consistent with the negative exponential distribution that is typical of uneven-aged stands (Table 1).

Table 1. Number of trees that were sampled at the six sites by diameter class at breast height (DBH, 1.3 m) and tree vigor class

DBH (cm)	M	S	C	R	Total
20–29.9	29	13	39	29	110
30–39.9	16	20	28	28	92
40–49.9	15	16	21	20	72
50–59.9	6	9	7	2	24
Total	66	58	95	79	298

In each sampling year, we recorded the diameter at breast height (DBH; ±0.1 cm), the vigor class, the total height and the height at the base of the live crown (±0.1 m) of each tree. We also measured crown radii (±0.01 m) along the four cardinal directions (N, E, S, and W) from the centre of the bole to the edge of the crown using vertical sighting [27]. Increment cores that were sampled at 1.3 m above ground level were then air-dried and progressively sanded down to allow a clear identification of growth rings. Ring widths were measured using a Velmex micrometer (±0.002 mm) and were corrected to take into account radial shrinkage during drying.

2.3. Tree Vigor Classification

We evaluated tree vigor with a four-class system, which is assumed to be related to the probability of tree mortality over the next 25-year period [28]. As described by Delisle-Boulianne et al. [1], the distinction among classes is based upon the presence and severity of certain types of defects. These are grouped into eight categories: (1) conks and stromata; (2) cambial necrosis; (3) stem deformations and injuries; (4) stem base and root defects; (5) stem and bark cracks; (6) woodworms and sap wells; (7) crown decline; and (8) forks and pruning defects. Dichotomous keys that are based upon these various defects can be used to assign any tree to one of the four following classes: Class M corresponds to trees with major defects that will likely die during the next 25-year period; Class S refers to trees with major defects that are thought to not compromise their likelihood of survival in the medium term; Class C includes trees with some minor defects and that are still growing; and Class R is composed of trees that are almost defect-free that should be preserved for the future.

2.4. Vigor Indices

To examine relationships between the vigor classes that were based on apparent stem defects and quantitatively measured indices, we used four growth indices of which one was based on tree growth per unit leaf area, while the remaining three were solely based on tree diameter measurements.

2.4.1. Growth Efficiency Index

The growth efficiency index is defined as the annual production of stemwood per unit leaf area (Waring 1980) [11]. The average stemwood mass that was produced annually by each tree was calculated with the following equation, which has been adapted from Lambert et al. [29]

$$\Delta W_s = \frac{\beta_1 (D_t^{\beta_2} - D_{t-5}^{\beta_2})}{5} \tag{1}$$

where ΔW_s is the dry mass of stemwood produced annually (kg), D_t is tree DBH in the year of tree vigor class determination (cm), D_{t-5} is tree DBH five years prior to tree vigor class determination (cm), which was determined from increment cores, the parameter β_1 takes the empirical value of 0.1315 for sugar maple and 0.1932 for yellow birch, while the value of β_2 is 2.3129 for sugar maple and 2.1569 for yellow birch [29]. The leaf area of each tree was estimated with an equation that was developed by Moreau et al. [30]

$$LA = \beta_1 CSA^{\beta_2} \tag{2}$$

where LA is tree leaf area (m^2), CSA is crown surface area (m^2), which corresponds to the area of the geometric shape of sugar maple and yellow birch crowns (calculated with crown height and mean quadratic radius—Moreau et al. [30]), β_1 is 1.121 for sugar maple and 1.021 for yellow birch, and β_2 is 0.981 for sugar maple and 1.035 for yellow birch. The average stemwood mass produced annually (ΔW_s) and tree leaf area (LA) were then used to calculate the growth efficiency index (GE) of each tree as

$$GE = \Delta W_s / LA \tag{3}$$

2.4.2. Indices Solely Based upon Tree Diameter Measurements

All indices that were solely based upon tree diameter measurements were calculated to correspond to the two years of tree vigor class determination, and for periods Δt of 3, 5, 7, and 10 years prior to the year of tree vigor class determination. We tested three index types and each of them was calculated for the four Δt periods, for a total of 12 calculated indices. The first index type corresponds to the basal area increment (BAI) of trees that was calculated for each Δt. The second index type is a relative BAI, i.e., the ratio between the BAI of a tree for a given Δt and the basal area at the beginning of the period under consideration. The third index type is a ratio between the BAI of a tree for a given Δt and the BAI for an equivalent Δt that preceded the first Δt under consideration. This last index type indicates whether BAI is increasing (>1) or decreasing (<1). For example, if tree vigor class was determined in 2016 and Δt was three years, the first index type would correspond to the BAI that was calculated from 2013 to 2015, the second index type to the ratio between the BAI calculated from 2013 to 2015 and the tree basal area in 2012, and the third index type to the ratio between the BAI calculated from 2013 to 2015 and the BAI from 2010 to 2012.

2.5. Statistical Analysis

All analyses were performed in the R statistical environment (version 3.3.2, R Development Core Team 2016, Vienna, Austria). Model assumptions (homogeneity of variance and normality of residuals) were validated using Levene and Shapiro–Wilk tests, together with graphical analysis of the residuals.

To test the first hypothesis, which relates the growth efficiency index to tree vigor classes, we fitted a linear mixed model using the function *lme* in the *nlme* package [31]. This model included tree species as a fixed effect and site as a random effect. The growth efficiency index was ln-transformed to achieve normality of the residuals. The analysis was performed using all trees that were sampled from the six sites.

For the second hypothesis, which predicts that decreases in tree vigor classes are associated with similar decreases in growth-based indices, we used four sites for which two determinations of tree vigor class were performed, i.e., Duchesnay and the three Domtar sites. In a first step, we identified which growth-based index was most closely related to the growth efficiency index to use it as the best growth-based predictor to represent tree vigor as a continuous variable. This was performed using linear mixed models with the function *lme* of the *nlme* package. We compared the models with the Akaike information criterion (AIC), which is a measure of the loss of information that results from using a model to explain a particular variable [32]. As suggested by Mazerolle [33], we compared the models using differences in AIC between each model and the model with the lowest AIC value (Δ_i). We also computed the AIC weight (W_{t_i}), which is the probability that a model is the best among those models being compared [32], and the coefficient of determination (R^2). In a second step, we used a linear mixed ANOVA model to determine which growth-based index was best related to tree vigor classes (i.e., a categorical variable). The growth-based index that was best related to both growth efficiency index and tree vigor classes was then used to quantify its temporal changes between two qualitative determinations of tree vigor class.

In a third step, we submitted the previously calculated temporal changes in the growth-based index as an explanatory variable in a logistic regression model that was fitted with the function *glm* of the *stats* package. The dependent binary variable of this model (dummy variable) was formed using only vigorous trees that were determined at the first vigor evaluation, which either remained vigorous at the second evaluation (coded value = 1) or became non-vigorous at the second evaluation (coded value = 0). Consistent with the results of Delisle-Boulianne et al. [1], we separated individuals into two classes of tree vigor (i.e., vigorous and non-vigorous trees) by grouping tree vigor classes M and S to represent non-vigorous trees, and tree vigor classes C and R to represent vigorous trees. We used a cut-point approach to convert the continuous probabilities that were produced by the logistic regression into dichotomous results, i.e., non-vigorous vs. vigorous trees. We considered a tree as non-vigorous at the second vigor evaluation if the model predicted a probability that was lower than a predetermined threshold. This threshold was determined at a value where precision and recall were equal [34]. At this cut-point, positive and negative predictions are made in the same proportions as the prevalence of vigorous and non-vigorous trees in the calibration dataset [6].

To test the third hypothesis, which specifies that tree and stand state variables are adequate predictors of growth efficiency index, we fitted linear mixed models using the function *lme* of the *nlme* package [31]. These models were fitted to the entire tree sample from the six sites. The considered state variables included tree DBH, height, relative height, crown diameter, and crown surface area as well as stand basal area and density. The models were compared based upon values of AIC, Δ_i, W_{t_i}, and R^2.

3. Results

3.1. Relationship between the Growth Efficiency Index and Tree Vigor Classes

We found no statistical differences in growth efficiency index between sugar maple and yellow birch ($p = 0.8829$), which were grouped together for subsequent analyses. Even though the growth efficiency index values tended to increase with increasing tree vigor class (Figure 2), this relationship was not statistically significant ($p = 0.4800$).

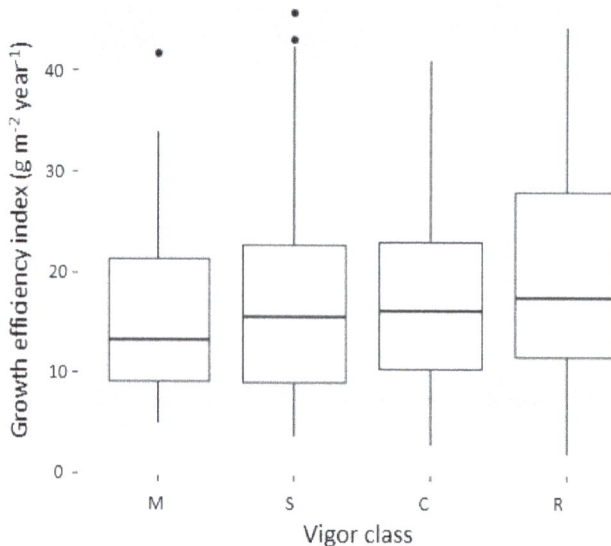

Figure 2. Box-and-whisker plots of the growth efficiency index as a function of tree vigor classes for the combined responses of sugar maple and yellow birch. The bold horizontal line within the boxes corresponds to the median (50th percentile), while the lower and upper limits of the boxes (interquartile range, IQR) represent the 25th and 75th percentiles, respectively. The whiskers are upper and lower values of 1.5 × IQR, and the individual points are outliers.

3.2. Growth-Based Indices and Changes in Tree Vigor Class

The basal area increment that was calculated over a five-year period (BAL_{1-5}) was the growth-based index that was most closely related to the growth efficiency index (Table 2), and was among the best indices that were related to tree vigor classes (Table 3). This index value was significantly different between the lowest and highest tree vigor classes (i.e., M and R classes) at both vigor evaluations ($p = 0.0192$ and $p = 0.0009$, respectively), and between the M and the C classes at the second vigor evaluation ($p = 0.0007$). Therefore, we included the temporal changes of BAL_{1-5} between the two vigor evaluations as an explanatory variable in the logistic regression model to predict the binary tree vigor classes. Validation of this model indicates that 66% of the trees (i.e., 107 of 161) were correctly classified as vigorous or non-vigorous trees (Table 4). Yet, the validation procedure also indicates that only 16% of trees (i.e., 5 of 32) that became non-vigorous at the second vigor evaluation were correctly classified. The weak capacity of the model to classify correctly vigorous trees that later became non-vigorous is illustrated by the substantial overlapping of BAL_{1-5} between observed vigorous and non-vigorous trees (Figure 3). This considerable overlap may explain the small range of predicted values of probability that trees remain vigorous, which are all close to the probability threshold (0.78) separating vigorous and non-vigorous trees (Figure 3).

Table 2. Statistics for the best linear mixed models that related the growth efficiency index to different growth-based indices. All models also include an intercept, tree species (the fixed effect), and a random site effect.

Growth-Based Index	AIC	Δ_i	W_{t_i}	R^2
BAL_{1-5}	2232.1	0	9.0×10^{-1}	0.35
BAL_{1-7}	2236.5	4.5	9.6×10^{-2}	0.34
BAL_{1-3}	2249.3	17.3	1.6×10^{-4}	0.31
BAL_{1-10}	2251.0	19.9	4.2×10^{-5}	0.31
BAL_{1-3}/BA_4	2268.2	36.2	1.3×10^{-8}	0.27
BAL_{1-5}/BA_6	2275.6	43.5	3.6×10^{-10}	0.26

Note: BAI is the basal area increment, BA is the basal area, AIC is the Akaike information criterion, Δ_i is the difference in AIC between each model and the model with the lowest AIC, W_{t_i} is the probability that a model is the best among those being compared, and R^2 is the coefficient of determination. The subscripts that are associated with growth-based indices correspond to the years during which the variables were computed, with year 0 corresponding to the time of tree vigor evaluation.

Table 3. Results of the analyses of variance (one-way ANOVAs) relating the most relevant growth-based indices to tree vigor classes. All models also included an intercept, the tree species (fixed effect), and a random site effect.

	First Vigor Evaluation ($n = 225$)		Second Vigor Evaluation ($n = 298$)	
Growth-Based Index	F	p-Value	F	p-Value
BAL_{1-7}	10.68	<0.0001	5.83	0.0007
BAL_{1-5}	9.56	<0.0001	5.83	0.0007
BAL_{1-10}	11.63	<0.0001	5.73	0.0008
BAL_{1-3}	7.86	<0.0001	4.65	0.0034
BAL_{1-7}/BA_8	3.18	0.0249	5.90	0.0006
BAL_{1-5}/BA_6	2.76	0.0430	5.70	0.0008

Note: The two tree vigor evaluations were conducted at intervals of 8 to 10 years on the same trees.

Table 4. Validation of the model predicting non-vigorous (0) and vigorous (1) trees based upon differences in BAL_{1-5} between the two tree vigor evaluations.

	Predicted = 0	Predicted = 1
Observed = 0	5	27
Observed = 1	27	102

Note: All trees were vigorous at the first period of evaluation. A tree was considered to be non-vigorous at the second period of evaluation if the model predicted a probability $p < 0.78$. A total of 161 trees were used for the validation operation.

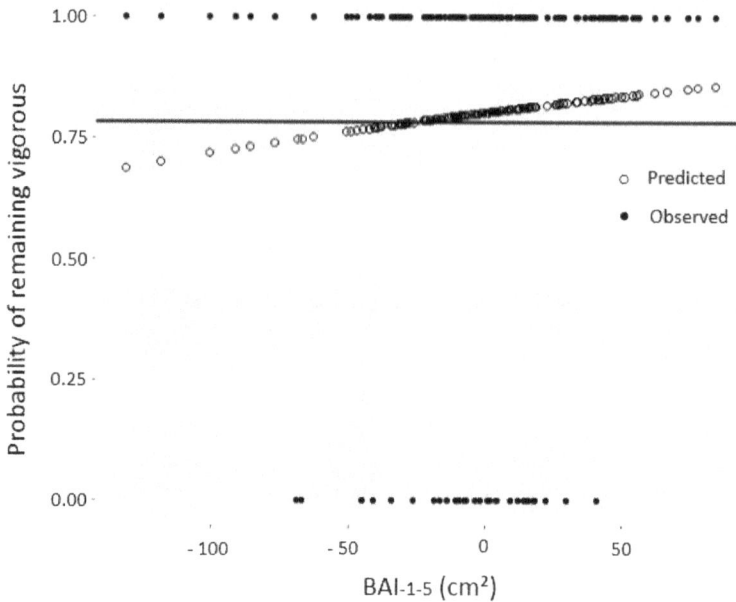

Figure 3. Probability that a vigorous tree at the first evaluation period either becomes non-vigorous (0) or remains vigorous (1) at the second evaluation period as a function of the growth-based index BAL_{1-5}. The horizontal line represents the probability threshold (0.78) below which a tree was considered as non-vigorous.

3.3. Predicting the Growth Efficiency Index

Based upon AIC values that were calculated from models including various state variables at tree and stand levels, the most parsimonious model predicting the growth efficiency index included only tree relative height, i.e., the height of the tree of interest divided by the height of the tallest tree in a stand (Table 5). However, the predictive ability of this model was rather poor, with a R^2 of only 0.05. When the same state variables were related to the three most relevant growth-based indices, the most parsimonious models included tree crown surface area and relative height. Although, the predictive abilities of these models were higher than that of the preceding one, with R^2 varying from 0.37 to 0.39, both parameters were never significantly ($\alpha = 0.05$) different from zero (results not shown), suggesting weak relationships.

Table 5. Statistics of linear mixed models relating the growth efficiency index (GE) and three growth-based indices to state variables at tree and stand levels. All models also included an intercept, tree species (fixed effect), and a random site effect.

Dependent Variables	Independent Variables	AIC	Δ_i	W_{t_i}	R^2
GE	RH	2327.5	0.0	9.5×10^{-1}	0.05
	CSA + RH	2333.2	5.7	5.4×10^{-2}	0.11
BAL_{1-5}	CSA + RH	2943.9	0.0	1.0×10^{-1}	0.39
	CSA	2958.4	14.5	6.9×10^{-4}	0.38
BAL_{1-7}	CSA + RH	3139.9	0.0	1.0×10^{-1}	0.39
	CSA	3157.8	17.9	1.3×10^{-4}	0.37
BAL_{1-10}	CSA + RH	3320.1	0.0	1.0×10^{-1}	0.37
	CSA	3339.8	19.7	5.5×10^{-5}	0.36

Note: RH is tree relative height, CSA is the crown surface area, DBH is the diameter at breast height (1.3 m), and BA is stand basal area.

4. Discussion

4.1. Relationships between Quantitative and Qualitative Tree Vigor Indices

The principal objective of this study was to establish relationships between a system of tree vigor classification based upon apparent stem defects and some quantitative tree vigor indices to evaluate the ability of the former to identify low-vigor trees. A first relationship involved the growth efficiency index that expresses the amount of wood produced annually per unit leaf area. Because the allocation priority to stemwood is low in normally growing trees [12], a high value of this index suggests that tree carbon resources far exceed priority allocations, implying strong overall tree vigor [11]. Accordingly, this index successfully predicted the vigor and mortality probabilities of trees affected by defoliating and boring insects [15–17,19]. In our case, the growth efficiency index was not significantly related to the vigor classes that were based upon apparent stem defects, which is a first indication of the lack of reliability of the classification system for tree vigor evaluation.

To lend support to the preceding result, we used BAL_{1-5}, which is a growth-based index that is closely related to tree vigor classes (Table 3), to determine whether its temporal changes were related to changes in tree vigor classes. In focusing only upon tree vigor changes that were observed over time for the same trees, this analysis allowed us to remove a portion of the between-tree variation from the vigor class determinations, which can notably be influenced by tree hierarchical position [35,36] and age [37]. Even when this inter-tree variation was removed, only 16% of vigorous trees that became non-vigorous 8–10 years later could be correctly classified. Nevertheless, significant differences in BAL_{1-5} were detected between the two extreme classes (i.e., M and R classes), while the intermediate classes were not distinguished from the extreme classes. These results agree with those of Hartmann et al. [10] and Guillemette et al. [9], who significantly differentiated only classes M and R using a growth-based index to predict the mortality probability of sugar maple trees.

Overall, these results indicate that the apparent stem defects that were used in the tree classification system under study are weakly related, at best, to the actual vigor of sugar maple and yellow birch trees. Interestingly, Power and Havreljuk [38] observed a strong relationship between these vigor classes and stem quality classes that are commonly used in northeastern Canada [39]. This observation is likely related to the fact that the same stem defects are used in both classifications. Thus, it seems that despite being weak indicators of tree vigor, some of these apparent stem defects are useful for estimating the potential monetary value of hardwoods [2].

The tree vigor system under study is currently applied in northern hardwood stands to identify trees that would be harvested during partial cut operations and, consequently, to improve the vigor of future stands. Nevertheless, some observations suggest that the selection rules that are used by this classification system may not achieve the expected benefit. For example, Nolet et al. [40] observed that the proportion of moribund trees (M class) that was determined by the classification system increased with increasing site quality, while tree mortality that was measured 10 years after tree vigor class determination indicated the opposite trend. These conflicting results underscore the need that such a classification system for evaluating tree vigor should include variables other than those that represent only apparent stem defects. Additional potential variables could reflect past stand management practices, site quality, tree social status, stand density, and tree bark characteristics, among others [41]. Further, our results and those of Hartmann et al. [10] and Guillemette et al. [9] indicate that the use of intermediate vigor classes seems ambiguous in determining the vigor or the mortality probability of trees. Therefore, it is possible that simplifying the tree vigor system by eliminating intermediate classes could improve the tree marking process without affecting its quality. To this end, Cecil-Cockwell and Caspersen [42] proposed a parsimonious, three-class system that takes into account tree vigor and stem quality while eliminating superfluous classes that do not influence the value of sugar maple trees. Indeed, tree mortality of hardwood species was significantly related to a binary classification of tree vigor, whereas a two-class stem product system also contributed to tree mortality predictions [43], further suggesting that criteria for stem quality and tree vigor are frequently confounded.

4.2. Possible Drawbacks of the Growth Efficiency Index

The classification system under study reflects the probability of tree mortality during the upcoming cutting cycle, i.e., a 25-year period [28]. This includes mortality risks that are related to stem structural defects, such as leaning trees, stem base injuries, fork cracks and stem cracks [7]. These defects, especially when they are associated with decay, increase the risks of stem breakage that is caused by wind. Yet these defects do not necessarily affect the diameter growth and the vigor of trees [10], implying that growth efficiency and growth-based indices could be rather insensitive to these defects. Therefore, the lack of correspondence between these sources of mortality risk and tree diameter measurements could explain, in part, the weak relationship between vigor classes and the quantitative vigor indices.

The growth efficiency index has been used by several authors to evaluate the vigor of hardwood species (e.g., [44–46]), but some limitations regarding the use of this index should be considered for deciduous species. For instance, annual wood production of sugar maple seems less closely related to leaf area than what has been observed for most conifers [18]. This can be explained by the fact that unlike most conifers, hardwood species must produce new foliage each spring to activate the photosynthetic process. Because the production of this new foliage is related to the quantity of carbohydrates that is stored during previous years, wood production of hardwood species is more strongly dependent upon the growing conditions that prevailed during these years than are conifers [47]. Therefore, environmental factors such as drought, low temperature, or insect defoliation can influence tree growth for several years [18], especially in less vigorous trees [10]. These effects may thus contribute to explaining the weak relationship that was observed between the quantitative vigor indices and the vigor classes based upon apparent stem defects. To reduce these effects, it should be noted that tree diameter growth was averaged over a five-year period to avoid introducing effects of episodic events into the vigor indices.

4.3. Estimating the Growth Efficiency Index from State Variables

Because tree diameter growth is rarely available from usual forest inventory data, low-vigor trees are difficult to identify using the growth efficiency index or growth-based indices, which would benefit from being estimated from tree and stand state variables. Candidate variables that can be related to the growth efficiency index include stand density [15,16], tree social status [11,36], tree crown size [48], and tree diameter [49]. Yet, none of these variables allowed us to estimate adequately either the growth efficiency index or growth-based indices of sugar maple and yellow birch trees in the uneven-aged hardwood stands under study. The low predictive ability of our relationships may be partly explained by the exclusion from our sample of all trees with a DBH less than 20 cm and those that were neither dominant nor codominant. Nevertheless, better predictions of the growth efficiency index could be obtained by considering other state variables such as tree bark characteristics. Indeed, old trees often have rough bark [50], which could explain the significant relationship between bark appearance and diameter growth of sugar maple and yellow birch trees [41]. Between two trees of similar diameter, the one with the smoothest bark would be younger and faster growing than the one with the roughest bark. The possible relationship between bark appearance and the growth efficiency index is worth testing to help identify low-vigor trees more easily during tree marking operations in northern hardwood stands.

5. Conclusions

This study demonstrates that tree vigor classes that were deduced from apparent stem defects were weakly related to the growth efficiency index and other indices solely based upon tree diameter measurements. Since the development of a new vigor classification system is a long process, systems that are based upon apparent stem defects should continue to be used over the medium-term in order to improve the vigor of degraded northern hardwood stands. However, these systems would benefit from being simplified because their current complexity does not provide a gain in precision. In addition, a promising quantitative estimator of tree vigor, the growth efficiency index, was poorly related to usual characteristics of sugar maple and yellow birch trees. Further studies should be conducted in an attempt to relate the growth efficiency index to other tree characteristics, such as bark appearance, to increase our ability to distinguish vigorous from non-vigorous trees, and to identify the defects that actually reflect tree vigor and growth.

Author Contributions: E.M. performed the field and laboratory data collection, analyzed the data, and contributed to the study design and to the paper writing; S.B. provided past measurements from the Duchesnay experimental site, and contributed to the study design and to the paper writing; G.M. contributed to the fieldwork and to the paper writing; D.P. conceived the study, provided the funding of the project, and contributed to the data analysis and to the paper writing.

Funding: This research was funded by a Discovery Grant from the Natural Sciences and Engineering Research Council (NSERC) of Canada awarded to David Pothier.

Acknowledgments: We are grateful to Sharad Kumar Baral, Eloïse Dupuis, Alexandre Morin-Bernard, Félix Poulin, Corryne Vincent, Marie-Laure Lusignan, Émilie St-Jean, Pierre Racine, and Martine Lapointe for their help with the fieldwork and in the laboratory. We also thank William F.J. Parsons for reviewing and making valuable comments on the manuscript.

Conflicts of Interest: The authors declare no conflict of interest.

References

1. Delisle-Boulianne, S.; Fortin, M.; Achim, A.; Pothier, D. Modelling stem selection in northern hardwood stands: Assessing the effects of tree vigour and spatial correlations with a copula approach. *Forestry* **2014**, *87*, 607–617. [CrossRef]
2. Havreljuk, F.; Achim, A.; Auty, D.; Bédard, S.; Pothier, D. Integrating standing value estimations into tree marking guidelines to meet wood supply objectives. *Can. J. For. Res.* **2014**, *44*, 750–759. [CrossRef]
3. Pothier, D.; Fortin, M.; Auty, D.; Delisle-Boulianne, S.; Gagne, L.-V.; Achim, A. Improving tree selection for partial cutting through joint probability modeling of tree vigor and quality. *Can. J. For. Res.* **2013**, *43*, 288–298. [CrossRef]
4. Nyland, R.D. Selection system in northern hardwoods. *J. For.* **1998**, *96*, 18–21.
5. Guillemette, F.; Bédard, S.; Fortin, M. Evaluation of a tree classification system in relation to mortality risk in Québec northern hardwoods. *For. Chron.* **2008**, *84*, 886–899. [CrossRef]
6. Laliberté, J.; Pothier, D.; Achim, A. Adjusting harvest rules for red oak in selection cuts of Canadian northern hardwood forests. *Forestry* **2016**, *89*, 402–411. [CrossRef]
7. Boulet, B.; Landry, G. *La Carie des Arbres: Fondements, Diagnostic et Application*, 3rd ed.; Les Publications du Québec: Québec, QC, Canada, 2015; p. 372.

8. Fortin, M.; Bédard, S.; Guillemette, F. *Estimation par Simulation Monte Carlo de la Probabilité de Mortalité Quinquennale de L'érable à Sucre, du Bouleau Jaune et du Hêtre à Grandes Feuilles en Peuplements de Feuillus en Fonction de la Classification MSCR*; Avis Technique; Gouvernement du Québec, Ministère des Ressources Naturelles et de la Faune, Direction de la Recherche Forestière: Québec, QC, Canada, 2008; p. 11.

9. Guillemette, F.; Bédard, S.; Havreljuk, F. *Probabilités de Mortalité des Feuillus Selon le Classement de la Priorité de Récolte*; Direction de la Recherche Forestière: Québec, QC, Canada, 2015; p. 11.

10. Hartmann, H.; Beaudet, M.; Messier, C. Using longitudinal survival probabilities to test field vigour estimates in sugar maple (*Acer saccharum* Marsh.). *For. Ecol. Manag.* **2008**, *256*, 1771–1779. [CrossRef]

11. Waring, R.H.; Thies, W.G.; Muscato, D. Stem growth per unit of leaf area: A measure of tree vigor. *For. Sci.* **1980**, *26*, 112–117.

12. Waring, R.H. Characteristics of trees predisposed to die. *Bioscience* **1987**, *37*, 569–574. [CrossRef]

13. Christiansen, E.; Waring, R.H.; Berryman, A.A. Resistance of conifers to bark beetle attack: Searching for general relationships. *For. Ecol. Manag.* **1987**, *22*, 89–106. [CrossRef]

14. Waring, R.H.; Newman, K.; Bell, J. Efficiency of tree crowns and stemwood production at different canopy leaf densities. *Forestry* **1981**, *54*, 129–137. [CrossRef]

15. Mitchell, R.G.; Waring, R.H.; Pitman, G.B. Thinning lodgepole pine increases tree vigor and resistance to mountain pine beetle. *For. Sci.* **1983**, *29*, 204–211.

16. Waring, R.H.; Pitman, G.B. Modifying lodgepole pine stands to change susceptibility to mountain pine beetle attack. *Ecology* **1985**, *66*, 889–897. [CrossRef]

17. Coops, N.C.; Waring, R.H.; Wulder, M.A.; White, J.C. Prediction and assessment of bark beetle-induced mortality of lodgepole pine using estimates of stand vigor derived from remotely sensed data. *Remote Sens. Environ.* **2009**, *113*, 1058–1066. [CrossRef]

18. Reed, D.D.; Pregitzer, K.S.; Liechty, H.; Burton, A.J.; Mroz, G.D. Productivity and growth efficiency in sugar maple forests. *For. Ecol. Manag.* **1994**, *70*, 319–327. [CrossRef]

19. Coyea, M.R.; Margolis, H.A. The historical reconstruction of growth efficiency and its relationship to tree mortality in balsam fir ecosystems affected by spruce budworm. *Can. J. For. Res.* **1994**, *24*, 2208–2221. [CrossRef]

20. Boivin-Dompierre, S.; Achim, A.; Pothier, D. Functional response of coniferous trees and stands to commercial thinning in eastern Canada. *For. Ecol. Manag.* **2017**, *384*, 6–16. [CrossRef]

21. Pedersen, B.S. Modeling tree mortality in response to short and long-term environmental stresses. *Ecol. Model.* **1998**, *105*, 347–351. [CrossRef]

22. Yao, X.; Titus, S.J.; MacDonald, S.E. A generalized logistic model of individual tree mortality for aspen, white spruce, and lodgepole pine in Alberta mixedwood forests. *Can. J. For. Res.* **2001**, *31*, 283–291. [CrossRef]

23. Bigler, C.; Bugmann, H. Growth-dependent tree mortality models based on tree rings. *Can. J. For. Res.* **2003**, *33*, 210–221. [CrossRef]

24. Duchesne, L.; Ouimet, R.; Morneau, C. Assessment of sugar maple health based on basal area growth pattern. *Can. J. For. Res.* **2003**, *33*, 2074–2080. [CrossRef]

25. Saucier, J.-P.; Grondin, P.; Robitaille, A.; Gosselin, J.; Morneau, C.; Richard, P.J.H.; Brisson, J.; Sirois, L.; Leduc, A.; Morin, H.; et al. Chapitre 4: Écologie forestière. In *Manuel de Foresterie*; Éditions MultiMondes: Montréal, QC, Canada, 2009; pp. 165–316.

26. Grondin, P.; Jean, N.; Hotte, D. *Intégration de la Végétation et de ses Variables Explicatives à des Fins de Classification et de Cartographie D'unités Homogènes du Québec Méridional*; Direction de la Recherche Forestière: Québec, QC, Canada, 2007; p. 62.

27. Russell, M.B.; Weiskittel, A.R. Maximum and largest crown width equations for 15 tree species in Maine. *North. J. Appl. For.* **2011**, *28*, 84–91.

28. Boulet, B. *Défauts Externes et Indices de la Carie des Arbres*; Les Publications du Québec: Québec, QC, Canada, 2007; p. 291.

29. Lambert, M.-C.; Ung, C.-H.; Raulier, F. Canadian national tree aboveground biomass equations. *Can. J. For. Res.* **2005**, *35*, 1996–2018. [CrossRef]

30. Moreau, E.; Bédard, S.; Baral, S.K.; Pothier, D. Evaluating electrical resistivity tomography to estimate sapwood area and leaf area of sugar maple and yellow birch. *Ecohydrology* **2018**, e2014. [CrossRef]

31. Pinheiro, J.; Bates, D.; DebRoy, S.; Sarkar, D.; R Core Team. Nlme: Linear and Nonlinear Mixed Effects Models. R Package Version 3.1-137. 2018. Available online: https://CRAN.R-project.org/package=nlme (accessed on 21 September 2018).

32. Burnham, K.P.; Anderson, D.R. *Model Selection and Multimodel Inference: A Practical Information-Theoretic Approach*, 2nd ed.; Springer: New York, NY, USA, 2002; p. 488.

33. Mazerolle, M.J. Improving data analysis in herpetology: Using Akaike's information criterion (AIC) to assess the strength of biological hypotheses. *Amphibia-Reptilia* **2006**, *27*, 169–180. [CrossRef]

34. Fawcett, T. *ROC Graphs: Notes and Practical Considerations for Researchers*; Technical Report HPL-2003-4; HP Laboratories: Palo Alto, CA, USA, 2004.

35. Maguire, D.A.; Brissette, J.C.; Gu, L.H. Crown structure and growth efficiency of red spruce in uneven-aged, mixed-species stands in Maine. *Can. J. For. Res.* **1998**, *28*, 1233–1240. [CrossRef]

36. Ryan, M.G.; Phillips, N.; Bond, B.J. The hydraulic limitation hypothesis revisited. *Plant Cell Environ.* **2006**, *29*, 367–381. [CrossRef] [PubMed]

37. Seymour, R.S.; Kenefic, L.S. Influence of age on growth efficiency of *Tsuga canadensis* and *Picea rubens* trees in mixed-species, multiaged northern conifer stands. *Can. J. For. Res.* **2002**, *32*, 2032–2042. [CrossRef]

38. Power, H.; Havreljuk, P. Predicting hardwood quality and its evolution over time in Quebec's forests. *Forestry* **2018**, *91*, 259–270. [CrossRef]

39. Petro, F.J.; Calvert, W.W. *La Classification des Billes de Bois Francs Destinées au Sciage*; Forintek Canada Corp.: Ottawa, ON, Canada, 1990.

40. Nolet, P.; Hartmann, H.; Bouffard, D.; Doyon, F. Predicted and observed sugar maple mortality in relation to site quality indicators. *North. J. Appl. For.* **2007**, *24*, 258–264.

41. Gauthier, M.-M.; Guillemette, F. Bark type reflects growth potential of yellow birch and sugar maple at the northern limit of their range. *Plant Ecol.* **2017**, *219*, 381–390. [CrossRef]

42. Cecil-Cockwell, M.J.L.; Caspersen, J.P. A simple system for classifying sugar maple vigour and quality. *Can. J. For. Res.* **2015**, *45*, 900–909. [CrossRef]

43. Fortin, D.; Bédard, S.; DeBlois, J.; Meunier, S. Predicting individual tree mortality in northern hardwood stands under uneven-aged management in southern Québec, Canada. *Ann. For. Sci.* **2008**, *65*, 205. [CrossRef]

44. Allen, C.B.; Will, R.E.; Jacobson, M.A. Production efficiency and radiation use efficiency of four tree species receiving irrigation and fertilization. *For. Sci.* **2005**, *51*, 556–569.

45. Binkley, D.; Stape, J.L.; Bauerle, W.L.; Ryan, M.G. Explaining growth of individual trees: Light interception and efficiency of light use by *Eucalyptus* at four sites in Brazil. *For. Ecol. Manag.* **2010**, *259*, 1695–1703. [CrossRef]

46. Voelker, S.L.; Lachenbruch, B.; Meinzer, F.C.; Kitin, P.; Strauss, S.H. Transgenic poplars with reduced lignin show impaired xylem conductivity, growth efficiency and survival. *Plant Cell Environ.* **2011**, *34*, 655–668. [CrossRef] [PubMed]

47. Lane, C.J.; Reed, D.D.; Mroz, G.D.; Liechty, H.O. Width of sugar maple (*Acer saccharum*) tree rings as affected by climate. *Can. J. For. Res.* **1993**, *23*, 2370–2375. [CrossRef]

48. Zarnoch, S.J.; Bechtold, W.A.; Stolte, K.W. Using crown condition variables as indicators of forest health. *Can. J. For. Res.* **2004**, *34*, 1057–1070. [CrossRef]

49. Binkley, D.; Stape, J.L.; Ryan, M.G.; Barnard, H.R.; Fownes, J.H. Age-related decline in forest ecosystem growth: An individual-tree, stand-structure hypothesis. *Ecosystems* **2002**, *5*, 58–67. [CrossRef]

50. Clausen, K.E.; Godman, R.M. *Bark Characteristics Indicate Age and Growth Rate of Yellow Birch*; Research Note NC-75; U.S. Department of Agriculture, Forest Service, North Central Forest Experimental Station: St. Paul, MN, USA, 1969.

forests

Article

Short-Term Vegetation Responses to Invasive Shrub Control Techniques for Amur Honeysuckle (*Lonicera maackii* [Rupr.] Herder)

Graham S. Frank [1,2,*], Michael R. Saunders [1,2] and Michael A. Jenkins [1,2]

[1] Department of Forestry and Natural Resources, Purdue University, 715 W. State St., West Lafayette, IN 47907, USA; msaunder@purdue.edu (M.R.S.); jenkinma@purdue.edu (M.A.J.)
[2] Hardwood Tree Improvement and Regeneration Center (HTIRC), 715 W. State St., West Lafayette, IN 47907, USA
* Correspondence: graham.frank@oregonstate.edu or frankg@purdue.edu

Received: 27 August 2018; Accepted: 27 September 2018; Published: 30 September 2018

Abstract: Invasive shrubs in forest understories threaten biodiversity and forest regeneration in the eastern United States. Controlling these extensive monotypic shrub thickets is a protracted process that slows the restoration of degraded forest land. Invasive shrub removal can be accelerated by using forestry mulching heads, but evidence from the western United States indicates that mulching heads can promote exotic species establishment and mulch deposition can reduce native plant species abundance. We compared the effectiveness of the mulching head and the "cut-stump" method for controlling the invasive shrub Amur honeysuckle (*Lonicera maackii*), as well as their impacts on native plant community recovery, in mixed-hardwood forests of Indiana. After two growing seasons, mulching head treatment resulted in greater *L. maackii* regrowth and regeneration. The recovery of native plant abundance and diversity following shrub removal did not differ between the two methods. However, mulch deposition was associated with increased abundance of garlic mustard (*Alliaria petiolata*), an invasive forb. Increasing mulching head treatment depth reduced *L. maackii* regrowth, but additional study is needed to determine how it affects plant community responses. The mulching head is a promising technique for invasive shrub control and investigating tradeoffs between reducing landscape-scale propagule pressure and increased local establishment will further inform its utility.

Keywords: invasive plants; forest restoration; soil disturbance; herbicide effects; forest regeneration; floristic quality index; species composition

---

1. Introduction

Throughout the eastern United States, invasive woody shrubs are nearly ubiquitous in hardwood forest understories and pose a serious threat to successful forest management and restoration [1]. Examples of these invasive shrubs—mostly introduced in the late 18th to late 19th centuries—include numerous species of bush honeysuckle (*Lonicera* spp.) as well as Japanese barberry (*Berberis thunbergii* DC.), burning bush (*Euonymus alatus* [Thunb.] Siebold), and others. Proliferations of these invasive shrubs can change forest structure by forming dense monotypic thickets in the understory [2]. These thickets pose a threat to forest health by inhibiting native tree regeneration [3,4] and reducing the cover and diversity of herbaceous plants [5]. Structural change in the understory vegetation also affects wildlife, such as promoting seed predation, with implications for forest regeneration and biodiversity [6–8]. Controlling dense, well-established populations of invasive shrubs is both expensive and time consuming. For example, between 1994 and 2005, workers in Great Smoky Mountains National Park spent over 17,000 h controlling woody invasive plants [1]. Minimizing time and resources allocated to such efforts will depend on developing techniques to expedite the control of mature populations of invasive shrubs that have reached the saturation phase of invasion.

Controlling invasive shrubs is challenging because it not only requires the mechanical cutting of stems but also retreating the vigorous stump sprouting that occurs after cutting [9]. Clearing saws are frequently used to cut invasive shrubs due to their affordability, ease of use, and limited collateral damage. However, the follow-up treatment of resprouts with herbicides can be both time consuming and cost prohibitive [10]. Mulching treatments conducted with rotary mulching heads (i.e., forestry heads, mastication heads, or wood shredders; hereafter "mulching heads") are an alternative method that may accelerate shrub removal in dense thickets. Mulching heads are capable of shattering root collars (MRS pers. obs.), which can reduce resprouting [11] and may in turn reduce the need for herbicide treatments. According to Ward et al. [12], mulching head treatments were slightly more effective than clearing saws with no stump treatment for controlling the thicket-forming invasive shrub Japanese barberry, though comparisons with clearing saw removal should include the widely-recommended herbicide application to cut stumps (henceforth "cut-stump" method; [1]). For controlling Chinese privet (*Ligustrum sinense* Lour.), an invasive evergreen shrub found primarily in riparian areas of the southeastern United States, both techniques resulted in rapid regrowth the following year [13].

In addition to effectiveness and cost-efficiency, impacts on the native understory community must be considered when evaluating removal techniques [14]. Effective restoration requires control techniques that efficiently remove shrubs without degrading native plant populations or promoting further invasive species establishment [15–17]. Because the effectiveness of mulching heads in reducing shrub survival depends on shattering the root collar, using this equipment may result in substantial soil disturbance and damage to the roots and rhizomes of native perennials. While Luken et al. [18] found that experimental soil disturbance within small gaps (5 m diameter) created in honeysuckle thickets did not affect understory plant communities, large-scale control efforts may increase light availability to a greater extent, compounding the effects of substrate disturbance.

Mulching head treatments also deposit a layer of woody debris on the forest floor. This mulch layer may inhibit native plant cover and diversity through physical effects on understory species [19] but may also reduce future reinvasion of invasive shrubs. Thick layers of litter have been shown to limit the establishment of Amur honeysuckle (*Lonicera maackii* [Rupr.] Herder (Caprifoliaceae); hereafter "honeysuckle") as well as garlic mustard (*Alliaria petiolata* [M. Bieb.] Cavara & Grande), an invasive biennial forb frequently promoted by honeysuckle removal [17,20–22]. However, previous research has not determined whether inputs of mulch act similarly to thick litter layers in inhibiting the establishment of invasive shrubs in eastern hardwood forests.

Few studies have examined the effects of mulching head treatments on understory plant communities or compared its effects to those of cut-stump treatments, and none have distinguished the individual and interactive effects of substrate disturbance and mulch deposition from this technique on plant community recovery. Existing studies found greater non-native herbaceous cover following cut-stump removal of Japanese barberry [12] and greater re-establishment of Chinese privet after mulching head treatment [16] but otherwise similar plant community recovery between the two treatments. While these studies provide valuable information about plant community response to different treatments, additional evidence from systems with different dominant invasives, climate regimes, and native plant communities is necessary to tailor control techniques for specific restoration prescriptions.

Amur honeysuckle is a strong candidate for examining the efficacy of mulching head treatments and their effects on understory flora due to its tendency to form dense thickets [23]. This species is highly invasive in eastern hardwood forests, where it is widely dispersed by birds and deer [24,25], readily establishes along forest edges and in young stands [26], and can enter the expansion phase of population growth within 10 to 15 years [27], posing a challenge for control efforts. Honeysuckle is a strong competitor, in part due to its efficient resource use and long leaf phenology relative to native counterparts [28,29].

The primary objective of this study was to examine how understory plant communities respond to honeysuckle removal using mulching head treatments as compared to the currently recommended cut-stump method. Within this primary objective, we aimed to: (1) compare the effectiveness of the two methods at reducing honeysuckle cover and densities; (2) compare the effects of each method on native herbaceous plants, native woody seedlings, recruitment of new honeysuckle seedlings, and other exotic plant species; and (3) isolate the effects of mulch deposition from those of soil disturbance on the community structure of understory flora. We expected that both honeysuckle removal methods would increase understory diversity and cover but that herbaceous cover would be reduced by mulch deposition and that increases in herbaceous diversity would be primarily driven by exotic and ruderal species. We also expected that soil disturbance from the mulching head treatment would increase honeysuckle seedling densities and garlic mustard cover but that these effects would be mitigated by mulch deposition.

2. Methods

2.1. Study Area

Study sites were situated in mature secondary mixed-hardwood forests within the Central Till Plain region of Indiana at two locations near West Lafayette, Indiana—(1) Purdue University, Richard G. Lugar Forestry Farm (hereafter LF) and (2) Purdue University, Wildlife Area (hereafter PWA). Both locations consist of natural forest tracts intermixed with tree plantations and/or tallgrass prairie and are surrounded by an agricultural matrix. Mature *L. maackii* invasions formed dense thickets in the understories at all four sites. Prior to this study, none of the study sites had been treated for invasive plants. Estimated *L. maackii* establishment at each site was 30, 37, 33, and 25 years prior to the commencement of the study at the three sites at LF (LF1, LF2, LF3), and the site at PWA, respectively, based on annual growth rings from stem cross-sections collected at ground height from the five largest individuals at each site. Pretreatment density *L. maackii* stems >1.37 m tall ranged from 2104 stems ha^{-1} at PWA to 6729 stems ha^{-1} at LF2. Mean ($\pm$standard error) *L. maackii* cover in the 1–5 m stratum at each site ranged from 57.7% $\pm$ 6.8% at LF3 to 67.7% $\pm$ 5.3% at PWA. We selected sites that each had a component of *Quercus* L. spp. (oaks) and *Carya* Nutt. spp. (hickories) in the overstory, and each site also included *Acer* L. spp. (maples) and *Prunus serotina* Ehrh. (black cherry) in the overstory (Table 1). All sites were flat to gently sloping. Soils at LF sites are primarily Miami silt loams with some Kalamazoo and Ockley silt loams; PWA soils are Miami and Rainsville silt loams [30].

Table 1. Mean basal area (m^2 ha^{-1}) of overstory tree species at each site. Three sites were at the Richard G. Lugar Forestry Farm (LF1–3) and a fourth was at the Purdue Wildlife Area (PWA) in West Lafayette, IN. Data were collected using a variable radius plot (basal area factor = 2.296 m^2 ha^{-1}) during summer 2016 in treatment areas and winter 2016–2017 in reference areas.

Species	Common Name	Basal Area			
		LF1	LF2	LF3	PWA
Acer saccharinum L.	silver maple	–	–	–	1.53
Acer saccharum Marshall	sugar maple	2.30	1.15	3.83	–
Carya spp. Nutt.	hickory species	0.77	5.36	5.74	8.04
Celtis occidentalis L.	common hackberry	0.77	–	2.30	0.38
Cornus florida L.	flowering dogwood	0.38	–	–	–
Crataegus spp. L.	hawthorn species	0.38	0.77	–	–
Fraxinus americana L.	white ash	1.91	2.30	3.44	–
Juglans nigra L.	black walnut	2.30	3.06	1.53	–
Liriodendron tulipifera L.	tulip tree	0.77	–	–	–
Morus alba L.	white mulberry	–	–	–	0.77
Picea rubens Sarg.	red spruce	–	–	2.30	–

Table 1. *Cont.*

Species	Common Name	Basal Area			
		LF1	LF2	LF3	PWA
Platanus occidentalis L.	American sycamore	–	0.38	–	–
Prunus serotina Ehrh.	black cherry	3.44	0.77	1.91	6.12
Quercus alba L.	white oak	–	3.06	–	2.30
Quercus imbricaria Michx.	shingle oak	–	1.91	0.38	1.15
Quercus rubra L.	northern red oak	4.21	–	0.38	1.53
Quercus velutina Lam.	black oak	0.38	1.53	–	1.91
Robinia pseudoacacia L.	black locust	–	0.77	0.38	–
Sassafras albidum (Nutt.) Nees	sassafras	3.06	–	–	2.68
Tilia americana L.	American basswood	0.38	–	0.38	–
Ulmus americana L.	American elm	0.77	0.38	0.77	–
Ulmus rubra Muhl.	slippery elm	0.77	–	0.77	–
Total Site Basal Area		22.58	21.43	24.11	26.40

2.2. Experimental Design

Each study site contained a total of 24 vegetation monitoring plots arranged on a grid, consisting of 16 treatment plots and eight untreated reference plots. Plots were grouped into clusters of four, which would all receive the same honeysuckle removal treatment, in order to provide a buffer between different removal methods. Within groups of four plots, plot center points were placed 8 m apart. The nearest adjacent plot center points between separate four-plot groups were 11 m apart (Figure 1). Initial plot placement was randomly selected, with the constraint that the entire grid would be at least 10 m from any forest edge or previously-treated honeysuckle removal areas. Reference plots at each site were grouped together to avoid unintentional disturbance during honeysuckle removal but were randomly assigned to one side or the other of the experimental plots.

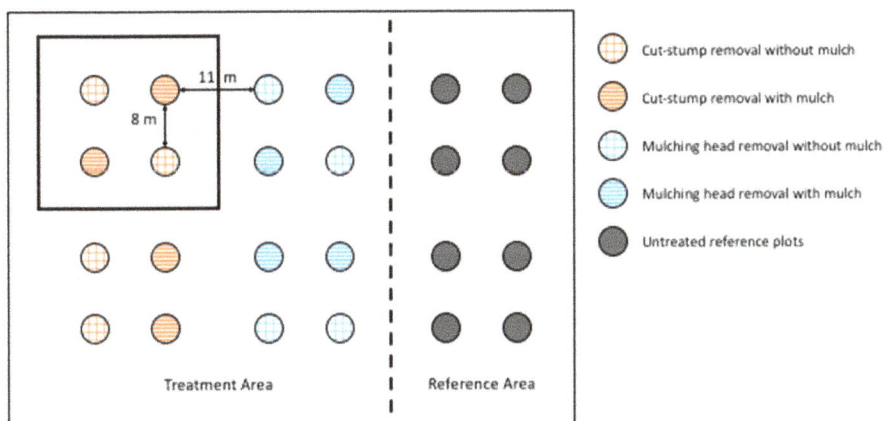

Figure 1. Schematic of an example site layout. Circles represent plot center points, with fill colors indicating different removal methods (cut-stump or mulching head). Removal methods were randomly assigned to each four-plot grouping (e.g., black square top-left) to avoid incidental impacts of the mulching head on non-target plots. Mulch cover (presence/absence) is represented by different fill patterns and was randomly assigned to two plots within each four-plot group. Circles with solid gray fill indicate reference plots (to the right of the dashed line). Spacing between plots in the same group was 8 m and spacing between plots of different groups was 11 m.

We randomly assigned one of two honeysuckle removal treatments—either "cut-stump" or "mulching head" removal—to each four-plot group such that each site had eight plots of each removal treatment. The cut-stump treatment consisted of cutting shrubs at the base with a gas-powered clearing saw and treating the resulting stumps with a volume-based herbicide mix of 15% triclopyr (Garlon 4 Ultra®; DowAgroSciences, Indianapolis, IN, USA), 3% imazypyr (Stalker®; BASF, Florham Park, NJ, USA), and 82% bark oil (Ax-It®; Townsend Corporation, Muncie, IN, USA) using a hand sprayer. We removed honeysuckle slash to avoid altering the light environment or physically obstructing browsing. Mulching head removal treatments were accomplished using a skidsteer-mounted forestry mulching head (Bull Hog®; Fecon Inc., Lebanon, OH, USA), attempting to clear honeysuckle shrubs as thoroughly as possible while avoiding incidental damage to native saplings and overstory trees. We removed any residual honeysuckle stems missed by the mulching head due to their proximity with native trees using a clearing saw but did not treat these with herbicide. Honeysuckle removal took place between 13 November and 4 December 2015.

To isolate the effects of mulch deposition from those of substrate disturbance by the mulching head, we added or removed mulch at two randomly selected plots within each four-plot treatment group. We took four measurements of mulch thickness at each mulching head treatment plot prior to mulch manipulations and attempted to replicate the highest site average depth of 2.5 cm across all mulched plots. Mulch additions were made primarily with woodchips from honeysuckle removed from cut-stump plots but included a small proportion of black cherry to imitate residual mulch from mulching head treatments.

2.3. Vegetation Sampling

We monitored understory plant communities at each plot prior to honeysuckle removal in 2015 and following removal in both 2016 and 2017 to directly estimate changes in community composition. We sampled spring perennials in late April to early May, which were defined as native perennial forbs that flower primarily from April to June [31]. We then conducted a thorough sampling of all plant taxa present from late July to early August (summer sampling). At each plot, we sampled all vegetation cover below 1 m ("herb-layer") within four 1-m² quadrats placed 1 m (for summer sampling) or 2 m (for spring sampling) from plot center in each cardinal direction. Within each quadrat, we estimated percent cover to the nearest integer for each plant taxon and counted woody stems by taxon, classifying the latter into two height classes: less than 0.5 m or 0.5 to 1.37 m.

We also estimated honeysuckle midstory cover (1.01 to 5 m) using cover class midpoints [32] in 4-m² quadrats oriented randomly, but consistently between years, at each plot. Shrubs and saplings greater than 1.37 m in height and less than 10 cm diameter at breast height (DBH; 1.37 m) were counted within 40 m² circular plots at two plots per treatment block and were classified into one of two size classes: (1) less than 5 cm DBH or (2) 5 to 9.9 cm DBH. We estimated percent canopy cover using a spherical crown densiometer (Forestry Suppliers Inc., Jackson, MS, USA), averaging four readings at each plot center point. We determined overstory composition using a variable radius plot (basal area factor 2.296 m² ha⁻¹) originating at the center of each block.

2.4. Soil Temperature

We collected data on soil temperature for 60 consecutive days from 12 March through 10 May 2017 by burying an iButton® data logger (Maxim Integrated, San Jose, CA, USA) approximately 1–2 cm below the surface of the mineral soil at the center of each plot. iButtons were coated in a thin layer of a spray-on rubber coating (PlastiDip®; Plasti Dip International, Blaine, MN, USA) before burying to protect against water damage and were programed to record temperature to the nearest 0.5 °C every hour. These data were summarized by calculating the mean, maximum, and minimum for each day at each plot.

2.5. Mulching Head Treatment Intensity

To determine how mulching head treatment intensity (i.e., depth) affects resprouting, we conducted an additional experiment at three other sites (two at LF and one at PWA), where eight plots were established on a grid with 8 × 8 m spacing. We determined depth of mulching head disturbance using wooden stakes pounded into the ground in 4-m intervals on a grid at each site, with the top 10 cm aboveground. While many stakes broke below ground during the treatment, a minimum of three of the nine stakes within 6 m of each plot were measurable (median 4.5) and we used these to calculate average treatment intensity at each plot.

On a single day the following spring, we counted the number of resprouting stumps in 40-m² circular plots around the centers of the eight plots at each site and measured the height, crown size (length and width), and number of stems of each resprout. Volume of each resprout was estimated by modelling the shrub as an ellipsoid with

$$volume = \frac{4}{3}\pi \left(\frac{length \times width \times height}{2} \right) \tag{1}$$

and the sum of these values was calculated to estimate total resprout volume per plot.

2.6. Data Preparation

To determine plot-level estimates of percent cover for each taxon in the herb-layer, we calculated the mean across four 1-m² quadrats from each plot. Woody seedling counts were summed across quadrats and multiplied by 25 to provide plot-level density estimates in seedlings 100 m⁻². Shrub and sapling counts within each 40-m² plot were multiplied by 250 to provide estimates in stems ha⁻¹. We also calculated total herbaceous cover and the total density of native seedling species at each plot.

For spring perennials, summer herbaceous plants, and woody seedlings, we calculated three measures of species diversity: taxonomic richness (S), the number of taxa present; Shannon's Diversity Index (H′), calculated as:

$$H' = -\sum_{i=1}^{S} p_i \ln p_i \tag{2}$$

where p_i is the proportion of the *i*th taxon in the dataset using cover for herbaceous plants and count for woody stems; and Pielou's Evenness Index (J′), calculated as:

$$J' = \frac{H'}{H'_{max}} \tag{3}$$

For the summer herb-layer plant community, we also calculated a floristic quality index (FQI) for each plot by assigning a coefficient of conservation (C) to each taxon according to values for Indiana [33] and then using the following equation [34]:

$$FQI = C_{ave} \times \sqrt{S} \tag{4}$$

This metric has been used previously to monitor the quality of understory community recovery after *Lonicera morrowii* removal [35]. C values for taxa identified to a pair of possible species (e.g., *Viola pubescens* or *V. striata*) were averaged but never differed by more than two points. Exotic species were automatically assigned a C value of 0. Grasses and sedges, which were only identified to family and genus, respectively, and unidentified taxa were excluded from FQI calculations.

2.7. Data Analysis

We analyzed whether changes in the response variables throughout the course of the study differed between treatments using linear mixed-effects models in the R package lme4 [36]. These models included fixed effects for "treatment"—a composite factor for removal method and mulch presence/absence—sampling year, and the interaction between sampling year and treatment. Random effects in these models included a random intercept for each site and a random year effect for each plot within a site. This longitudinal mixed-effects model for the response of garlic mustard

(*Alliaria petiolata*) failed to converge, so garlic mustard responses were analyzed with two separate models for the changes from 2015 to 2016 and from 2015 to 2017. The linear mixed-effects model for soil temperature included a fixed effect for treatment, random day effect for each plot, and random intercepts for each site and day.

To analyze differences between treatments in honeysuckle resprouting responses, we selected the most parsimonious subset of variables from a saturated model that included treatment, the pretreatment densities of each honeysuckle size class, and the interactions between treatment and these size classes. Models from this analysis were selected using Akaike's Information Criterion corrected for small sample sizes (AICc; [37]) in the R package MuMIn [38] and were constrained to include the fixed effect for treatment and a random intercept for each site.

We used square root or natural log transformations, when necessary, to improve normality and homogeneity of variance of residuals. If the model term of interest (treatment or sampling year × treatment interaction) was significant ($\alpha = 0.05$) or approaching significance ($\alpha = 0.10$), we tested planned comparisons for this term between: reference treatment and zero (i.e., did the response change in reference areas over the course of the study?); reference treatment and each operational treatment (i.e., cut-stump without mulch added—"Cut$^{[-]}$"—and mulching head treatment without mulch removed—"MH$^{[+]}$"); shrub removal with mulch and without; cut-stump removal and mulching head removal; and between the two operational treatments. Degrees of freedom were obtained using Satterthwaite's approximation in R package lmerTest [39] and are presented to the nearest integer for ease of interpretation. We tested planned comparisons using R package multcomp [40], and present single-step, adjusted *p*-values based on the joint normal or *t*-distribution of the linear function. Original data and R code for all analyses and figures are available in Supplementary Materials.

3. Results

3.1. Honeysuckle Responses

Both removal methods were initially successful at removing sapling-layer honeysuckle (>1.37 m tall), but stump sprouts quickly reentered the sapling layer. Sapling-layer honeysuckle were initially reduced from 4005 ± 307 shrubs ha^{-1} (mean ± standard error) across all pretreatment plots in 2015 to 94 ± 45 shrubs ha^{-1} in cut-stump removal plots and 110 ± 51 shrubs ha^{-1} in mulching head treatment plots in 2016. As resprouting stumps continued to grow from 2016 to 2017, the density of sapling-layer shrubs increased to 391 ± 143 stems ha^{-1} in cut-stump plots and to 781 ± 196 stems ha^{-1} in mulching head plots, though changes throughout the study period were not statistically different for the two removal treatments ($p > 0.99$). Sapling-layer honeysuckle densities changed little in reference areas, from 3734 ± 457 stems ha^{-1} in 2015 to 4016 ± 552 stems ha^{-1} in 2017 ($p > 0.99$).

Resprouting responses differed between the treatments, and these responses were influenced by pretreatment honeysuckle density (Table 2). Planned comparisons revealed that the presence of mulch negatively affected the height of resprouts ($p = 0.03$) but that height did not vary between removal methods. The mulching head treatment was associated with more stems per stump than cut-stump removal ($p < 0.01$), but mulch itself did not affect stems per stump. Significant interactions between treatment method and pretreatment honeysuckle densities for each resprout metric indicated that the degree to which pretreatment honeysuckle densities influenced resprouting depended on the treatment method (Table 2). The effect of small (<5 cm DBH) sapling-layer honeysuckle on densities of resprouts was greater in plots treated with the mulching head than with cut-stump removal ($p < 0.01$) and there was a weaker effect of large (>5 cm DBH) sapling layer honeysuckle on stems per stump following mulching head removal than cut-stump removal ($p < 0.01$). Midstory honeysuckle cover decreased more in each operational treatment than in reference areas (both $p < 0.01$; Table 3), but these changes were not different between operational treatments ($p > 0.99$).

Table 2. Honeysuckle resprouting responses to cut-stump and forestry mulching head removal treatments. Models were selected using Akaike's Information Criterion corrected for small sample sizes (AICc). Resprouting data were collected in 28.27 m^2 circular plots in late-April to early-May 2016 following shrub removal treatments the previous November–December at four mature secondary hardwood forest sites near West Lafayette, IN, USA.

	Cut-Stump		Mulching Head		Model Predictors	Test Statistic	p-Value
	No Mulch	With Mulch	No Mulch	With Mulch			
Resprouting stump density 100 m^{-2}	80 ± 23	39 ± 24	95 ± 24	95 ± 25	Treatment	$F_{3,51} = 0.77$	0.52
					Shrubs(lg)	$F_{1,50} = 27.54$	<0.01
					Shrubs(sm)	$F_{1,51} = 0.60$	0.44
					Seedlings(sm)	$F_{1,53} = 2.80$	0.10
					Treatment × Shrubs(sm)	$F_{3,51} = 3.75$	0.02
Height of resprouting stumps (cm)	31 ± 2	26 ± 3	25 ± 2	25 ± 2	Treatment	$F_{3,56} = 2.70$	0.054
					Seedlings(lg)	$F_{1,56} = 2.42$	0.13
					Treatment × Seedlings(lg)	$F_{3,56} = 2.85$	0.046
Stems per resprouting stump	11 ± 1	8 ± 1	9 ± 1	10 ± 1	Treatment	$F_{3,54} = 3.85$	0.01
					Shrubs(lg)	$F_{1,4} = 8.61$	0.04
					Shrubs(sm)	$F_{1,53} = 4.81$	0.03
					Treatment × Shrubs(lg)	$F_{3,54} = 3.43$	0.02

Shrubs(lg): Pretreatment density of honeysuckle individuals >1.37 m tall and >5 cm diameter, Shrubs(sm): >1.37 m tall and <5 cm diameter, Seedlings(lg): 0.5–1.37 m tall, Seedlings(sm): <0.5 m tall.

Table 3. Percent cover (mean ± standard error) of understory honeysuckle, midstory honeysuckle, and overall canopy in response to cut-stump and mulching head honeysuckle control. Understory, midstory, and canopy cover were estimated in four 1-m^2 quadrats, one 4-m^2 quadrat, or four spherical densiometer readings, respectively, at each plot. F and p-values are from linear mixed models testing whether each variable varied between treatments. Satterthwaite's approximation for degrees of freedom was used to estimate p-values and p-values are presented to the nearest integer. Data were collected in late July to early August each year at four mature secondary hardwood forest sites near West Lafayette, IN, USA.

		Ref.	Cut-Stump		Mulching Head		F Value [df]	p-Value
		—	No Mulch	With Mulch	No Mulch	With Mulch		
Canopy cover (%)	2015 (Pretreatment)	95 ± 0.3	94 ± 0.6	94 ± 0.6	94 ± 0.4	94 ± 0.4	1.26 [4,88]	0.29
	2015–2016 Change	−2 ± 0.3	−4 ± 0.9	−3 ± 0.9	−5 ± 0.9	−4 ± 0.9	4.55 [4,88]	<0.01
	2015–2017 Change	1 ± 0.4	−2 ± 0.8	−3 ± 0.9	−3 ± 0.7	−3 ± 0.8	8.13 [4,88]	<0.01
Midstory (1–5 m) L. maackii cover (%)	2015 (Pretreatment)	68 ± 4.5	55 ± 6.7	60 ± 6.6	76 ± 6.1	54 ± 8.7	2.01 [4,91]	0.10
	2015–2016 Change	9 ± 2.5	−54 ± 6.8	−60 ± 6.6	−76 ± 6.1	−54 ± 8.7	44.12 [4,90]	<0.01
	2015–2017 Change	4 ± 3.4	−52 ± 7.5	−59 ± 6.6	−75 ± 6.2	−52 ± 8.6	33.93 [4,91]	<0.01
Herb-layer (0–1 m) L. maackii cover (%)	2015 (Pretreatment)	18 ± 2.7	27 ± 3.7	29 ± 4.0	23 ± 3.8	24 ± 4.4	3.50 [4,88]	0.01
	2015–2016 Change	−5 ± 1.3	−17 ± 3.2	−20 ± 3.9	−4 ± 4.3	−11 ± 4.5	7.86 [4,88]	<0.01
	2015–2017 Change	−4 ± 1.7	−12 ± 3.1	−15 ± 4.2	0 ± 3.8	−7 ± 3.5	4.50 [4,88]	<0.01

We tested the role of mulching head treatment depth in determining honeysuckle resprouting at a separate set of sites. Plot-averaged mulching head treatment intensity ranged from 2.2 cm above soil surface to 1.5 cm below soil surface. Higher mulching head treatment intensity showed weak but statistically significant negative associations with the density of resprouting stumps ($p = 0.03$, $R^2 = 0.069$) and the size of resprouting shrubs ($p = 0.04$, $R^2 = 0.112$). However, the relationship between treatment intensity (i.e., depth) and the total estimated volume of resprouting shrubs on a plot was considerably stronger ($p < 0.01$, $R^2 = 0.311$; Figure 2). Total shrub volume combines both shrub size and stem density into a single value and is likely the most informative response metric for management that frequently includes foliar herbicide applications to resprouting stems.

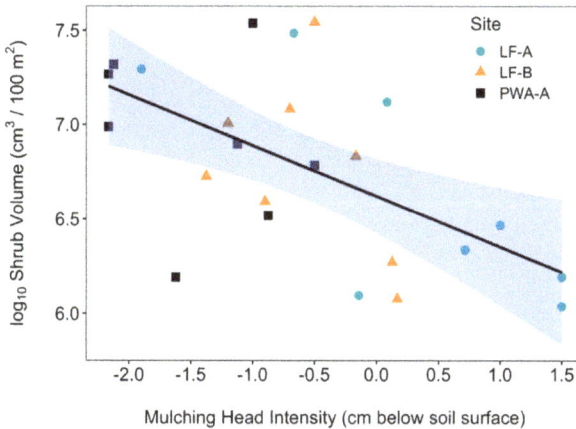

Figure 2. Relationship between mulching head treatment depth and the volume of resprouting shrubs. $R^2 = 0.311$, $p < 0.01$. Different symbols correspond to each site and shading indicates the 95% confidence interval. Shrub removal treatments and measurements of resprouting shrubs were conducted in winter and spring 2017, respectively, at three separate sites, near West Lafayette, IN, USA. Sites were located in the same areas (LF and PWA) but were distinct from sites used in comparing mulching head and cut-stump treatments. Negative x-axis values indicate distance above soil surface.

Overall, cut-stump removal reduced herb-layer honeysuckle cover (below 1 m) more than the mulching head treatment ($p < 0.01$), though the comparison between the operational treatments, Cut[−] and MH[+], was less pronounced (Table 4). The cut-stump control method resulted in lower herb-layer honeysuckle cover the first summer following removal, following reductions in percent cover of 17.2 ± 3.2 and 20.3 ± 3.9 in Cut[−] and Cut[+] plots and 4.2 ± 4.3 and 10.5 ± 4.5 in MH[−] and MH[+] plots, respectively, despite being higher initially (Figure 3 and Table 3). The following year, herb-layer honeysuckle cover increased similarly in both removal treatments (Figure 3).

Honeysuckle seedlings responded strongly to shrub removal initially but declined in all treatments the following year. Seedlings of the smaller size class (<0.5 m) increased in density by 1988 ± 429 seedlings 100 m^{-2} in Cut[−] plots and by 5542 ± 1853 seedlings 100 m^{-2} in MH[+] plots, as compared to a mean increase of 1266 ± 232 seedlings 100 m^{-2} in reference areas (Figure 3b). Despite initial differences, densities of small seedlings were more similar between treatments the following year (Figure 3b), resulting in no overall sampling year effect between treatments ($p > 0.10$, Table 4). Densities of large honeysuckle seedlings (0.5–1.37 m) decreased in the first year following shrub removal and increased the next year in all shrub removal treatments (Figure 3c); there was no difference in sampling year effects for this size class ($p > 0.10$, Table 4).

Table 4. Linear mixed effects model results for understory community response metrics to different shrub removal treatments, including richness (S), Shannon's diversity index (H'), and Pielou's evenness (J') each subset of understory taxa, floristic quality index (FQI) for the entire understory community, and cover of herbaceous plant groups based on life history strategy. Treatments included cut-stump (C) or forestry mulching head (M), with mulch (I^+) or without (I^-), and an untreated reference (R). Spring perennial cover data were collected late-April to early-May and all other data were collected mid-July to early-August from 2015 to 2017 at four mature secondary hardwood forest sites near West Lafayette, IN, USA.

	Treatment ANOVA		Treatment × Year ANOVA		Planned Comparisons (z Values)					
	df	F	df	F	R vs. 0	R vs. C[I^-]	R vs. M[I^+]	C[I^-] vs. M[I^+]	C vs. M	[I^-] vs. [I^+]
L. maackii										
Herb-Layer Cover (<1 m)	4, 81 ·	2.25 ·	4, 91	3.91 **	−1.51	−2.37 ·	−0.23	1.85	3.46 **	−0.84
Seedlings <0.5 m	4, 69 ·	2.40 ·	4, 132	0.63	−	−	−	−	−	−
Seedlings 0.5–1.37 m	4, 105	0.44	4, 187	1.30	−	−	−	−	−	−
Herb-Layer Native Plants [††]										
Cover	4, 71	0.31	4, 91	14.37 **	−1.80	5.84 **	4.91 **	−0.81	−0.10	−1.05
FQI	4, 89	0.43	4, 91	2.91 **	0.71	2.64 *	2.15	−0.43	−0.64	0.04
S	4, 98	1.77	4, 145	29.06 **	0.36	7.73 **	8.10 **	−2.03	2.48 ·	0.32
H'	4, 80	4.06	4, 91	7.36 **	1.70	3.45 **	4.02 **	0.50	2.26	−1.55
J'	4, 80	2.11 ·	4, 85	1.49	−	−	−	−	−	−
Spring Perennials										
Cover	4, 87 ·	2.43 ·	4, 91	1.57	−	−	−	−	−	−
S	4, 88	9.41	4, 91	0.39	−	−	−	−	−	−
H'	4, 88	0.78	4, 91	2.17 ·	2.71 *	1.25	1.01	−0.21	1.63	−1.93
J'	4, 73	2.67 *	4, 82	3.66 **	0.48	−0.31	2.50 ·	2.37 ·	3.31 **	0.16
Summer Herbs										
Cover	4, 76	0.45	4, 91	21.52	−1.01	5.97 **	6.59 **	0.54	1.83	−1.06
S	4, 114	1.65	4, 147	18.69 **	0.88	6.10 **	6.34 **	0.21	2.13	−1.83
H'	4, 82	1.57	4, 91	9.88 **	0.93	4.22 **	4.98 **	0.66	2.11	−1.19
J' [‡]	4, 67	1.70	4, 68	0.73	−	−	−	−	−	−
Annual Natives	4, 127	0.86	4, 99	2.20 ·	−0.01	0.69	0.98	−0.33	−0.49	−0.59
Biennial Natives	4, 87	0.69	4, 91	2.92 *	0.25	0.64	0.80	0.38	−0.05	0.24
Perennial Natives	4, 86	0.20	4, 91	14.33 **	0.16	1.97	2.74 *	−1.97	1.62	−0.25
Native Vines	4, 85	1.43	4, 91	8.13 **	−2.25	3.10 **	3.14 **	0.27	0.81	0.77
Native Graminoids	4, 96	0.60	4, 176	5.87 **	0.25	0.14	1.73	−0.87	2.37 ·	1.06
Exotics	4, 172	15.90	4, 169	2.18 ·	−0.28	0.19	0.03	−1.10	−0.06	−0.82
Native Seedlings										
Cover	4, 84	0.63	4, 182	5.03 **	−2.51 ·	3.89 **	2.18	−1.49	−0.56	−1.54
Density	4, 56	0.78	4, 91	2.44 ·	−2.17	2.85 *	1.15	−1.48	−1.93	−0.16
S	4, 85	1.55	4, 91	9.79 **	−1.36	4.48 **	5.04 **	0.49	0.62	0.07
H'	4, 87	1.69	4, 91	8.30 **	−1.27	4.25 **	4.41 **	0.14	0.69	−0.49
J'	4, 77	1.28	4, 75	3.08 *	−1.54	2.27	2.31 ·	0.01	1.28	−1.26

· $p < 0.10$; * $p < 0.05$; ** $p < 0.01$; [††] Includes all native woody and herbaceous species from summer sampling period; does not include spring perennials; [‡] Site PWA excluded due to an excessive number of zero values prior to shrub removal.

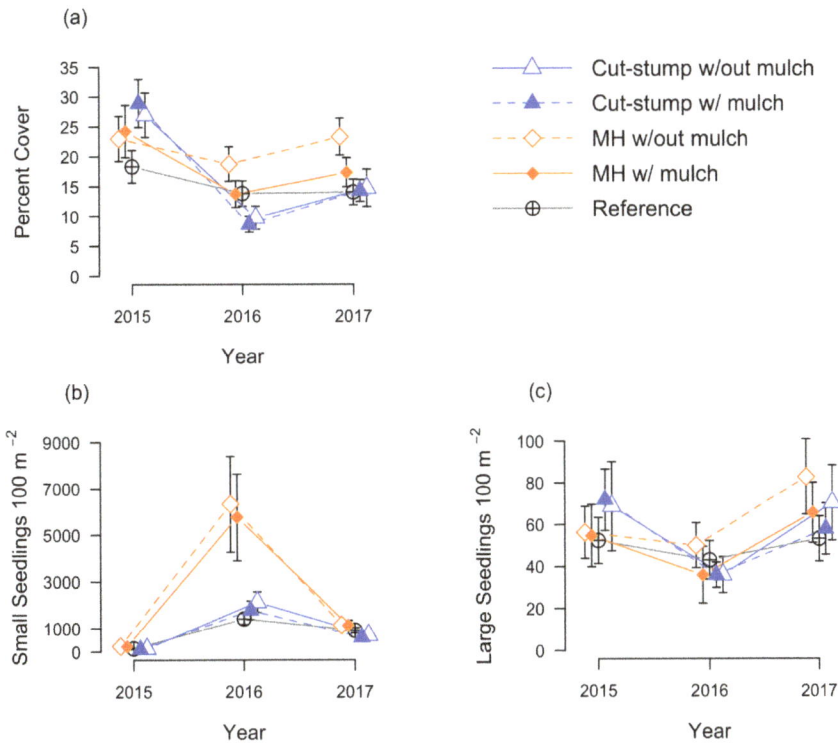

Figure 3. Honeysuckle responses to removal treatments conducted in November and December 2015, including percent cover below 1 m (**a**), density of honeysuckle seedlings <0.5 m tall (**b**), and density of honeysuckle seedlings 0.5–1.37 m tall (**c**). Honeysuckle removal was accomplished using cut-stump or forestry mulching head ("MH") treatments, with mulch removed/absent or added/present. Honeysuckle were not removed in reference areas for comparison. Dashed lines indicate management techniques with mulch manipulated (added/removed) and solid lines represent operational (no mulch manipulation) management techniques. Values are means ± standard error. Data were collected from mid-July to late August at four mature secondary hardwood forest sites near West Lafayette, IN, USA.

3.2. Native Seedling Responses

Decreases in native woody seedling densities between 2015 and 2016 were lower in Cut$^{[-]}$ plots (-3 ± 79 seedlings 100 m^{-2}) than in MH$^{[+]}$ plots (-220 ± 72 seedlings 100 m^{-2}), although plots randomly assigned to the mulching head treatment had higher native seedling densities prior to removal, and in 2016 the densities of native seedlings were similar in Cut$^{[-]}$ (406 ± 70 seedlings 100 m^{-2}) and MH$^{[+]}$ (416 ± 92 seedlings 100 m^{-2}) plots (Figure 4a). Native seedling densities decreased in reference areas throughout the study but increased in all removal treatments between 2016 and 2017 (Figure 4a), and there was no difference between the two operational treatments, Cut$^{[-]}$ and MH$^{[+]}$, in sampling year effects on native seedling densities (Table 4).

The most abundant native woody seedlings prior to shrub removal were white ash (*Fraxinus americana* L.) and sugar maple (*Acer saccharum* Marshall), which accounted for 45.7% ± 3.8% and 25.4% ± 3.3% of pretreatment native seedlings, respectively. Taken together, the pretreatment percentage of these shade tolerant species did not differ between treatment assignments ($p = 0.35$).

A longitudinal linear mixed-effects model indicated a negative sampling year effect on the percentage of these two species ($p = 0.04$), but this effect did not differ between treatments ($p = 0.16$).

Native seedling S and H′ increased similarly following both removal treatments (Figure 4b,c), and these changes were no different between removal treatments (Table 4). Native seedling J′ decreased in reference areas over the course of the study and increased following shrub removal treatments (Figure 4d). However, while there was an interaction between treatment and sampling year ($p = 0.02$), we were unable to detect any differences in the change in native seedling J′ between individual operational treatments and the reference (Table 4).

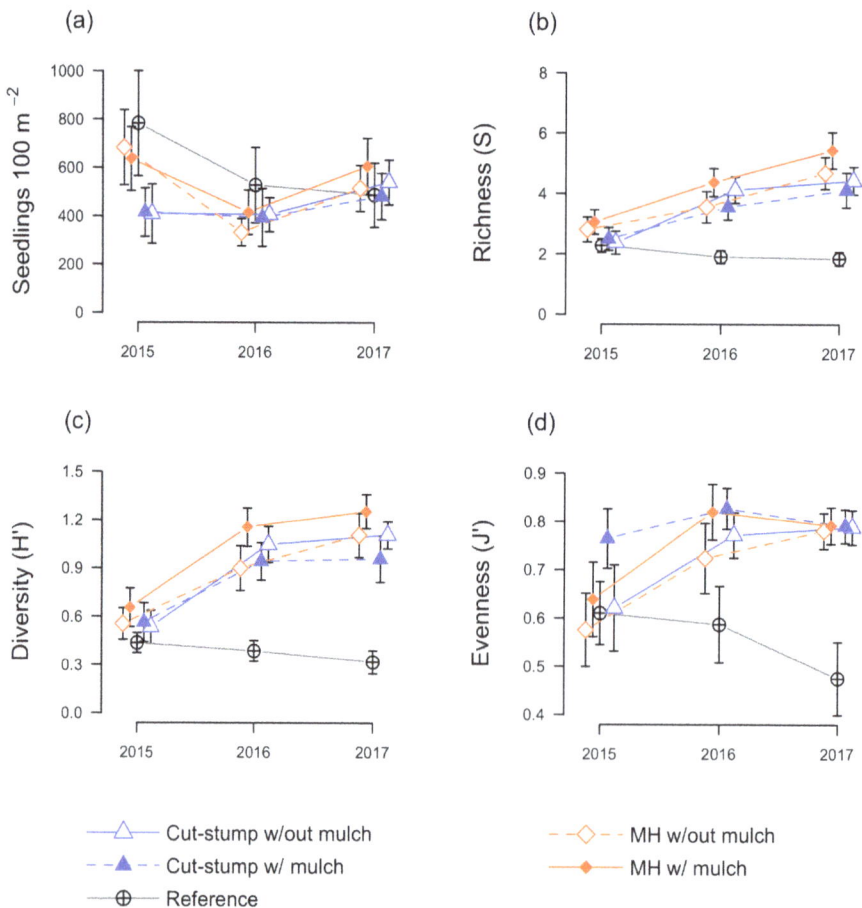

Figure 4. Densities (**a**) and diversity indices (**b–d**) of native woody species prior to invasive shrub removal (2015) and following shrub removal (2016–2017) accomplished using cut-stump or forestry mulching head ("MH") treatments, with mulch removed/absent or added/present. Invasive shrubs were not removed in reference areas for comparison. Dashed lines indicate management techniques with mulch manipulated (added/removed) and solid lines represent operational (no mulch manipulation) management techniques. Values are means ± standard error. Data were collected from mid-July to late August at four mature secondary hardwood forest sites near West Lafayette, IN, USA.

3.3. Herbaceous Plant Responses

Cover of spring perennials changed little over the course of the study and was generally low across treatments (Figure 5a). Slight changes in spring perennial cover were not different between treatments (Table 4). Spring perennial S and H′ increased across all treatments, including reference areas, over the course of the study, though H′ decreased slightly in both treatments with mulch in the first spring following shrub removal (Figure 5b,c). Spring perennial J′ pretreatment values varied between treatments (Table 4) but increased more following the mulching head treatment than cut-stump (Figure 5d).

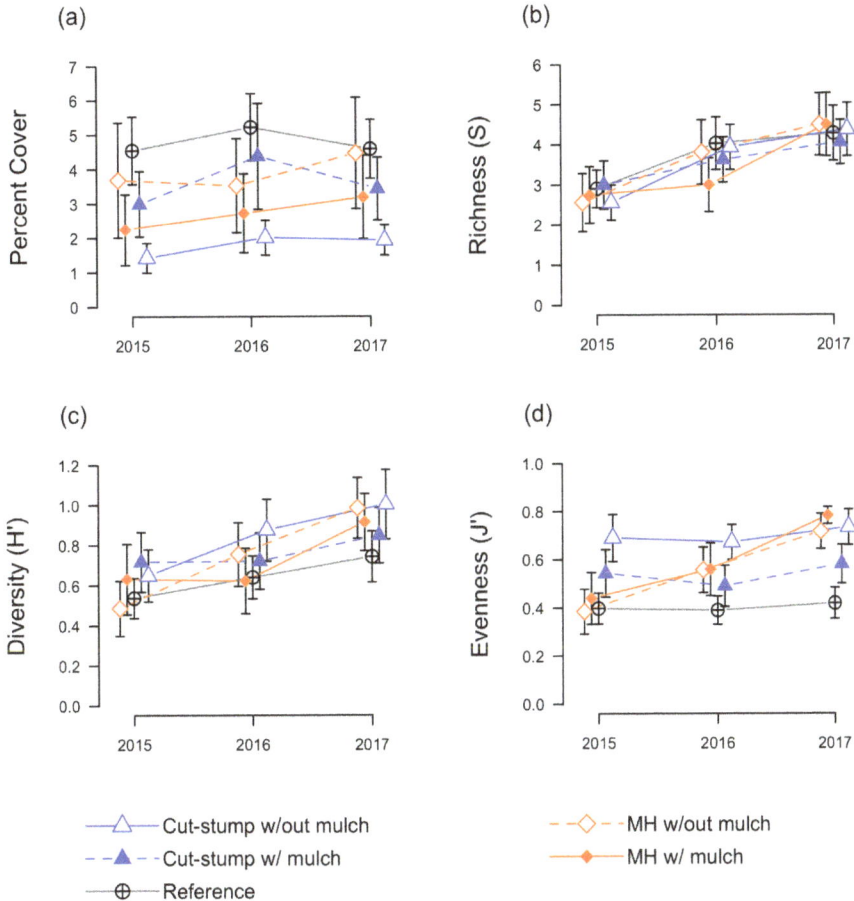

Figure 5. Percent cover (**a**) and diversity indices (**b–d**) of spring perennial herbaceous plants prior to invasive shrub removal (2015) and following shrub removal (2016–2017) accomplished using cut-stump or forestry mulching head ("MH") treatments, with mulch removed/absent or added/present. Invasive shrubs were not removed in reference areas for comparison. Dashed lines indicate management techniques with mulch manipulated (added/removed) and solid lines represent operational (no mulch manipulation) management techniques. Values are means ± standard error. Data were collected from mid-July to late August at four mature secondary hardwood forest sites near West Lafayette, Indiana, USA.

Total summer cover of herbaceous plants increased following all shrub removal treatments, and cover increases continued in the second post-treatment growing season (Figure 6a). Pretreatment mean percent cover of herbaceous plants ranged from 0.5 ± 0.2 at PWA to 6.1 ± 1.6 at site LF2 (grand mean 3.9 ± 0.6). Increases in percent herbaceous cover were greater in each operational treatment than in reference sites (both $p < 0.01$) and were similar across removal treatments ($p > 0.10$, Table 4). Herbaceous percent cover increased 9.2 ± 1.2 across cut-stump and mulching head treatments over the course of the study, as compared to a 0.8 ± 0.5 decrease in reference areas (Figure 6a).

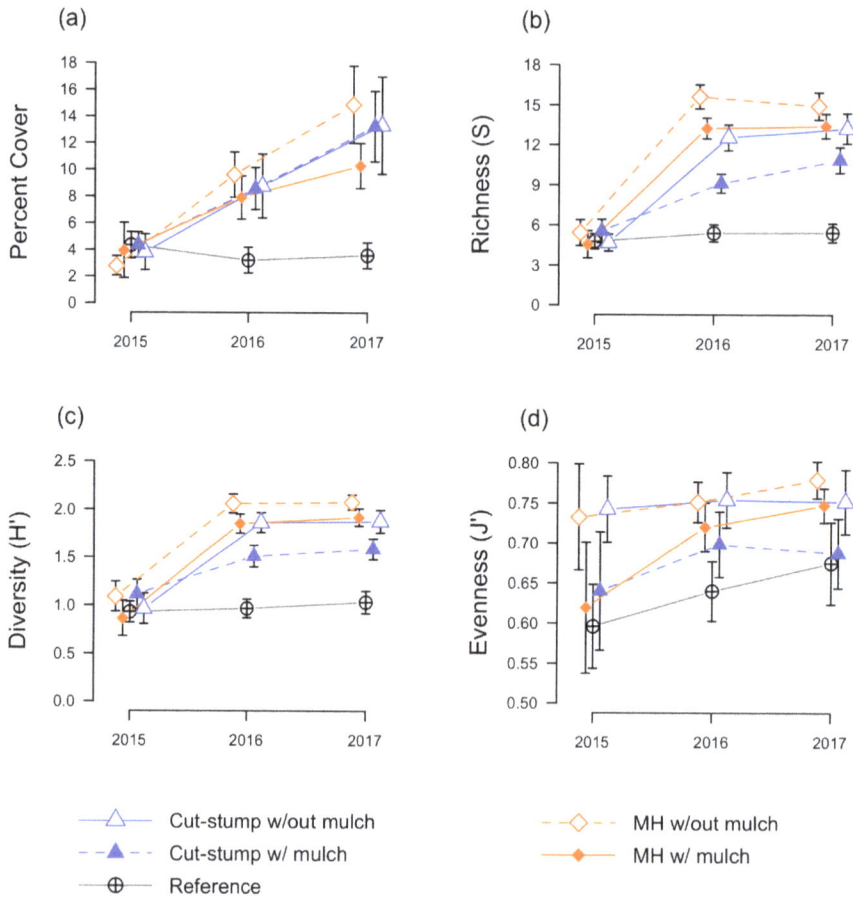

Figure 6. Percent cover (**a**) and diversity indices (**b–d**) of understory herbaceous plant species prior to invasive shrub removal (2015) and following shrub removal (2016–2017) accomplished using cut-stump or forestry mulching head ("MH") treatments, with mulch removed/absent or added/present. Invasive shrubs were not removed in reference areas for comparison. Dashed lines indicate management techniques with mulch manipulated (added/removed) and solid lines represent operational (no mulch manipulation) management techniques. Values are means ± standard error. Data were collected from mid-July to late August at four mature secondary hardwood forest sites near West Lafayette, IN, USA.

Summer herbaceous S and H′ both increased following shrub removal, but these increases were sharper initially than between the first and second post-treatment growing seasons and did not differ between operational removal treatments (both $p > 0.10$, Figure 6b,c). Mean herbaceous S increased by 8.1 ± 0.6 in shrub removal areas, as compared to an increase of 0.7 ± 0.4 in reference areas (Table 4). Mean herbaceous H′ increased 0.9 ± 0.2 in Cut[−] and 1.1 ± 0.2 in MH[+] plots, as compared to 0.1 ± 0.1 in reference areas (both $p < 0.01$, Table 4). Herbaceous J′ was highly variable and there were no differences in the change in herbaceous J′ between treatments (Figure 6d and Table 4).

Cover of native perennials increased more following MH[+] treatment than in reference areas ($p = 0.03$), and cover of native vines increased more in both operational treatments than in reference areas (both $p = 0.01$), but removal treatments did not differ in the responses of any herbaceous groups, including exotics (Table 4). Percent cover of garlic mustard responded similarly to both operational treatments in the first year, increasing from 0.1 ± 0.0 prior to removal treatments up to 1.1 ± 0.5 after Cut[−] treatment and 1.1 ± 0.4 after MH[+]. While garlic mustard cover was generally lower during the 2017 sampling period, decreasing to 0.3 ± 0.2 in Cut[−] plots and to 0.2 ± 0.1 in MH[+], two-year change in garlic mustard cover was higher following Cut[−] treatment than MH[+] treatment ($p = 0.06$). However, the number of fruiting garlic mustard individuals in 2017 was greater in MH[+] than Cut[−] treatment plots. Differences between operational treatments were driven by mulch ($p < 0.01$) rather than removal method ($p = 0.26$).

3.4. Floristic Quality

FQI calculated at the plot scale (4-m^2) for herb-layer species changed little in treatment plots following shrub removal but responded positively the following year, whereas FQI in reference plots declined slightly (Figure 7). Overall, FQI increased 0.95 ± 0.36 per year more in Cut[−] plots and 0.77 ± 0.36 per year more in MH[+] plots compared to reference areas ($p = 0.04$; $p = 0.14$), in which FQI did not change over the course of the study ($p = 0.92$). Mean C_{ave} across all plots prior to shrub removal was 2.5 ± 0.1, decreasing the following year to 1.9 ± 0.1 and 1.8 ± 0.1 in Cut[−] and MH[+] treatments, then rebounding somewhat to 2.2 ± 0.2 and 2.1 ± 0.1 in those same treatments, respectively.

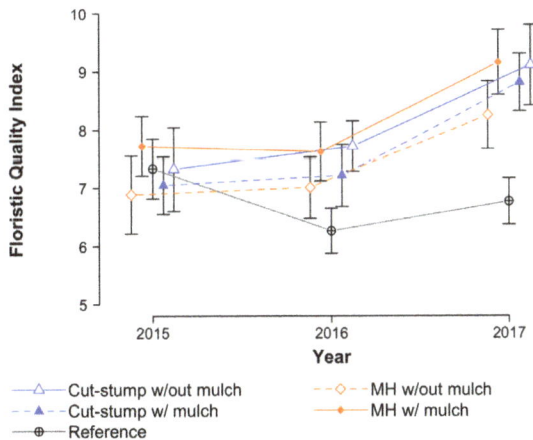

Figure 7. Floristic quality index (FQI) for understory communities, including exotic species, before (2015) and after (2016 and 2017) invasive shrub removal was accomplished using cut-stump or forestry mulching head ("MH") treatments, with mulch removed/absent or added/present. Invasive shrubs were not removed in reference areas for comparison. Dashed lines indicate management techniques with mulch manipulated (added/removed) and solid lines represent operational (no mulch manipulation) management techniques. Values are mean FQI ± standard error. Data were collected from four mature secondary hardwood forest sites near West Lafayette, IN, from mid-July to early August each year.

3.5. Soil Temperature

Honeysuckle removal treatments elevated soil temperatures, and mulch reduced daily temperature variability within removal areas. Daily mean soil temperature at 1 cm depth was $0.56° \pm 0.08°$ higher in removal plots without mulch than in removal areas and was $0.57° \pm 0.08°$ higher in removal plots with mulch than in removal areas (both $p < 0.01$). Across both removal methods, mulch did not affect mean temperature ($p > 0.99$). Removal plots with mulch had higher daily minimum soil temperatures and lower daily maximum temperatures than those without mulch (min.: $p < 0.01$, max.: $p = 0.05$). This buffering effect of mulch on temperature extremes resulted in no difference in daily soil temperature ranges between mulched removal plots and untreated reference plots ($p = 0.31$), whereas temperatures in removal plots without mulch ranged $0.32° \pm 0.07°$ more than in reference areas on a daily basis ($p < 0.01$).

4. Discussion

4.1. Treatment Effectiveness

In this comparison of two methods for invasive shrub removal, the mulching head was generally less effective than cut-stump removal at controlling invasive bush honeysuckle, as indicated by more rapid regrowth of herb-layer honeysuckle cover. This difference in honeysuckle cover was likely driven by differences in the responses of resprouting stumps we observed between removal methods. Without treating cut stumps, previous work on other invasive shrub taxa found no differences in regrowth between clearing saw and mulching head methods [12], suggesting that differences in resprouting between the two techniques may be driven by stump treatments. We saw faster regrowth of honeysuckle cover following both treatments compared to barberry responses observed by Ward et al. [12]. Conducting shrub removal in late fall, as opposed to early spring, may be responsible for more prolific resprouting, as non-structural carbohydrate stores in roots are highest in the fall just before leaf abscission [35].

Honeysuckle removal in nearby sites the following year indicated that increasing the depth of the mulching head treatment can mitigate resprouting (Figure 2), suggesting that a relatively shallow treatment depth in this study may be partially responsible for the relatively vigorous honeysuckle resprouting we observed in mulching head treatment areas. However, native plant community impacts of the mulching head along a gradient of treatment intensity have not yet been investigated, and one of the reasons for similar native community responses between control methods in this study may be that we limited soil disturbance associated with the mulching head treatment. This balance between reducing resprouting and protecting belowground plant structures requires additional study as best management practices are developed for this relatively new control technique.

Although we did not examine the phenology of treated stumps, one explanation for shorter resprouting shrubs in mulched plots is that mulch buffered soil against cold winter temperatures, and shrubs in these plots may have taken longer to meet the chilling requirement required to break endodormancy, though the chilling requirement for Amur honeysuckle is minimal [41]. Interactions between treatment method and pretreatment stem density effects on resprouting suggest that the mulching head treatment was less effective at controlling the resprouting of small sapling-layer shrubs but more effective than the cut-stump method for controlling large sapling layer shrubs, though we were unable to track individual stems due to the nature of the mulching head treatment.

4.2. Responses of Exotic Flora

Greater honeysuckle percent cover in mulching head-treated plots in 2016 compared to cut-stump plots is also attributable to the strong response of honeysuckle seedlings in the first year, which is consistent with rapid honeysuckle seedling responses in other removal experiments [17]. Canopy openness was no different between removal methods (Table 3), so the much greater density of honeysuckle seedlings following mulching head treatment is more likely the result of soil disturbance, which has been implicated in the germination of other invasive shrubs [42]. Re-establishment of Chinese privet was also higher in areas treated with a mulching head in riparian forests in Georgia, USA [16]. The mulching head likely disrupts the litter layer on the forest floor more than the cut-stump method, which can increase the germination and establishment of both honeysuckle and garlic mustard seedlings [20]. We expected that mulch would limit the density of emergent honeysuckle seedlings, but any such effect was dwarfed by the difference between removal methods (Figure 3b).

We also expected that garlic mustard, an exotic invasive herb that regularly colonizes sites after honeysuckle removal [15,17,21], would be promoted more by the mulching head treatment, but suppressed by residual mulch, and this hypothesis was partially supported. Increases in garlic mustard cover were similar between treatments, despite greater honeysuckle cover in mulching head plots. Recent work suggests that direct interactions with other components of the understory plant community may not be as strong a determinant of garlic mustard abundance as ungulate herbivory pressure [43]. Despite these similar initial increases in garlic mustard cover between treatments, there were more fruiting garlic mustard individuals in mulched plots the second year. Due to the biennial life history strategy of garlic mustard, with basal rosettes that overwinter, severe winter conditions can limit garlic mustard survival [44]. Insulated soil temperatures beneath mulch may have contributed to higher winter survival rates, resulting in more fruiting individuals the following growing season.

The senescence of a large proportion of honeysuckle seedlings after the first year post-treatment is also consistent with previous work on honeysuckle control [15]. Self-thinning of woody seedlings can be a density-dependent response to soil pathogens, even in temperate forests [45], suggesting that without the strong resprouting response observed in this study, more of these honeysuckle germinants may have persisted. While honeysuckle invasion reduces conspecific seed viability [46], honeysuckle seedlings may grow better in soils previously conditioned by invasive shrubs [47]. Moreover, invasive shrubs may be less susceptible to fungal pathogens due to enemy release [48], but some evidence suggests that Amur honeysuckle may be susceptible to a leaf blight fungus that affects native *Lonicera* species [15]. Testing understory responses to the mulching head in combination with strategies to further mitigate regrowth, including greater treatment intensities, treating in the spring, and applying follow-up control methods, will help to elucidate how the pulse of honeysuckle regeneration after mulching head treatments responds to lower densities of resprouting mature individuals.

4.3. Native Plant Community Recovery

Generally, native plant community recovery was similar between both treatment methods, despite greater honeysuckle regrowth following the mulching head treatment. In this region, woodlands flat enough to permit mulching head utilization have likely been subject to numerous severe disturbances during the last 200 years, including row cropping, grazing pressure, and high grading. This history of anthropogenic disturbance, coupled with approximately 30 years of honeysuckle invasion, may have had a filtering effect on the existing plant communities such that species more sensitive to substrate disturbance had been excluded long before the initiation of this study [49]. Several recovery metrics exhibited greater differences between treatments the first year after shrub removal than the trajectory over the course of the entire study would suggest. Had the study only followed shrub removal effects for one post-treatment growing season, mulching head impacts on native seedling densities, stronger responses of honeysuckle seedling densities in mulching head plots, and negative effects of mulch on spring perennial diversity would have indicated that the mulching head was less suitable for achieving restoration goals than the cut-stump method. However, the lack of these differences over a two-year study period emphasizes the necessity for multi-year monitoring of restoration objectives.

While total cover of summer herbaceous plants responded positively to all removal treatments, the cover of spring perennials did not recover during the course of this study, in contrast to other studies reporting rapid responses of spring herbs to invasive shrub removal [17,50]. However, the pretreatment cover of spring perennial herbs observed by Shields et al. [17] was far higher than in this study, and highly variable spring perennial cover between plots in this study may have been driven by the distribution of mayapple (*Podophyllum peltatum* L.) across study sites. Mayapple is relatively broadleaved compared to the other spring perennials in this study and accounted for 35.6% of all spring perennial cover recorded during the study but occurred at only 28 of 96 plots in 2017. Where it did occur, mayapple accounted for 54.3 ± 5.9% of spring perennial cover, indicating that this particular species may have driven the high variability in spring perennial cover and masked responses in the rest of the spring flora.

We observed a delay in the recovery of the FQI during the first year post-treatment, despite immediate increases in species richness, corresponding with concurrent declines in C_{ave}. FQI values two years after removal in this study were slightly lower than those calculated after *Lonicera morrowii* removal in Pennsylvania [35], likely due to the smaller plot size used in this study (4 m^2 vs. 5 m^2). However, C_{ave} values in removal areas in 2017 were lower than comparable reference sites in the Central Till Plain (e.g., mature upland forest: C_{ave} = 3.3 [34]), indicating that initial honeysuckle removal treatments at these sites did not completely restore the quality of the understory community to that of a similar uninvaded habitat.

In addition to focusing on Amur honeysuckle, two features of this study set it apart from previous comparisons of the effectiveness of forestry mulching heads and cut-stump treatments for invasive shrub removal and impacts on understory community [12,16]. Namely, we did not apply any sort of follow-up treatment to control regrowth from stumps during the 21-month period following initial removal covered by this study and we specifically separated the effects of mulch deposition from physical effects of the two removal methods. We decided against follow-up foliar herbicide applications in order to inform how initial shrub control treatments would influence the vigor of regrowth and assess the necessity for follow-up treatments. Waiting until the second post-treatment growing season to apply follow-up treatments, typically foliar herbicide applications, is likely to require greater herbicide volumes and time investment due to an increase in the number of resprouting stumps that attain heights 0.5–1.37 m (Figure 3c) or even reach the sapling layer (>1.37 m). This increase in larger honeysuckle individuals corresponded with greater honeysuckle cover in the second year post-treatment, despite steep declines in the number of smaller honeysuckle seedlings. We observed increases in native plant cover and diversity despite invasive shrub cover rebounding quickly, which is consistent with other findings that follow-up treatments do not impact native community recovery in the short term [12].

Mulch deposition had small effects on honeysuckle resprout height and soil temperature variability. While mulch deposition had no statistically significant effects on native plant community recovery in this study, there were consistent trends of mulch having effects counter to those of the mulching head removal method in isolation, which may be responsible for the similar responses observed between operational Cut[−] and MH[+] treatments (Table 4).

5. Conclusions

This study indicates that forestry mulching heads are a promising tool for expediting the removal of heavy shrub invasions from eastern hardwood forests and thereby restoring the native forest understory community. Native plant community responses were similar between the mulching head and cut-stump treatments and increases in Amur honeysuckle seedlings following mulching head treatment were largely temporary. Although the mulching head treatment was slightly less effective than the cut-stump method at preventing honeysuckle regrowth in this study, our results indicate that greater treatment depth with the mulching head can reduce post-treatment shrub volume. Furthermore, its ability to remove mature invasive shrubs rapidly over large areas may help to reduce recolonization from adjacent seed sources when used in a management context. This potential for reduced propagule pressure at the landscape scale, and possible implications for invasion dynamics, should be the subject of future research. Floristic quality indices showed a relatively weak response to any shrub removal technique in this study, suggesting that these communities are dominated by disturbance-tolerant species that may have relatively low conservation value. Therefore, removal of invasive woody plants is only one of a suite of approaches that managers may need to employ to restore many degraded hardwood forests.

Supplementary Materials: Original data and R code used in analyses are available online at http://www.mdpi.com/1999-4907/9/10/607/s1..

Author Contributions: Conceptualization: G.S.F., M.R.S., and M.A.J.; Data Curation: G.S.F.; Formal Analysis: G.S.F.; Funding Acquisition: M.A.J.; Investigation: G.S.F. and M.A.J.; Methodology: M.R.S. and M.A.J.; Project Administration: M.A.J.; Resources: M.R.S. and M.A.J.; Software: G.S.F.; Supervision: M.A.J.; Validation: G.S.F., M.R.S., and M.A.J.; Visualization: G.S.F.; Writing—Original Draft: G.S.F.; Writing—Review and Editing: M.R.S. and M.A.J.

Funding: Funding for this study was provided by the Hardwood Tree Improvement Center at Purdue University with funds from the Fred M. van Eck Foundation for Purdue University. Additional funding was provided by the USDA National Institute of Food and Agriculture, McIntire Stennis Cooperative Forestry Research Program (project IND011533MS).

Acknowledgments: We gratefully acknowledge the assistance of Brian Beheler, Michael Loesch-Fries, Charlotte Owings, Ivy Widick, and Will Zak in implementing treatments and collecting field data.

Conflicts of Interest: The authors declare no conflicts of interest.

References

1. Webster, C.R.; Jenkins, M.A.; Jose, S. Woody invaders and the challenges they pose to forest ecosystems in the eastern United States. *J. For.* **2006**, *104*, 366–374. [CrossRef]
2. Mascaro, J.; Schnitzer, S.A. *Rhamnus cathartica* L. (common buckthorn) as an ecosystem dominant in southern Wisconsin forests. *Northeast. Nat.* **2007**, *14*, 387–402. [CrossRef]
3. Hartman, K.M.; McCarthy, B.C. Restoration of a forest understory after the removal of an invasive shrub, Amur honeysuckle (*Lonicera maackii*). *Restor. Ecol.* **2004**, *12*, 154–165. [CrossRef]
4. Schulte, L.A.; Mottl, E.C.; Palik, B.J. The association of two invasive shrubs, common buckthorn (*Rhamnus cathartica*) and Tartarian honeysuckle (*Lonicera tatarica*), with oak communities in the midwestern United States. *Can. J. For. Res.* **2011**, *41*, 1981–1992. [CrossRef]
5. Collier, M.H.; Vankat, J.L.; Hughes, M.R. Diminished plant richness and abundance below *Lonicera maackii*, an invasive shrub. *Am. Midl. Nat.* **2002**, *147*, 60–71. [CrossRef]
6. Meiners, S.J. Apparent competition: An impact of exotic shrub invasion on tree regeneration. *Biol. Invasions* **2007**, *9*, 849–855. [CrossRef]
7. Schmidt, K.A.; Whelan, C.J. Effects of exotic *Lonicera* and *Rhamnus* on songbird nest predation. *Conserv. Biol.* **1999**, *13*, 1502–1506. [CrossRef]
8. Watling, J.I.; Hickman, C.R.; Orrock, J.L. Invasive shrub alters native forest amphibian communities. *Biol. Conserv.* **2011**, *144*, 2597–2601. [CrossRef]
9. Luken, J.O.; Mattimiro, D.T. Habitat-specific resilience of the invasive shrub Amur honeysuckle (*Lonicera maackii*) during repeated clipping. *Ecol. Appl.* **1991**, *1*, 104–109. [CrossRef] [PubMed]
10. Bailey, B.G.; Saunders, M.R.; Lowe, Z.E. A cost comparison of five midstory removal methods. In Proceedings of the 17th Central Hardwood Forest Conference, Lexington, KY, USA, 5–7 April 2010; pp. 535–543.
11. Ducrey, M.; Turrel, M. Influence of cutting methods and dates on stump sprouting in Holm oak (*Quercus ilex* L) coppice. *Ann. For. Sci.* **1992**, *49*, 449–464. [CrossRef]
12. Ward, J.S.; Williams, S.C.; Worthley, T.E. Comparing effectiveness and impacts of Japanese barberry (*Berberis thunbergii*) control treatments and herbivory on plant communities. *Invasive Plant Sci. Manag.* **2013**, *6*, 459–469. [CrossRef]
13. Hanula, J.L.; Horn, S.; Taylor, J.W. Chinese privet (*Ligustrum sinense*) removal and its effect on native plant communities of riparian forests. *Invasive Plant Sci. Manag.* **2009**, *2*, 292–300. [CrossRef]
14. Flory, S.L.; Clay, K. Invasive plant removal method determines native plant community responses. *J. Appl. Ecol.* **2009**, *46*, 434–442. [CrossRef]
15. Boyce, R.L. Recovery of native plant communities in southwest Ohio after *Lonicera maackii* removal. *J. Torrey Bot. Soc.* **2015**, *142*, 193–204. [CrossRef]

16. Hudson, J.R.; Hanula, J.L.; Horn, S. Impacts of removing Chinese privet from riparian forests on plant communities and tree growth five years later. *For. Ecol. Manag.* **2014**, *324*, 101–108. [CrossRef]

17. Shields, J.M.; Saunders, M.R.; Gibson, K.D.; Zollner, P.A.; Dunning, J.B.; Jenkins, M.A. Short-term response of native flora to the removal of non-native shrubs in mixed-hardwood forests of Indiana, USA. *Forests* **2015**, *6*, 1878–1896. [CrossRef]

18. Luken, J.O.; Kuddes, L.M.; Tholemeier, T.C. Response of understory species to gap formation and soil disturbance in *Lonicera maackii* thickets. *Restor. Ecol.* **1997**, *5*, 229–235. [CrossRef]

19. Miller, E.M.; Seastedt, T.R. Impacts of woodchip amendments and soil nutrient availability on understory vegetation establishment following thinning of a ponderosa pine forest. *For. Ecol. Manag.* **2009**, *258*, 263–272. [CrossRef]

20. Bartuszevige, A.M.; Hrenko, R.L.; Gorchov, D.L. Effects of leaf litter on establishment, growth and survival of invasive plant seedlings in a deciduous forest. *Am. Midl. Nat.* **2007**, *158*, 472–477. [CrossRef]

21. Cipollini, K.; Ames, E.; Cipollini, D. Amur honeysuckle (*Lonicera maackii*) management method impacts restoration of understory plants in the presence of white-tailed deer. *Invasive Plant Sci. Manag.* **2009**, *2*, 45–54. [CrossRef]

22. Wilson, H.N.; Arthur, M.A.; Schörgendorfer, A.; Paratley, R.D.; Lee, B.D.; McEwan, R.W. Site characteristics as predictors of *Lonicera maackii* in second-growth forests of central Kentucky, USA. *Nat. Areas J.* **2013**, *33*, 189–198. [CrossRef]

23. Luken, J.O. Population structure and biomass allocation of the naturalized shrub *Lonicera maackii* (Rupr.) Maxim. in forest and open habitats. *Am. Midl. Nat.* **1988**, *119*, 258–267. [CrossRef]

24. Bartuszevige, A.M.; Gorchov, D.L. Avian seed dispersal of an invasive shrub. *Biol. Invasions* **2006**, *8*, 1013–1022. [CrossRef]

25. Castellano, S.M.; Gorchov, D.L. White-tailed deer (*Odocoileus virginianus*) disperse seeds of the invasive shrub, amur honeysuckle (*Lonicera maackii*). *Nat. Areas J.* **2013**, *33*, 78–80. [CrossRef]

26. Flory, S.L.; Clay, K. Invasive shrub distribution varies with distance to roads and stand age in eastern deciduous forests in Indiana, USA. *Plant Ecol.* **2006**, *184*, 131–141. [CrossRef]

27. Shields, J.M.; Jenkins, M.A.; Saunders, M.R.; Zhang, H.; Jenkins, L.H.; Parks, A.M. Age distribution and spatial patterning of an invasive shrub in secondary hardwood forests. *For. Sci.* **2014**, *60*, 830–840. [CrossRef]

28. Fridley, J.D. Extended leaf phenology and the autumn niche in deciduous forest invasions. *Nature* **2012**, *485*, 359–362. [CrossRef] [PubMed]

29. Heberling, J.M.; Fridley, J.D. Functional traits and resource-use strategies of native and invasive plants in Eastern North American forests. *New Phytol.* **2013**, *200*, 523–533. [CrossRef] [PubMed]

30. Web Soil Survey. Soil Survey Staff, Natural Resources Conservation Service, United States Department of Agriculture. Available online: https://websoilsurvey.sc.egov.usda.gov (accessed on 15 October 2015).

31. Yatskievych, K. *Field Guide to Indiana Wildflowers*; Indiana University Press: Bloomington, IN, USA, 2000; ISBN 0-253-21420-3.

32. Peet, R.K.; Wentworth, T.R.; White, P.S. A flexible, multipurpose method for recording vegetation composition and structure. *Castanea* **1998**, *63*, 262–274.

33. Rothrock, P.E. *Floristic Quality Assessment in Indiana: The Concept, Use, and Development of Coefficients of Conservatism*; Final Report; ARN A: Indianapolis, IN, USA, 2004.

34. Rothrock, P.E.; Homoya, M.A. An evaluation of Indiana's floristic quality assessment. *Proc. Indiana Acad. Sci.* **2005**, *114*, 9–18.

35. Love, J.P.; Anderson, J.T. Seasonal effects of four control methods on the invasive morrow's honeysuckle (*Lonicera morrowii*) and initial responses of understory plants in a southwestern Pennsylvania old field. *Restor. Ecol.* **2009**, *17*, 549–559. [CrossRef]

36. Bates, D.; Maechler, M.; Bolker, B.; Walker, S. Fitting linear mixed-effects models using lme4. *J. Stat. Softw.* **2015**, *67*, 1–48. [CrossRef]

37. Sugiura, N. Further analysts of the data by akaike's information criterion and the finite corrections. *Commun. Stat. Theory Methods* **1978**, *7*, 13–26. [CrossRef]

38. Barton, K. MuMIn: Multi-Model Inference. R Package Version 1.15.6. Available online: https://CRAN.R-project.org/package=MuMIn (accessed on 10 January 2016).

39. Kuznetsova, A.; Brockhoff, P.B.; Christensen, R.H.B. lmerTest: Tests in linear mixed effects models. *J. Stat. Softw.* **2017**, *82*, 13. [CrossRef]

40. Hothorn, T.; Bretz, F.; Westfall, P. Simultaneous inference in general parametric models. *Biom. J.* **2008**, *50*, 346–363. [CrossRef] [PubMed]

41. Polgar, C.; Gallinat, A.; Primack, R.B. Drivers of leaf-out phenology and their implications for species invasions: Insights from Thoreau's Concord. *New Phytol.* **2014**, *202*, 106–115. [CrossRef] [PubMed]

42. De Villalobos, A.E.; Vázquez, D.P.; Martin, J.L. Soil disturbance, vegetation cover and the establishment of the exotic shrub *Pyracantha coccinea* in southern France. *Biol. Invasions* **2010**, *12*, 1023–1029. [CrossRef]

43. Kalisz, S.; Spigler, R.B.; Horvitz, C.C. In a long-term experimental demography study, excluding ungulates reversed invader's explosive population growth rate and restored natives. *Proc. Natl. Acad. Sci. USA* **2014**, *111*, 4501–4506. [CrossRef] [PubMed]

44. Biswas, S.R.; Wagner, H.H. A temporal dimension to the stress gradient hypothesis for intraspecific interactions. *Oikos* **2014**, *123*, 1323–1330. [CrossRef]

45. Packer, A.; Clay, K. Soil pathogens and spatial patterns of seedling mortality in a temperate tree. *Nature* **2000**, *404*, 278–281. [CrossRef] [PubMed]

46. Orrock, J.L.; Christopher, C.C.; Dutra, H.P. Seed bank survival of an invasive species, but not of two native species, declines with invasion. *Oecologia* **2012**, *168*, 1103–1110. [CrossRef] [PubMed]

47. Kuebbing, S.E.; Classen, A.T.; Call, J.J.; Henning, J.A.; Simberloff, D. Plant-soil interactions promote co-occurrence of three nonnative woody shrubs. *Ecology* **2015**, *96*, 2289–2299. [CrossRef] [PubMed]

48. DeWalt, S.J.; Denslow, J.S.; Ickes, K. Natural-enemy release facilitates habitat expansion of the invasive tropical shrub *Clidemia hirta. Ecology* **2004**, *85*, 471–483. [CrossRef]

49. Dorrough, J.; Scroggie, M.P. Plant responses to agricultural intensification. *J. Appl. Ecol.* **2008**, *45*, 1274–1283. [CrossRef]

50. Miller, K.E.; Gorchov, D.L. The invasive shrub, *Lonicera maackii*, reduces growth and fecundity of perennial forest herbs. *Oecologia* **2004**, *139*, 359–375. [CrossRef] [PubMed]

forests

MDPI

Article

The Potential Distribution of Tree Species in Three Periods of Time under a Climate Change Scenario

Pablo Antúnez, Mario Ernesto Suárez-Mota *, César Valenzuela-Encinas and
Faustino Ruiz-Aquino

División de Estudios de Postgrado-Instituto de Estudios Ambientales, Universidad de la Sierra Juárez,
Avenida Universidad S/N, Ixtlán de Juárez, 68725 Oaxaca, México; pantunez4@gmail.com (P.A.);
cesarvzlae@gmail.com (C.V.-E.); ruiz.aquino@unsij.edu.mx (F.R.-A.)
* Correspondence: mesuarez@unsij.edu.mx; Tel.: +55-951-5536362

Received: 1 September 2018; Accepted: 9 October 2018; Published: 11 October 2018

Abstract: Species distribution models have become some of the most important tools for the assessment of the impact of climatic change, and human activity, and for the detection of failure in silvicultural or conservation management plans. In this study, we modeled the potential distribution of 13 tree species of temperate forests distributed in the Mexican state Durango in the Sierra Madre Occidental, for three periods of time. Models were constructed for each period of time using 19 climate variables from the MaxEnt (Maximum Entropy algorithm) modelling algorithm. Those constructed for the future used a severe climate change scenario. When comparing the potential areas of the periods, some species such as *Pinus durangensis* (Martínez), *Pinus teocote* (Schiede ex Schltdl. & Cham.) and *Quercus crassifolia* (Bonpl.) showed no drastic changes. Rather, the models projected a slight reduction, displacement or fragmentation in the potential area of *Pinus arizonica* (Engelm.), *P. cembroides* (Zucc), *P. engelmanni* (Carr), *P. leiophylla* (Schl), *Quercus arizonica* (Sarg), *Q. magnolifolia* (Née) and *Q. sideroxila* (Humb. & Bonpl.) in the future period. Thus, establishing conservation and reforestation strategies in the medium and long term could guarantee a wide distribution of these species in the future.

Keywords: Bioclimatic niche; Durango; Mexican tree species; MaxEnt; non-parametric correlation

1. Introduction

In conservation biology, estimating areas of potential distribution of species through the modeling of an ecological niche, has had a variety of applications for learning the current and future state of species [1]. Yet, human activities have produced significant changes in the distribution of species in the different ecosystems of the world [2], with the temperate forests of Mexico being one of the most affected, due to the excessive extraction of some species of commercial interest, mainly of the *Pinus* genus like *Pinus durangensis* Martínez, *P. arizonica* Engelm., *P. cooperi* C. E. Blanco and *P. engelmannii* Carr in the Mexican northwest [3,4]. On the other hand, the changes in land-use and activities related to the extraction of timber species, are among the main causes of the disturbance of habitat and the loss of diversity in the temperate forests of Mexico [5–8]. Also, climate change is another phenomenon to be considered when studying the distribution of plants since many habitats have undergone severe changes. Even so, the survival of some plant species is at risk because of this [9]. Projecting the potential distribution of species using climatic variables becomes very important to evaluate and foresee any alteration of either natural or anthropogenic origin [1,10,11].

One way of estimating the potential distribution of a species for a future period is by modeling their ecological niche considering that the geographical distribution of a given species is not random, but obeys environmental factors such as altitude, topographic position, temperature, humidity and precipitation, among others. Likewise, the distribution of plants is associated with physiographic (aspect and slope) and edaphic factors [12–14].

One of the most used tools to model the distribution of species in geographic space and their environmental tolerances, is the Maximum Entropy algorithm (MaxEnt), whose predictions start with the principle of maximum probability of occurrence of the species from the presence data and the bioclimatic variables associated with each sampling site [15].

The MaxEnt method has been used successfully in studies related to the analysis of biodiversity, the conservation of the species niche, the identification of priority conservation areas and the implementation of actions focused on preventing the establishment of invasive species in a locality [16–18]. The maps of potential distribution projected by MaxEnt and the spatial analysis by means of a GIS have proved their effectiveness by providing a reasonable approximation of the niche of the species, even with a small sample size [19–22].

An effective way to measure the effect of environmental variations on the geographic distribution of high-interest plants is to evaluate any change manifested when going from one period of time, with specific climatic characteristics, to another, such as the period of the most recent glaciation and the contemporary period. In this way, it is possible to predict a possible displacement, fragmentation or reduction of the potential area in the face of future climate change scenarios.

This study had two objectives: (i) to identify if there is a significant change in the potential distribution of 13 tree species (highly valued both economically and ecologically) native to the temperate forests of Durango, northwestern Mexico, as a function of 19 climatic variables, considering three periods of time; and (ii) to identify species with risks of decreasing their potential distribution area in the future (2080 to 2100), under a scenario of severe climate change. In this study the spatial resolution of the data was 1 km^2. The hypothesis was that the potential distribution of each species does not change significantly between one period of time and another.

2. Materials and Methods

2.1. Study Region

The modeled region corresponds to a portion of the mountain system of the Sierra Madre Occidental (SMO) in the State of Durango (northwest of Mexico), between geographical coordinates (WGS 84) 26°50′ and 22°17′ N and 107°09′ and 102°30′ W, covering an area of about 6.33 million ha (Figure 1). In this region, forests of oak pine, pine-oak, oak or pine, together with other species typical of the temperate climate, are mainly distributed. Portions of temperate mesophytic as well as tropical deciduous and semi-deciduous forests are also found on the western slope of the region [23]. The annual rainfall fluctuates from 250 to 1444 mm and the mean annual temperature from 8.3 to 26.2 °C [24].

2.2. Obtaining Data

A total of 13 of the most representative species of the study area were analyzed, with eight of the genus *Pinus* (*P. arizonica* Engelm. var. stormiae Martínez, *P. strobiformis* Engelm., *P. cembroides* Zucc, *P. cooperi* C. E. Blanco, *P. durangensis* Martínez, *P. engelmannii* Carr, *P. leiophylla* Schiede ex Schltdl. et Cham., and *P. teocote* Schiede ex Schltdl. et Cham) and five of the genus *Quercus* (*Q. arizonica* Sarg, *Q. crassifolia* Humb. & Bonpl., *Q. grisea* Liebm, *Q. magnoliifolia* Née and *Q. sideroxyla* Humb. & Bonpl.) selected.

The presence records of the 13 species were collected using two data sources. The first data-set was obtained from on-line herbarium specimens provided by The National Herbarium (MEXU http://www.ib.unam.mx/botanica/herbario/). The second data-set was obtained directly in the field, from 1804 sampling plots, distributed systematically every five kilometers in the study area. These plots were established by the National Forestry Commission (CONAFOR) for the National Forest and Soil

Inventory 2004–2009 [25]. These presence records were converted to geographical coordinates and finally a single database was generated by combining both subsets of data.

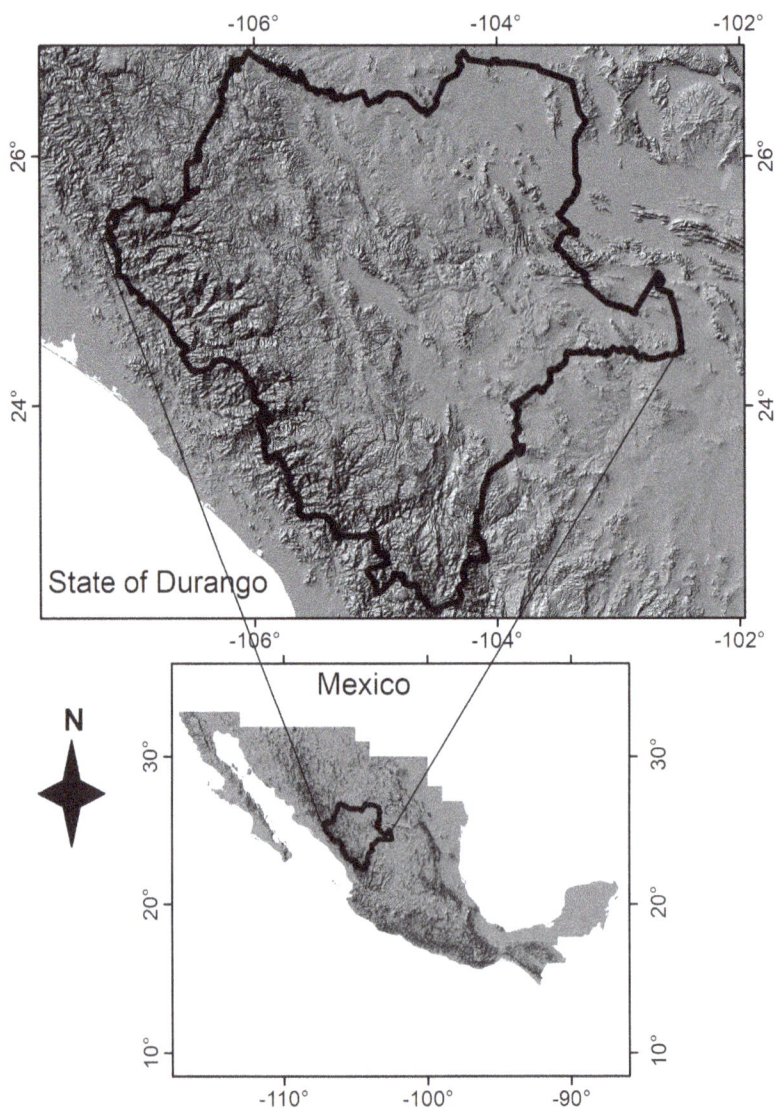

Figure 1. Study area, located in northwest of Mexico. The axes are geographic coordinates.

It should be mentioned that the temperate forests of the SMO are under forest management and it is possible that anthropogenic activities have altered the natural distribution of the species studied. For example, some conifer species such as *Pinus durangensis* Martínez, *P.cooperi* C. E. Blanco, *P. arizonica* Engelm. var. stormiae, *P. teocote* Schiede ex Schltdl. et Cham and *P. engelmannii* Carr, are extracted more than *Pinus strobiformis* Engelm. or *Pinus leiophylla* Schiede ex Schltdl. et Cham. due to having desirable phenotypic characteristics for the forest industry [26]. But since there is no precise quantification of the impact of anthropogenic activities, this factor was not incorporated into the models.

2.3. Distribution Modeling

The potential distribution of each species was modeled as a function of 19 environmental variables (Table 2) which were obtained from the WorldClim database (http://www.worldclim.org/ version 2) [27,28], with a spatial resolution of 1 km. To accomplish this, we used the Maximum Entropy algorithm (MaxEnt) [15], following the methodology used by several authors [17,29,30]. Potential distribution models for three periods of times were generated: (i) period of the most recent glaciation (21,000 years ago), (ii) present period and (iii) the future, corresponding to the period from 2080 to 2100. For this last period, the general circulation model used was the NIES 99 (http://www.ipcc.ch/ipccreports/sres/emission/index.php?idp=35) under an A2A scenario, a very severe scenario [31]. The MaxEnt method has proven its effectiveness in making predictions based on information from presence data [32–35] and whose results express the value of habitat suitability for each species as a function of probability according to environmental variables. A high value of the distribution function in a given cell suggests very favorable conditions for the presence of an analyzed species [15].

The results obtained from the MaxEnt method, were changed to Boolean layers (presence-absence data) using ArcMap software package 10.1 (http://desktop.arcgis.com/en/arcmap/), considering a cutoff threshold equal to 10% omission errors [30,36]. In each model, the variables that explained 70% or more of the data variability in the principal component analysis (PCA) were included [30,37]. The PCA analyses were carried out with R software [38].

For each species, presence records were divided into 75% (randomly selected) for model training and 25% for model validation, using 100 replicates in 500 iterations with different random partitions (cross-validation method).

In order to measure the degree of association or similarity between the relative abundance of one species (presence-absence data per cell of 1 km^2) between one period of time and another, the Phi coefficient was used [39]. Finally, to identify if there are significant differences between the potential areas of one period of time and another, the average area of the 13 species was compared, using the Kruskal-Wallis test, testing the hypothesis that the average area of a period of time is similar to the other periods. The level of significance used in all the analyses was 0.05.

2.4. Model Evaluation

Models were evaluated using the area under the curve (AUC) described by Phillips et al. [15], whose values are calculated from the Receiver Operating Characteristic (ROC). When AUC showed values equal to or greater than 0.9, we considered the models to be robust; values of AUC 0.7–0.9 were considered moderately robust; and values close to 0.5 were considered not robust [33]. In all cases, the logistic output format was used [34].

3. Results

Observing the AUC values, the models for the scenarios of the three periods were consistent and robust showing values higher than 0.90 for all species, except for *Quercus arizonica* and *Quercus sideroxyla* (Table 1). The climatic variables used to model the potential niche of the species are listed in Table 2, which mainly includes average, maximum and minimum temperatures and rainfall in specific periods. The areas projected for the three periods of time, are shown in Figures 2 and 3.

The Phi coefficients revealed a high and positive correlation between presence/absence records of the contemporary and future period with an average of 0.81 for 92% of the species studied. The highest Phi coefficients were observed between the contemporary period and the future for *P. durangensis* and *P. cooperi* with a value of 0.87 for each species, followed by *P. strobiformis* with a coefficient of 0.85. In contrast, the smallest absolute coefficients (three of them negative) were observed between the period of the most recent glaciation and the contemporary period, and between the period of the most recent glaciation and the future period for *Pinus cembroides* and *strobiformis*, whose Phi coefficients were less than 0.05 (Table 3).

Table 1. List of species studied and their records used in the potential distribution models.

Species	AUC Values		
	Past	Present	Future
Pinus arizonica Engelm. var. *stormiae* Martínez	0.960	0.961	0.962
Pinus strobiformis Engelm	0.927	0.945	0.945
Pinus cembroides Zucc	0.948	0.944	0.950
Pinus cooperi C.E.Blanco	0.961	0.965	0.966
Pinus durangensis Martínez	0.915	0.909	0.908
Pinus engelmanni Carr	0.926	0.929	0.908
Pinus leiophylla Schiede ex Schltdl. et Cham.	0.954	0.937	0.943
Pinus teocote Schiede ex Schltdl. et Cham.	0.951	0.942	0.947
Quercus arizonica Sarg	0.791	0.784	0.794
Quercus crassifolia Humb. & Bonpl.	0.966	0.964	0.967
Quercus grisea Liebm	0.942	0.939	0.939
Quercus magnolifolia Née	0.964	0.967	0.967
Quercus sideroxyla Humb. & Bonpl.	0.879	0.897	0.898

Table 2. List of climatic variables used in the models and their respective acronyms. The variables that showed the highest and significant correlation coefficients (highlighted in bold) were used to model the potential areas.

Acronyms	Description	PC1	PC2
bio_01	Mean Annual Temperature (°C)	0.34	0.06
bio_02	Mean Diurnal Range (Mean of monthly max. temp. min. temp.) (°C)	**0.93**	0.07
bio_03	Isothermality (Bio_02/Bio_07) (×100)	0.31	−0.41
bio_04	Temperature Seasonality (standard deviation × 100) (Coefficient of Variation)	**0.98**	0.19
bio_05	Max Temperature of Warmest Month (°C)	−0.31	0.13
bio_06	Min Temperature of Coldest Month (°C)	**0.72**	−0.03
bio_07	Temperature Annual Range (Bio_05–Bio_06) (°C)	**0.96**	0.13
bio_08	Mean Temperature of Wettest Quarter (°C)	0.19	0.13
bio_09	Mean Temperature of Driest Quarter (°C)	0.44	0.16
bio_10	Mean Temperature of Warmest Quarter (°C)	0.12	0.11
bio_11	Mean Temperature of Coldest Quarter (°C)	0.50	0.027
bio_12	Annual Precipitation (mm)	**0.88**	0.45
bio_13	Precipitation of Wettest Month (mm)	**0.88**	0.42
bio_14	Precipitation of Driest Month (mm)	0.22	0.49
bio_15	Precipitation Seasonality (Coefficient of Variation)	−0.33	−0.24
bio_16	Precipitation of Wettest Quarter (mm)	**0.88**	0.44
bio_17	Precipitation of Driest Quarter (mm)	0.52	0.45
bio_18	Precipitation of Warmest Quarter (mm)	**0.79**	0.42
bio_19	Precipitation of Coldest Quarter (mm)	**0.72**	0.43

Table 3. Phi correlation coefficients for the presence/absence values, comparing the degree of association among the three periods of time.

Species Studied	Periods of Time	Past	Present
P. arizonica	Present	0.75	-
	Future	0.67	0.79
P. strobiformis	Present	0.73	-
	Future	0.75	0.84
P. cembroides	Present	0.52	-
	Future	0.65	0.46
P. cooperi	Present	0.80	-
	Future	0.78	0.82
P. duranguensis	Present	0.72	-
	Future	0.73	0.82
P. engelmanni	Present	0.73	-
	Future	0.74	0.84
P. leiophylla	Present	0.67	-
	Future	0.71	0.78

Table 3. *Cont.*

Species Studied	Periods of Time	Past	Present
P. teocote	Present	0.72	-
	Future	0.67	0.81
Q. arizonica	Present	−0.04	-
	Future	0.02	0.71
Q. crassifolia	Present	−0.04	-
	Future	−0.05	0.85
Q. grisea	Present	0.79	-
	Future	0.74	0.87
Q. magnolifolia	Present	0.79	-
	Future	0.74	0.87
Q. sideroxyla	Present	0.75	-
	Future	0.69	0.69

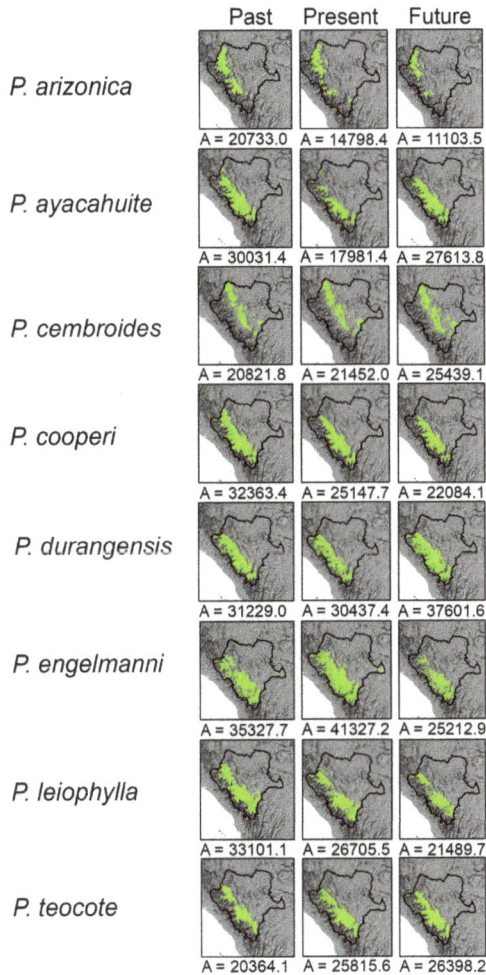

Figure 2. Potential distribution areas of *Pinus* species for the period of the most recent glaciation (21,000 years ago), present period and into the future (2080 to 2100).

When comparing the average values of the projected surfaces of the 13 species studied, no significant differences were found among the three periods of time, according to the Kruskal-Wallis test, with a significance level of 0.05. However, observing the areas of each species separately, several of them showed changes, as was the case of *Pinus arizonica*, *P. leiophylla*, *P. cooperi* and *Quercus grisea*, whose projected areas decreased as they passed from one period to another. This was more evident for the first two species between the period of the most recent glaciation and the contemporary period (Figure 4). The potential area of some species showed a change from one period to another. For example, *P. cembroides* projected an increase of 19,701 km² from the present to the future period and *P. strobiformis* projected an increase of 9633 km² (Figures 2 and 3).

In general, the projected potential areas were variable for each species, and although in most cases there were minor changes in the projected surface, slight displacements, a possible fragmentation of the habitats or discontinuity in the distribution of species were observed from one period of time to another (Figure 4). Habitat fragmentation was clearer for *P. arizonica*, *P. leiophylla*, *Q. arizonica*, *Q. grisea* and *Q. magnolifolia* (Figures 2 and 3).

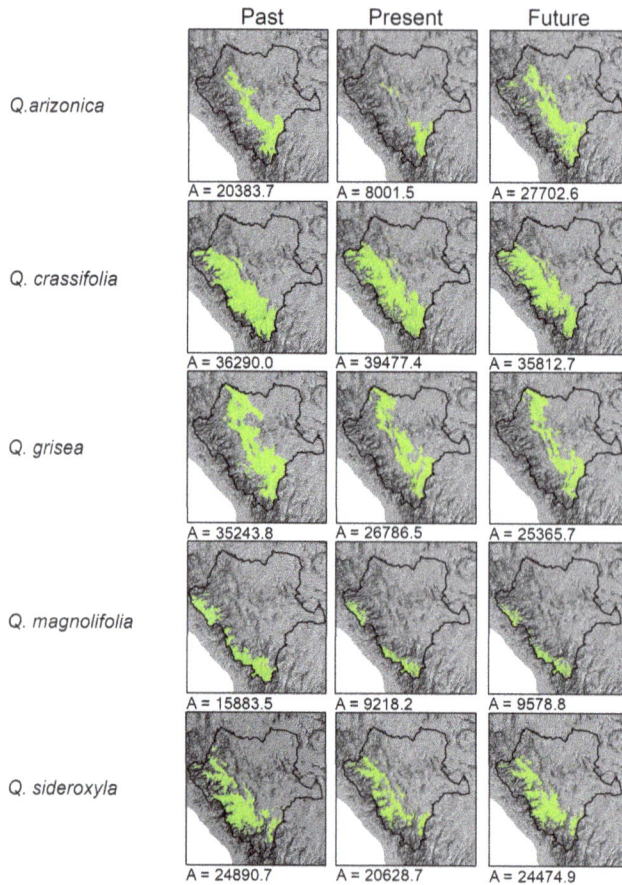

Figure 3. Potential distribution areas of *Quercus* species for the period of the most recent glaciation (21,000 years ago), present period and into the future (2080 to 2100).

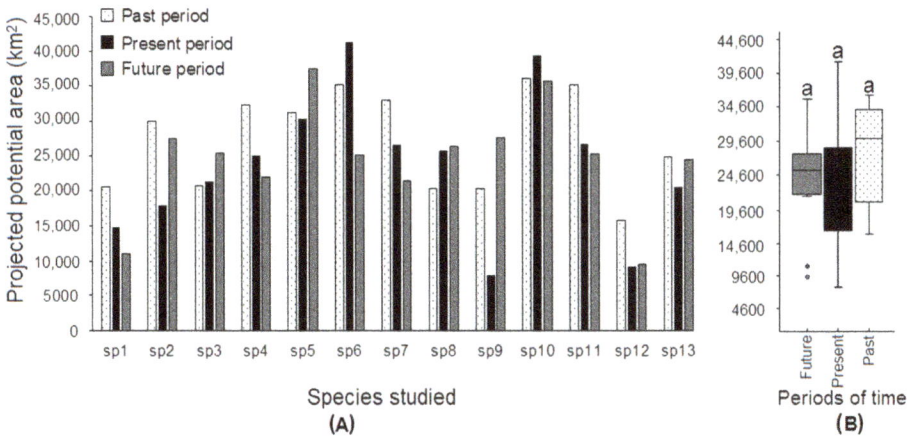

Figure 4. (A) Potential areas projected for each species for different periods of time. (B) Comparison
of means by the Kruskal-Wallis test, using the potential areas of the 13 species of each period; means
sharing a letter are not significantly different (*p* < 0.05). sp1: *P. arizonica*, sp2: *P. ayacahuite*, sp3:
P. cembroides, sp4: *P. cooperi*, sp5: *P. duranguensis*, sp6: *P. engelmanni*, sp7: *P. leiophylla*, sp8: *P. teocote*, sp9:
Q. arizonica, sp10: *Q. crassifolia*, sp11: *Q. grisea*, sp12: *Q. magnolifolia*, sp13: *Q. sideroxyla*.

4. Discussion

MaxEnt is a modeling instrument that has gained relevance in recent years for analyzing ecological
characteristics of species using presence records, which are usually collected from specimen records in
plant collection centers (e.g., MEXU) or field survey records. In this study, the areas projected (km²) by
models, apparently did not change from one period to another for most of the species (Figures 2 and 3).
Yet, the fragmentation or displacement of areas with a suitable climate was evident in the maps for
several species, as in the case of *Pinus arizonica*, *P. cembroides*, *P. engelmanni*, *P. leiophylla*, *Quercus arizonica*,
Q. magnolifolia and *Q. sideroxila* (Figures 3 and 4). Considering that the real interval of climate tolerance
(optimal interval) is variable between one species and another, although they cohabit in the same
region [26], it is possible that there were overestimates of the models. For this reason, the small
difference between the areas projected does not necessarily suggest that the species have broad climatic
tolerances or that the plants have a high capacity to adapt to contrasting climatic conditions.

The observations of which species show a fragmentation or discontinuity of the areas projected
are valuable findings since they can be indicators of a high sensitivity to changes in climate values.
In the end, a fragmentation can trigger a reduction in the real area of a species in the medium- or
long-term, a problem that has been observed in the last decades in the temperate forests of the SMO [40]
and in some Mexican terrestrial ecosystems [41,42]. Moreover, in the SMO, there are no studies on the
real anthropogenic effects, since they could accelerate the alteration of the natural distribution of some
species. For example, during the harvesting of roundwood, regeneration and other herbaceous species
are damaged in many zones of the study region due to the poor design of the road network used to
transport the roundwood. If the quantitative effects of anthropogenic activity were known, a specific
weight could be included and assigned when making the potential distribution models.

The Phi coefficients revealed different degrees of similarity between one period of time and
another. For example, when comparing the corresponding values of the contemporary period and the
future period of both *Quercus magnolifolia* and *Q. grisea*, a high correlation was observed (Phi = 0.87),
suggesting a high similarity in the pattern of distribution of these species in these two periods of time.
Conversely, a weak correlation was observed between the values of the past and present period of
both *Quercus crassifolia* and *Quercus arizonica* (Table 3).

Potential distribution models have been useful for making important decisions in the area of ecology [14,43,44], although they should be used with caution as several authors have warned [45–48], because this type of modeling has limitations, which include the iterative procedure of the MaxEnt algorithm [49] as well as the possible errors of commission and omission when using climatic factors to characterize the habitat of a species [48]. A commission error related to climatic variables could occur when the affinity of a species in a site is attributable to an event or situation that cannot be conditioned by climatic variables [48] whereas an error of omission is any error related to the underestimation of the climatic range [50] which can be translated as a false prediction of absence [48].

In some cases, the areas projected by the models are larger than the real distributions of the species studied, since the models do not consider biotic interactions [32] or the effects of human activities whose impact on the distribution pattern has increased significantly in recent years [2,51]. Given that there is no precise quantification of the magnitude of the impact of human activities on the presence/absence of the species, this factor was not included in the model. Likewise, the models also exclude the dynamic nature of the individuals studied, the alteration by pests or diseases, or other situations that, for a short period, could drastically alter the distribution and abundance of forest species [52].

On the other hand, changes in the fundamental niche of a species can occur as a result of the plasticity of traits of morphological, physiological or behavioral type or by the classification of genotypes. For example, they could occur by the spatial segregation of individuals with certain functional features [53–55]. Finally, the niche of a species can expand or move, after a long period of time, as a genetic response to the new environmental conditions [56,57].

In general, the statistical results suggest that the potential areas of most species are not experiencing drastic changes in the three modeled time periods. Nevertheless, establishing strategies for restoration, conservation, and reforestation in the medium- and long-term can guarantee not only the survival, but also the wide distribution, of the species of greatest economic and ecological interest, particularly those whose potential areas were reduced, displaced or fragmented, such as *Pinus engelmanni*, *P. arizonica*, *P. leiophylla*, *P. cooperi* and *Quercus grisea*, whose loss of potential area before the change from one period to another was evident (Figures 2 and 3). In these models, historical factors, biotic interactions or other limitations that affect the dispersion of the plants are not incorporated, since the magnitude of impact (quantitative values) of each of them on the distribution and abundance of organisms is unknown, and thus it is impossible to assign them an appropriate weight in the models. In future studies, we foresee incorporating physiographic variables, topographic variables and soil properties, as well as testing other analysis methods.

5. Conclusions

The projections reported here could be useful for decision makers. For example, for a preliminary assessment of risks of the decrease, displacement, or fragmentation of the spaces with ideal conditions for the species studied. The results could also be used to define the degree of change in the climatic domain for each of these species. Incorporating other variables like soil properties (pH, electrical conductivity, texture, etc.), physiographic variables, the annual deforestation index, or any variable that could modify the natural distribution of plants, could increase the certainty of the models. The potential areas for most of the species did not vary between one period and another. Still, establishing integral management plans including conservation and reforestation strategies, at a regional level can guarantee the continued wide distribution of the species studied. Likewise, it is highly recommended that strategies designed to minimize the impacts generated by the excessive extraction of some species, such as *Pinus durangensis* (cataloged as Near Threatened by the IUCN), are established.

Author Contributions: P.A. conceived and coordinated the research project, statistical analysis, and the primary writing of the text. M.E.S.-M. designed the models, analyzed the data, and helped in writing the text. C.V.-E. conducted formatting, and was the advisor and text editor. F.R.-A.: was the advisor and text editor.

Acknowledgments: We are grateful to the Secretaría de Educación Publica (SEP; IDCA: 24332 and Proyect PRODEP-UNSIJ-PTC-028). We acknowledge and wish to show our appreciation for the technical assistance provided by Irene Bautista Juárez, Teresa Martínez Martínez, Diana Nava Juárez and Marleny B. Ramírez Aguirre. We are grateful to Colin M. Gee and two anonymous reviewers for constructive comments on earlier versions of the manuscript.

Conflicts of Interest: The authors declare no conflict of interest.

References

1. Soberón, J.; Nakamura, M. Niches and distributional areas: Concepts, methods, and assumptions. *Proc. Natl. Acad. Sci. USA* **2009**, *106*, 19644–19650. [CrossRef] [PubMed]
2. Crowther, T.W.; Glick, H.B.; Covey, K.R.; Bettigole, C.; Maynard, D.S.; Thomas, S.M.; Tuanmu, M.N. Mapping tree density at a global scale. *Nature* **2015**, *525*, 201–205. [CrossRef] [PubMed]
3. Perosa, M.; Rojas, J.F.; Villagra, P.E.; Tognelli, M.F.; Carrara, R.; Alvarez, J.A. Distribución potencial de los bosques de *Prosopis flexuosa* en la Provincia Biogeográfica del Monte (Argentina). *Ecol. Austral.* **2014**, *24*, 238–248.
4. Martínez-Antúnez, P.; Hernández-Díaz, J.C.; Wehenkel, C.; López-Sánchez, C.A. Estimación de la densidad de especies de coníferas a partir de variables ambientales. *Madera Bosques* **2015**, *21*, 23–33. [CrossRef]
5. Sánchez-Cordero, V.; Cirelli, V.; Murguíal, M.; Sarkar, S. Place prioritization for biodiversity representation using species ecological niche modelling. *Biodivers. Inform.* **2005**, *2*, 11–23. [CrossRef]
6. Rodríguez, F.J.; Pereda, M.E. La dinámica espacial de los ecosistemas del estado de Durango. *Ra Ximhai* **2012**, *8*, 91–96.
7. Calderón-Aguilera, L.E.; Rivera-Monroy, V.H.; Porter-Bolland, L.; Martínez-Yrízar, A.; Ladah, L.B.; Martínez-Ramos, M.; Alcocer, J.; Santiago-Pérez, A.L.; Hernandez-Arana, H.A.; Reyes-Gómez, V.M.; et al. An assessment of natural and human disturbance effects on Mexican ecosystems: Current trends and research gaps. *Biodivers. Conserv.* **2012**, *21*, 589–617. [CrossRef]
8. Galicia, L.; Potvin, C.; Messier, C. Maintaining the high diversity of pine and oak species in Mexican temperate forests: A new management approach combining functional zoning and ecosystem adaptability. *Can. J. For. Res.* **2015**, *45*, 1358–1368. [CrossRef]
9. Thuiller, W.; Albert, C.; Araújo, M.B.; Berry, P.M.; Cabeza, M.; Guisan, A.; Sykes, M.T. Predicting global change impacts on plant species' distributions: Future challenges. *Perspect. Plant. Ecol.* **2008**, *9*, 137–152. [CrossRef]
10. Sykes, M.T.; Prentice, I.C.; Cramer, W. A bioclimatic model for the potential distributions of north European tree species under present and future climates. *J. Biogeogr.* **1996**, *23*, 203–233.
11. Kearney, M.; Porter, W.P. Mechanistic niche modelling: Combining physiological and spatial data to predict species' ranges. *Ecol. Lett.* **2009**, *12*, 334–350. [CrossRef] [PubMed]
12. Hanson, H.C.; Churchill, E.D. *The Plant Community*; Reinhold Publishing Corp.: New York, NY, USA, 1961; pp. 1–218.
13. Chapman, S.B. *Methods in Plant Ecology*; Blackwell Scientific: Oxford, UK, 1976; pp. 1–580.
14. Guisan, A.; Thuiller, W. Predicting species distribution: Offering more than simple habitat models. *Ecol. Lett.* **2005**, *8*, 993–1009. [CrossRef]
15. Phillips, S.J.; Anderson, R.P.; Schapire, R.E. Maximum entropy modeling of species geographic distributions. *Ecol. Model.* **2006**, *190*, 231–259. [CrossRef]
16. Anderson, R.P.; Peterson, A.T.; Gómez-Laverde, M. Using niche-based GIS modeling to test geographic predictions of competitive exclusion and competitive release in South American pocket mice. *Oikos* **2002**, *98*, 3–16. [CrossRef]
17. Suárez-Mota, M.E.; Villaseñor, J.L.; López-Mata, L. La región del Bajío, México y la conservación de su diversidad florística. *Rev. Mex. Biod.* **2015**, *86*, 799–808. [CrossRef]
18. Suárez-Mota, M.E.; Ortiz, E.; Villaseñor, J.L.; Espinosa-Garcia, F.J. Ecological niche modeling of invasive plant species according to invasion status and management needs: The case of *Chromolaena odorata* (Asteraceae) in South Africa. *Pol. J. Ecol.* **2016**, *64*, 369–383. [CrossRef]
19. Townsend, P.; Klusa, D.A. New distributional modelling approaches for gap analysis. *Animal Conservation. Zool. Soc. Lond.* **2003**, *6*, 47–54. [CrossRef]

20. Leal-Nares, O.; Mendoza, M.E.; Pérez-Salicrup, D.; Geneletti, D.; López-Granados, E.; Carranza, E. Distribución potencial del *Pinus martinezii*: Un modelo espacial basado en conocimiento ecológico y análisis multicriterio. *Rev. Mex. Biodivers.* **2012**, *83*, 1152–1170. [CrossRef]

21. Fourcade, Y.; Engler, J.O.; Rödder, D.; Secondi, J. Mapping species distributions with MAXENT using a geographically biased sample of presence data: A performance assessment of methods for correcting sampling bias. *PLoS ONE* **2014**, *9*, e97122. [CrossRef] [PubMed]

22. Chefaoui, R.M.; Hortal, J.; Lobo, J.M. Potential distribution modelling, niche characterization and conservation status assessment using GIS tools: A case study of Iberian *Copris* species. *Biol. Conserv.* **2005**, *122*, 327–338. [CrossRef]

23. Rzedowski, J. *Vegetación de México*; Limusa: México, D.F., Mexico, 1978; pp. 1–432.

24. Silva-Flores, R.; Pérez-Verdín, G.; Wehenkel, C. Patterns of tree species diversity in relation to climatic factors on the Sierra Madre Occidental, Mexico. *PLoS ONE* **2014**, *9*, 105034. [CrossRef] [PubMed]

25. CONAFOR (Comisión Nacional Forestal). Manual and Procedures for Field Sampling—National Forest and Soil Inventory. 2009. Available online: http://www.snieg.mx/contenidos/espanol/iin/Acuerdo_3_X/Manual_y_Procedimientos_para_el_Muestreo_de_Campo_INFyS_2004--2009.pdf (accessed on 17 November 2017).

26. Antúnez, P.; Wehenkel, C.; López-Sánchez, C.A.; Hernández-Díaz, J.C. The role of climatic variables for estimating probability of abundance of tree species. *Pol. J. Ecol.* **2017**, *65*, 324–338. [CrossRef]

27. Hijmans, R.J.; Cameron, S.E.; Parra, J.L.; Jones, P.G.; Jarvis, A. Very high resolution interpolated climate surfaces for global land areas. *Int. J. Clim.* **2005**, *25*, 1965–1978. [CrossRef]

28. Fick, S.E.; Hijmans, R.J. Worldclim 2: New 1-km spatial resolution climate surfaces for global land areas. *Int. J. Climatol.* **2017**, *37*, 4302–4315. [CrossRef]

29. Cuervo-Robayo, A.; Téllez-Valdés, O.; Gómez, M.A.; Venegas-Barrera, C.S.; Manjarrez, J.F.; Mártinez-Meyer, E. An update of high-resolution monthly climate surface for Mexico. *Int. J. Climatol.* **2013**, *34*, 2427–3437. [CrossRef]

30. Cruz-Cárdenas, G.; López-Mata, L.; Villaseñor, J.L.; Ortiz, E. Potential species distribution modeling and the use of principal component analysis as predictor variables. *Rev. Mex. Biodiver.* **2014**, *85*, 189–199. [CrossRef]

31. Nakicenovic, N.; Swart, R. *IPCC: Special Report on Emissions Scenarios*; Cambridge University Press: Cambridge, UK, 2000.

32. Elith, J.; Graham, J.C.H.; Anderson, P.; Anderson, R.P.; Dudík, M.; Ferrier, S.; Guisan, A.; Hijmans, R.J.; Huettmann, F.; Leathwick, J.R.; et al. Novel methods improve prediction of species' distributions from occurrence data. *Ecography* **2006**, *29*, 129–151. [CrossRef]

33. Peterson, A.T.; Soberón, J.; Pearson, R.G.; Anderson, R.; Martínez-Meyer, E.; Nakamura, M.; Araujo, M. *Ecological Niches and Geographic Distributions*; Princeton University Press: Princeton, NJ, USA, 2011; pp. 1–314.

34. Phillips, S.J.; Dudik, M. Modeling of species distributions with MaxEnt: New extensions and a comprehensive evaluation. *Ecography* **2008**, *31*, 161–175. [CrossRef]

35. Phillips, S.J. Transferability, sample selection bias and background data in presenceonly modelling: A response to Peterson et al. (2007). *Ecography* **2008**, *31*, 272–278. [CrossRef]

36. Pearson, R.G.; Raxworthy, C.J.; Nakamura, M.; Townsend Peterson, A. Predicting species distributions from small numbers of occurrence records: A test case using cryptic geckos in Madagascar. *J. Biogeogr.* **2007**, *34*, 102–117. [CrossRef]

37. Jolliffe, I.T. Principal component analysis and factor analysis. In *Principal Component Analysis*; Springer: New York, NY, USA, 1986; pp. 115–128.

38. R Core Team. R: A Language and Environment for Statistical Computing. R Foundation for Statistical Computing. 2018. Available online: https://www.r-project.org/ (accessed on 1 January 2017).

39. Kuhn, G.M. The phi coefficient as an index of ear differences in dichotic listening. *Cortex* **1973**, *9*, 450–457. [CrossRef]

40. González-Elizondo, M.S.; González-Elizondo, M.; González, L.R.; Enríquez, I.L.; Rentería, F.R.; Flores, J.T. Ecosystems and diversity of the Sierra Madre Occidental. In *Merging Science and Management in a Rapidly Changing World: Biodiversity and Management of the Madrean Archipelago III and 7th Conference on Research and Resource Management in the Southwestern Deserts, Tucson, AZ, USA, 1–5 May 2012*; Gottfried, G.J., Ffolliott, P.F., Gebow, B.S., Eskew, L.G., Collins, L.C., Eds.; Proceedings RMRS-P-67; US Department of Agriculture, Forest Service, Rocky Mountain Research Station: Fort Collins, CO, USA, 2013; pp. 204–211.

41. Toledo, V.M.; Ordóñez, M. The biodiversity scenario of Mexico: A review of terrestrial habitats. In *Biological Diversity of Mexico. Origins and Distribution*; Ramamoorthy, T.P., Bye, R., Lot, A., Eds.; Oxford University Press: New York, NY, USA, 1993; pp. 757–777.

42. Challenger, A. *Utilización y Conservación de los Ecosistemas Terrestres de México: Pasado, Presente y futuro*; Conabio-Instituto de Biología, UNAM-Agrupación Sierra Madre SC. Distrito Federal: Tlalpan, Mexico, 1998.

43. Guisan, A.; Zimmermann, N.E. Predictive habitat distribution models in ecology. *Ecol. Model.* **2000**, *135*, 147–186. [CrossRef]

44. Barve, N.; Barve, V.; Jiménez-Valverde, A.; Lira-Noriega, A.; Maher, S.P.; Peterson, A.T.; Villalobos, F. The crucial role of the accessible area in ecological niche modeling and species distribution modeling. *Ecol. Model.* **2011**, *222*, 1810–1819. [CrossRef]

45. Loiselle, B.A.; Howell, C.A.; Graham, C.H.; Goerck, J.M.; Brooks, T.; Smith, K.G.; Williams, P.H. Avoiding pitfalls of using species distribution models in conservation planning. *Conserv. Biol.* **2003**, *17*, 1591–1600. [CrossRef]

46. Araújo, M.B.; Pearson, R.G.; Thuiller, W.; Erhard, M. Validation of species–climate impact models under climate change. *Glob. Chang. Biol.* **2005**, *11*, 1504–1513. [CrossRef]

47. Carneiro, L.R.D.A.; Lima, A.P.; Machado, R.B.; Magnusson, W.E. Limitations to the use of species-distribution models for environmental-impact assessments in the Amazon. *PLoS ONE* **2016**, *11*, e0146543. [CrossRef] [PubMed]

48. Kadmon, R.; Farber, O.; Danin, A. A systematic analysis of factors affecting the performance of climatic envelope models. *Ecol. Appl.* **2003**, *13*, 853–867. [CrossRef]

49. Ward, G.; Hastie, T.; Barry, S.; Elith, J.; Leathwick, J.R. Presence-only data and the EM algorithm. *Biometrics* **2009**, *65*, 554–563. [CrossRef] [PubMed]

50. Walker, P.A.; Cocks, K.D. HABITAT: A procedure for modelling a disjoint environmental envelope for a plant or animal species. *Glob. Ecol. Biogeogr. Lett.* **1991**, *1*, 108–118. [CrossRef]

51. Hunter, P. The human impact on biological diversity. How species adapt to urban challenges sheds light on evolution and provides clues about conservation. *EMBO Rep.* **2007**, *8*, 316–318. [CrossRef] [PubMed]

52. Antúnez, P.; Hernández-Díaz, J.C.; Wehenkel, C.; Clark-Tapia, R. Generalized models: An application to identify environmental variables that significantly affect the abundance of three tree species. *Forests* **2017**, *8*, 59. [CrossRef]

53. Ackerly, D. Canopy gaps to climate change -extreme events, ecology and evolution. *New Phytol.* **2003**, *160*, 2–4. [CrossRef]

54. Ackerly, D.D. Community assembly, niche conservatism, and adaptive evolution in changing environments. *Int. J. Plant. Sci.* **2003**, *164*, 165–184. [CrossRef]

55. Eiserhardt, W.L.; Svenning, J.C.; Borchsenius, F.; Kristiansen, T.; Balslev, H. Separating environmental and geographical determinants of phylogenetic community structure in Amazonian palms (Arecaceae). *Bot. J. Linn. Soc.* **2013**, *171*, 244–259. [CrossRef]

56. Davis, M.B.; Shaw, R.G.; Etterson, J.R. Evolutionary responses to changing climate. *Ecology* **2005**, *86*, 1704–1714. [CrossRef]

57. Davis, M.B.; Shaw, R.G. Range shifts and adaptive responses to quaternary climate change. *Science* **2001**, *292*, 673–679. [CrossRef] [PubMed]

![forests logo] *forests*

MDPI

Article

Recent Trends in Large Hardwoods in the Pacific Northwest, USA

Jonathan W. Long [1,*], Andrew Gray [2] and Frank K. Lake [3]

[1]   USDA Forest Service Pacific Northwest Research Station, Davis, CA 95618, USA
[2]   USDA Forest Service Pacific Northwest Research Station, Corvallis, OR 97331, USA; agray01@fs.fed.us
[3]   USDA Forest Service Pacific Southwest Research Station, Arcata, CA 95521, USA; franklake@fs.fed.us
*   Correspondence: jwlong@fs.fed.us; Tel.: +1-530-759-1744

Received: 1 September 2018; Accepted: 10 October 2018; Published: 19 October 2018

Abstract: Forest densification, wildfires, and disease can reduce the growth and survival of hardwood trees that are important for biological and cultural diversity within the Pacific Northwest of USA. Large, full-crowned hardwoods that produce fruit and that form large cavities used by wildlife were sustained by frequent, low-severity fires prior to Euro-American colonization. Shifts in fire regimes and other threats could be causing declines in, large hardwood trees. To better understand whether and where such declines might be occurring, we evaluated recent trends in Forest Inventory and Analysis (FIA) data from 1991–2016 in California and southern Oregon. We included plots that lay within areas of frequent fire regimes during pre-colonial times and potential forest habitats for fisher, a rare mammal that depends on large live hardwoods. We analyzed changes in basal area for eight hardwood species, both overall and within size classes, over three time periods within ecoregions, and in public and private land ownerships. We found the basal area to generally be stable or increasing for these species. However, data for California black oak suggested a slight decline in basal area overall, and among both very large trees and understory trees; that decline was associated with fire mortality on national forest lands. In addition, mature trees with full crowns appeared to sharply decline across all species. Many trends were not statistically significant due to high variation, especially since more precise data from remeasured trees were only available for the two most recent time periods. Continued analysis of these indicators using remeasured trees will help to evaluate whether conservation efforts are sustaining large, full-crowned trees and their associated benefits.

Keywords: forest restoration; wildfire; biological diversity; cultural diversity; ecosystem services; monitoring; indicators; inventory; Native Americans; non-timber forest products

1. Introduction

Forests in the Pacific Northwest of USA are generally dominated by and renowned for conifer trees, but hardwoods are still important, distinctive components of forests in the region [1]. Several of these species, including California black oak, Oregon white oak, tanoak, canyon live oak, giant chinquapin, Pacific madrone, bigleaf maple, and California laurel, form large and old trees. The importance of oaks and other hardwood trees has been recognized in the forests of eastern North America, where hardwoods are often dominant [2]. Research has highlighted the importance of hardwood communities for sustaining biological and cultural diversity in Pacific Northwest, while suggesting that changes in forest management policies and practices may be needed to stave off their declines [3–5].

1.1. Background on the Potential for Hardwood Declines

Since Euro-American colonization, the forests in the region have changed significantly, due to timber harvest, fire suppression, introduced pathogens, and climate change [6–8]. During the past century, timber production on both public and private lands prioritized conifers over hardwood trees. In the mid-20th century, softwood producers poisoned large tanoak and other hardwoods using the "hack-and-squirt" technique to favor the regeneration of conifer trees [6,8]. Large hardwood trees may have already declined from the pre-colonial era as Euro-American settlers cut tanoak for tanneries, and oaks and chinquapins for fuel and lumber [6]. Climate vulnerability analyses have projected that hardwood species such as black oak, tanoak, and maple are likely to expand their ranges under the warmer and more fire-prone conditions that are expected [9,10]. However, recent studies have highlighted how changes in fire regimes, including fire suppression and reduction of Native American influence, combined with large, severe, and recurring wildfires, could trigger declines in large hardwoods in both woodland [11] and forest areas [12,13]. In addition, the introduced pathogen *Phytophthora ramorum*, has caused hardwood mortality (described as "sudden oak death") within coastal forests of Northern California and southwestern Oregon [14], although climatic constraints, eradication and containment efforts have helped to mitigate its spread.

Declines in large hardwoods may jeopardize benefits, also known as ecosystem services, particularly for several rare wildlife species [15] and for Native Americans. Tribes in the region value many of these hardwood species for food, medicine, implements, construction materials, and other applications [5,16,17]. Village sites are commonly associated with groves of these hardwood trees [18,19]. Large trees had special cultural and spiritual significance as gathering places [19] and sacred trees [20]. Native Americans used large cavities in many of these species as shelters and as closets to store ceremonial items, food, firewood, and tools [20,21]. Because of the importance of hardwoods for sustaining traditional cultural values, tribes have initiated efforts to restore groves of these species based upon tribal traditional practices (Figure 1).

Figure 1. The North Fork Mono Tribe has been using thinning, burning, and other treatments to restore meadows with large California black oaks on the Sierra National Forest, and to promote desired qualities including low, broad crowns to facilitate the production and gathering of acorns (Photo credit: Ron Goode).

Fruits of the eight hardwood species are consumed by a variety of wildlife species, and the hard nuts produced by oaks, tanoak, and chinquapin are particularly important because they can persist as a winter food source. Cavities and platforms in large hardwoods are also especially important as habitat for resting and raising young by fishers (*Pekania pennanti*), spotted owls (*Strix occidentalis caurina* and *S. o. occidentalis*), and great gray owls (*Strix nebulosa nebulosa* and *S. n. yosemitensis*) [22,23]. Research in many parts of the world have highlighted the importance of large trees that form cavities or "hollows" that are essential for some wildlife species [24].

High-intensity fires kill the above-ground stems of these hardwoods, even of mature trees, or otherwise causes damage that leads to their demise. Even in stands where hardwood trees are growing larger, which should afford greater resistance to fire, forest structure and fuel conditions may still be increasing their vulnerability to top-kill. As recent examples, the Poomacha Fire of 2007 led to the toppling of a giant canyon live oak on Palomar Mountain near San Diego within the following year [25], and the Soberanes Fire of 2016 felled the national champion madrone in the Big Sur area of coastal California [26].

These hardwood species are generally resilient to wildfire because top-killed trees typically resprout quickly. Consequently, high severity fires can facilitate a resurgence of hardwoods in areas where conifers have become dominant [27]. Because of this response, some proponents contend that active management is not needed to conserve hardwoods. For example, Baker [28] contended that shifts from open, hardwood-dominated conditions to more closed, conifer-dominated conditions reflects gradual forest recovery that occurs naturally over centuries, and then is reset by high-severity fires. Successional dynamics based upon infrequent, severe wildfires might be expected, particularly in areas that were sheltered from more frequent wildfires by chance, topography, and reduced Native American influence. Some wildlife biologists who are concerned with sustaining rare species, such as spotted owls and fishers, have also opposed the removal of conifers around oaks [8]. However, the combination of fire suppression punctuated by large, intense wildfires may be detrimental to sustaining legacy groves of large hardwoods and the services they provide [11]. Stand-replacing wildfire provides a slow and uncertain pathway for restoring mature hardwoods, particularly as some areas experience reburn well before hardwoods can mature [13].

Representing an alternative strategy, managers have proposed active management treatments to encourage the conservation and restoration of large legacy hardwood trees. Many scientists and tribes have recommended interventions to restore hardwood communities and to counteract the forest densification associated with fire suppression [8,29]. Such perspectives are rooted in the understanding that very frequent fires (reoccurring within 15 years) occurred prior to Euro-American colonization in many forest areas due to combinations of natural ignitions and burning by Native Americans [30,31]. Such a frequent fire regime killed small trees, resulting in more open forests with large individual trees, according to accounts by early scientists such as Jepson [6] and Native Americans [8,32] in the region.

1.2. Need for Current Trend Analysis

To understand whether declines in large hardwoods is a significant concern that warrants interventions, it is important to consider both recent and longer-term trends. Historical datasets tend to have many limitations for evaluating hardwoods in general, and large trees in particular [8]. However, in more recent decades, the USDA Forest Service's Forest Inventory and Analysis (FIA) program has provided standardized data across all ownerships in the United States. Studies in forests of eastern North America have found that alterations of frequent fire regimes, including the reduction of Native American influences, have led to declines in oaks [33]. An analysis of FIA data from the eastern USA highlighted concerns for oak decline from reports that intermediate-size oaks were not sufficient to replace dominant and co-dominant trees [2].

Several studies in the eastern United States have specifically used FIA data to evaluate the potential for nontimber forest products (NTFPs) that are important to tribes [34]; one study of ginseng evaluated changes in the volume of hardwood trees (>12.7 cm diameter at breast height (DBH)) [35], while another used specially collected measurements of birch bark to target the resource availability for tribal use [36]. However, the analysis of FIA data to evaluate NTFPs in the western United States has largely been limited to describing the occurrence of common understory plants of interest to harvesters [37], although a recent inventory report also noted that tanoak was the most abundant of NTFP species in California [38].

In California, studies have suggested that densities of hardwood trees, based upon recent FIA data, have not declined in general, compared to conditions reported in the 1930s surveys of forest vegetation led by Albert Wieslander. For example, McIntyre et al. [39] found that that period may have been a low point for hardwoods, which was followed by increases into the late 20th century. Dolanc et al. [40] found that madrone, maple, tanoak, and canyon live oak all increased in their frequencies of occurrence in plots from the 1930s to the early 2000s, while black oak slightly decreased. Among species that were more abundant, they found that tanoak, canyon live oak, and black oak had significantly increased in density, specifically among trees 10.2–60.9 cm DBH, while the density of larger trees was stable. Within Yosemite National Park, Lutz et al. [41] compared data from 1930s Wieslander plots to plots that were resampled by the National Park Service in the 1990s. They found that large (>31 cm DBH) black oaks and canyon live oaks were 57% and 98% more dense in the 1990s, although those differences were not statistically significant. Maple and laurel trees (>10 cm DBH) both declined during that period, but they were uncommon in both periods, as were any very large (>61 cm DBH) hardwoods. Recently, Zielinski and Gray [42] reported that the resting habitat for the fisher had not declined, based upon FIA data and two models of fisher habitat that considered hardwoods; their southern Sierra model included the mean DBH of all hardwoods [43], while their northwestern California model included the total hardwood basal area per unit area [44]. A summary of FIA data from 1981 to 2000 by Waddell and Barrett [45] found that in California, there were significant increases in net growth for black oak, white oak, maple, laurel, madrone, and tanoak. Altogether, these previous studies of inventory data did not point to hardwood declines.

However, a targeted study of potentially more sensitive indicators could reveal important patterns for particular species or within particular areas. Recently, Long et al. (2017) reported recent declines in black oak biomass on national forest lands in California, in association with fire mortality. Recognizing the importance of black oaks in terms of tribal use and for fishers, a draft revision of the management plan for the Sierra National Forest proposed the abundance of large and full-crowned black oaks as a monitoring indicator [46]. Those recent developments suggested a need to further explore trends in large hardwoods.

1.3. Objectives and Scope

Our overall research question was whether these hardwood species were exhibiting a decline in large trees that might reduce the benefits for Native American tribes and rare species such as the fisher. We addressed this question by analyzing trends in recent forest inventory data within a region of Southwest Oregon and California where declines might be expected due to shifts in fire regime and where results are most relevant to fisher conservation. Pure hardwood stands in lower elevation woodlands are important for tribes and wildlife, including owls, but not for fishers. Our analysis explores trends as new inventory data have become available. However, we did test the hypotheses that we identified, based upon a review of relevant literature.

2. Materials and Methods

2.1. Literature Review to Select Indicators

We briefly reviewed scientific literature that describes the values of large hardwoods for Native Americans and wildlife, as well as previous research into the threats and trends for such trees. Such a review serves to identify the indicators and geographic factors that are important to consider in evaluating the trends over this broad region. Table 1 summarizes the information that is gleaned from the review, including the maximum tree size and age, the typical age of fruit production, shade tolerance, fire tolerance, and vulnerability to *P. ramorum*. We focus on the eight hardwood species, because they can occur frequently as large trees (>20 m tall and >100 cm DBH) in mixed-conifer forests, and they can live for several centuries. We did not include species that are generally limited to small trees (<20 m tall) such as Pacific dogwood (*Cornus nuttallii*) and California hazelnut (*Corylus cornuta*). We also did not include tall species that are more typically short-lived, grow in riparian areas, and regenerate following floods or stand-replacing fires, such as black cottonwood (*Populus trichocarpa*), red alder (*Alnus rubra*), or aspen (*Populus tremuloides*).

Table 1. Hardwood species included in the analysis and attributes related to size, age, and disturbances.

Scientific Name	*Arbutus menziesii*	*Quercus garryana*	*Quercus kelloggii*	*Acer macrophyllum*	*Chrysolepis chrysophylla*	*Notho-lithocarpus densiflorus*	*Quercus chrysolepis*	*Umbellularia californica*
Species code	ARME	QUGA	QUKE	ACMA	CHCH	NODE	QUCH	UMCA
Common name	Pacific madrone	Oregon white oak	California black oak	bigleaf maple	giant chinquapin	tanoak	canyon live oak	California laurel
Upper DBH (cm)	122	246	274	338	244	277	330	404
Upper height (m)	38	37	40	49	46	63	30	53
Upper age (years)	400	500	500	300	500	250+	300	200+
Typical onset of fruiting (years)	30	20	30	10	<40	30	20	30–40
Shade tolerance	intermediate to low	low	low	high	intermediate	high	high	intermediate
Fire tolerance	1	2	2	1	1	1.5	0.5	0.5
Vulner-ability to *Phytophthora ramorum*	mortality	not affected	mortality	no mortality; infection is limited to leaf spots	mortality	mortality	mortality	no mortality; infection is limited to leaf spots

Notes: The ages of fruit production are referenced in McDonald and Tappeiner [47] and species profiles in the Fire Effects Information System (FEIS) [48–53]. Maximum age, height, and diameter at breast height (DBH) are from Niemiec et al. [54], except for canyon live oak, which is derived from Tollefson [49]; extraordinary specimens exceed such maximums. Fire tolerance codes, derived from the FEIS species profiles, are: 2 = moderate and large trees typically survive moderate-severity fires, 1.5 = large trees only usually survive moderate-severity fires, 1 = large trees sometimes survive moderate-severity fires, 0.5 = large trees rarely survive even moderate-severity fires without top-kill. Information on vulnerability to *Phytophthora ramorum*, which causes sudden oak death, is from Waddell and Barrett [45] and Frankel [55].

2.1.1. Species Distribution

The eight species occur throughout much of northwestern California and southwestern Oregon. Bolsinger [56] reported that all of these species except giant chinquapin were "major" hardwood species on timberlands in California. All eight occur in the Sierra Nevada, but four of them (giant chinquapin, white oak, madrone, and tanoak) have small and patchy distributions. On the other hand, black oak, canyon live oak, laurel, and maple are distributed throughout the Sierra Nevada and into southern California (Figure 2). Black oak and canyon live oak are two of the most dominant contributors of biomass to California forests [45].

2.1.2. Indicators for Fruit Production

The nut-bearing oaks and tanoaks do not typically produce fruits until at they are at least 20–30 years old, whereas maple and madrone bear fruit earlier (Table 1). However, at least several of these species are capable of producing relatively small quantities of fruit in resprouted stems within six years, as reported for black oak, tanoak, madrone, and chinquapin [47,59]. As a general rule for these species, trees that produce fruits in high quantity are many decades old (requiring more time to mature than many competing conifer species) and have full crowns due to growth in open conditions [6,47,60]. Some species may enter prime production after a century [47], although one study suggested that production in white oak may reach a plateau after only 60–80 years [60]. Acorn production in white oak is greatest in trees with full crowns and increased exposure to light, which may be described as a "mushroom-shape" that is marked by low, broad crowns [60,61] (Figure 1). Similar relationships between shape and nut productivity have been described for tanoak [6] and black oak [62]. Consequently, both basal area, as a proxy for age, and crown attributes are important indicators of productivity, while tree height and wood volume may be less important.

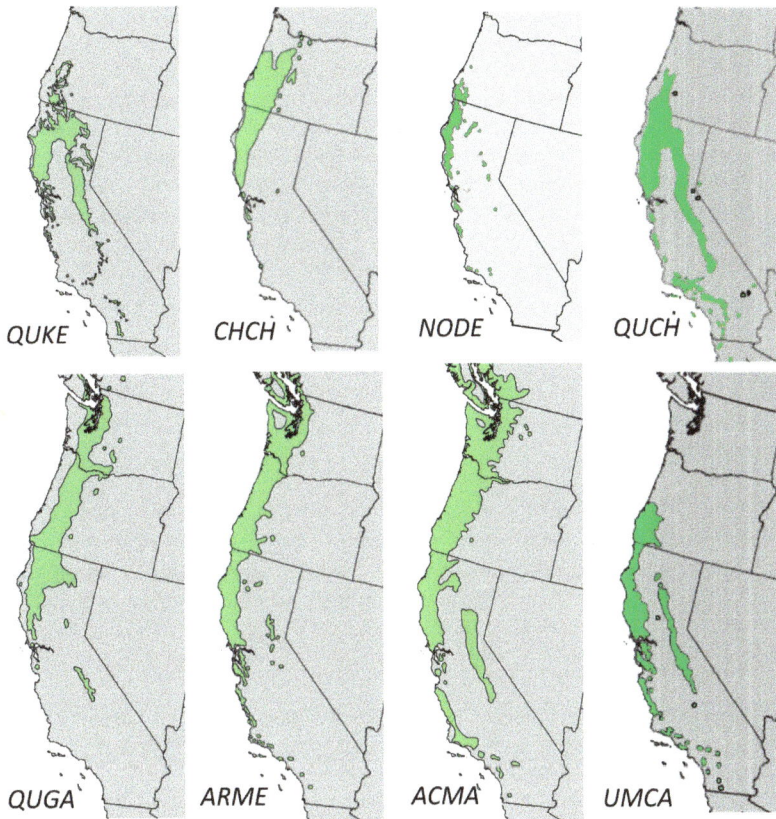

Figure 2. Distributions of the eight study species of hardwood trees within the far western USA from Little [57,58].

2.1.3. Indicators for Wildlife Habitat

Studies of fisher habitat in the Pacific Northwest of North America have found that all eight of the hardwood species we analyzed are used by fisher, along with black cottonwood and aspen [15,63]. Live trees harbor 91% of the fisher dens found in hardwoods [15]. Trees used by fisher are also very large, as Yaeger [63] (whose study included all of the species except laurel) reported that the median diameter at breast height (DBH) of hardwoods used, primarily for resting, was 77.2 cm at the Hoopa Valley Indian Reservation (January 1996 to June 1998) and 63.5 cm on the Shasta-Trinity National Forest in northwestern California. Accordingly, Aubry et al. [64] found that basal area of live hardwoods 51–100 cm DBH was a useful predictor of habitat quality across the range of the fisher.

Owls, including both northern and California spotted owls and great gray owls, also use cavities in large hardwoods. For example, North et al. [65] reported that 11% of the California spotted owls on the Sierra National Forest on the Sequoia-Kings Canyon National Park nested in cavities in very large and old black oaks (91 cm mean DBH, >139 years old within the study). They also found that spotted owls built nests on platforms in somewhat smaller (mean 52 cm DBH) canyon live oak trees. A recent review [23] reported that great gray owls used hardwoods (mostly black oaks) with a mean DBH of 102 cm, although six of the 16 oaks used for nesting were dead. Altogether, these studies suggest that basal area of large living hardwoods is an important indicator of wildlife habitat quality. While dead trees also have value for wildlife, sustaining large live hardwoods also ensures a supply of dead trees.

2.1.4. Tolerance of Shade, Fire, and *Phytophthora ramorum*

These species have complex interactions with each other and with conifers that are influenced by competition for light, fire effects, and disease. In the absence of disturbances such as wildfires, all of these hardwoods are succeeded by conifers, particularly white fir (*Abies concolor*) and Douglas fir (*Pseudotsuga menziesii*) [1,66]. The more shade-tolerant species, including tanoak, maple, canyon live oak, and giant chinquapin can become dominant in densifying forests that were formerly dominated by the more shade-intolerant species (Table 1). The eight species have variable resistance to fire depending upon their size (Table 1); fire generally kills small stems, but larger stems are able to survive low-intensity fires [67–69]. Lutz, van Wagtendonk and Franklin [41] noted that canyon live oak was much less common, and laurel was absent within burned plots in Yosemite National Park; such findings support the assignment of the lowest fire tolerance ratings for those species. The sudden oak death pathogen (*P. ramorum*) reportedly infects all of the species except white oak [45,55]. Laurel is very susceptible and facilitates spread of the disease; while chinquapin, madrone, canyon live oak, and black oak all have experienced mortality, the loss of large tanoak trees been one of the main impacts to ecosystems [70].

2.2. Overview of Forest Inventory Data

The FIA program measures environmental variables every 10 years in the western United States at sample points arranged in a hexagonal grid (with centers of hexagons 5.47 km apart for one point per 2400 ha) across all ownerships [71]. Individual trees are identified and measured on nested plots for different sized trees, and their status (live or dead), size, condition, and estimated cause of death are tracked over time [38]. When plots are sampled, disturbances that occurred between measurements are noted, including cutting, fire, cutting & fire, and "other", which includes insect, disease, beaver, porcupine, bear, ice, wind, flood, and landslides. The disturbances noted for each plot did not necessarily result in mortality of hardwood species; for example, cutting could have involved removal of only conifers.

2.3. Study Area

The study area (Figure 3) encompasses forested lands in California and Oregon with dominant vegetation that ranges from oak (*Quercus*) and juniper (*Juniperus*) species at the dry end, to Douglas fir and California mixed conifer in mesic areas, to true fir (*Abies*) forests in subalpine areas [72]. We grouped areas of broadly similar vegetation and disturbance regimes within contiguous ecoregions [73]. This study area includes part of the range for both subspecies of spotted owls and great gray owl. We selected plots in areas characterized by a "frequent, low severity" fire regime [74] to focus on areas where shifts in fire regimes would be most prominent. We also confined the study area to potential habitats for the fisher, which we defined by adding a buffer of 5 km around areas that are designated as being "selected for" or "intermediate" (meaning neutral rather than "selected against") in a fisher habitat model [75]. The buffer accounts for the fact that the habitat model attempted to select mature stands based on remote sensing analysis, while our objective was to describe conditions within zones of potential habitat, regardless of the current forest condition. Although both fisher habitat and areas of frequent fire regime extend further north into Washington State, we excluded plots that were north of 43.7 degree latitude, which approximates the northernmost extent of the Klamath ecoregion. There were 9.15 million ha of forest land that met the study criteria, represented by 3971 plots that ranged in elevation from 15 to 2890 m.

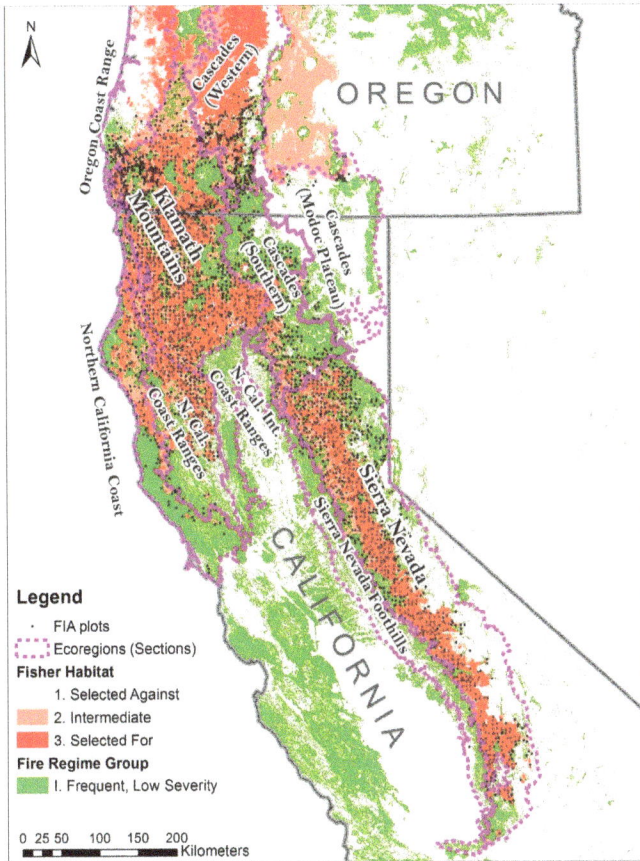

Figure 3. Map of the study area in California and Oregon, USA, showing ecoregions, areas of frequent, low severity fire regime, habitat for fishers, and the approximate location of the Forest Inventory and Analysis (FIA) plots that have at least one tree from the eight study species.

2.4. Measurement Periods

Three measurement periods were represented in the data. The first measurement period ("time 1") consisted of:

- the FIA periodic inventory of non-National Forest System (NFS) lands in California (1991–1994; mean 1992),
- the NFS inventory in California (1992 to 2000; mean 1998),
- the FIA periodic inventory of non-NFS lands in Oregon (1995–1998; mean 1997), and,
- the NFS inventory of Oregon (1993–1997; mean 1995).

The second remeasurement period ("time 2") installed an annual inventory on all lands where a tenth of the plots were expected to be measured every year in spatially-representative panels [71]. The first panel cycle for California and Oregon began in 2001 and ended in 2010, except in California on NFS lands, where the schedule was accelerated, to install all plots on NFS lands by 2006. The third period started in 2007 with remeasurements of plots that were established on the accelerated schedule on California NFS lands, while on other areas, remeasurements began a new panel cycle in 2011.

The last year of data that was available and compiled for this study was 2016. Inclusion of the accelerated annual plots on California NFS lands ensured a full remeasurement sample of those lands. Overall, the mean measurement years for plots in the three inventory assessments were 1996, 2004, and 2012, for mean remeasurement intervals of 7.8 years between times 1 and 2, and 8.8 years between times 2 and 3. The periodic and annual data are available in a public database [76].

We analyzed results from the full set of annual plots in the study area and also for the subset of those that were included in the initial inventory at time 1. The difference in basal area per unit area between those two sets at time 2 was added to the time 1 calculation. This adjustment is an estimate of the value at time 1 if we had measured the same set of plots as in the annual inventory. However, because the sample of other public lands was so different (there were no plots on other federal lands in California at time 1), the three-period comparisons were only made for NFS and private lands. Since the first period used a different plot design, comparisons across the three periods have higher variance and lower significance than the remeasurement from time 2 to time 3, which compared tree-to-tree change using the same plot design. The basal area per unit area is a better metric for comparing values across inventories than totals, because it avoids the confounding effects of definitional and sampling differences in the total area of forest land.

2.5. Metrics and Hypotheses

Based upon the review of potential indicators above, we hypothesized that declines in the basal area of large trees would be greatest in the shade-intolerant black oak and white oak, which depend on frequent fires. We expected madrone to follow that same pattern, while recognizing that it is less tolerant of fire and somewhat more tolerant of shade. We hypothesized that any declines would be even less in the five more shade-tolerant, fire-intolerant species—tanoak, canyon live oak, maple, giant chinquapin, and laurel—because those species may be favored by a longer fire regime and more shading. Furthermore, some of those species, such as canyon live oak, are often dominant on steep slopes that produce little fuel and burn less frequently than the surrounding landscape, so that they may be less affected by shifts in fire regimes [1].

Figure 4 summarizes the overall analysis approach and key variables used to filter and sort results. Because management goals and practices may differ among ownership categories, we stratified results between NFS lands, other public lands, and private lands (which includes tribal lands). Within public land areas, we considered both "reserved" areas, which have more restrictions on timber harvest (primarily wilderness, national and state parks, and national monuments), and "unreserved" areas. We expected more declines in the hardwood basal area due to cutting on private lands, and more declines due to natural disturbances, including fire, insects, and disease on public lands, particularly in "reserved" areas. However, we did not hypothesize that trends would necessarily be different across the ownership categories, particularly because such differences in management predate the inventory baseline. We also stratified results by ecoregions (Figure 4). Since we had already filtered the areas to include only areas of frequent fire regime, we did not have explicit expectations for variations among ecoregions, except that we expected that species vulnerable to mortality by *P. ramorum* would have declined in the California Coast ecoregion, where a quarantine had been imposed to constrain its spread [45].

We focused on live tree basal area per unit area as a primary response variable. We calculated means and variances using double sampling for post-stratification [77], and tested for differences based on the Z statistic, which is customary for inventory data where estimates based upon large sample sizes approach the normal distribution. We examined results with standard size classes ranging from sapling, small, medium, large, and very large trees (Figure 4). Based upon a simple linear regression between age and DBH reported for black oak in McDonald and Tappeiner [47], the breaks between these size classes would be equivalent to 30, 66, 149, and 185 years, respectively. Based upon the data for tanoaks from Sonoma County [78], those size class breaks would be equivalent to 46, 69, 114, and 137 years, respectively.

We considered trends in crown condition because trees that are suppressed are more prone to mortality and less productive [58]. We therefore examined the crown condition class, which describes both overstory trees (classified as open-grown, dominant, or co-dominant) and understory trees (classified as intermediate or suppressed). Shifts in basal area from open, overstory trees, to suppressed understory trees could be a leading indicator of overall declines. Accordingly, we hypothesized that the basal area in medium-sized or larger trees (>28 cm DBH) with full crowns (classified as open-grown or dominant), may have declined as forests became more dense. We also analyzed change in basal area within overstory versus understory trees within plots that were undisturbed, to help evaluate trends in general forest succession.

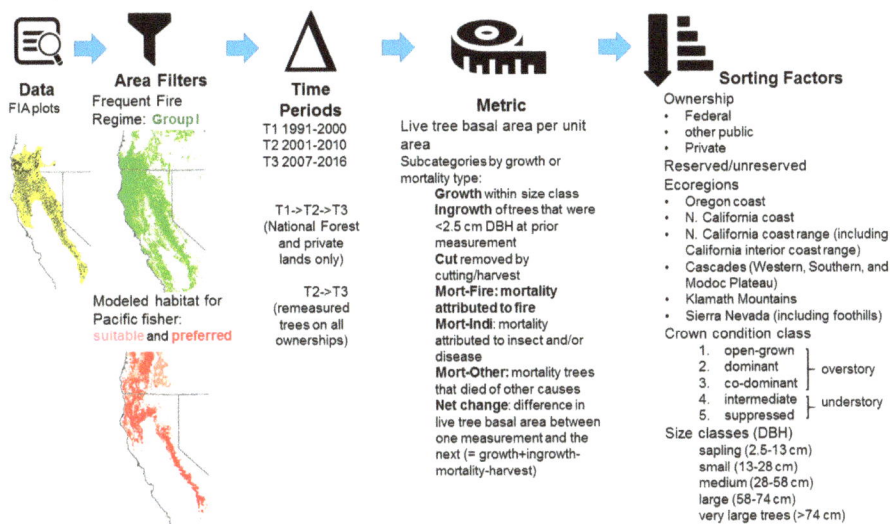

Figure 4. Schematic showing the factors considered in the trend analysis of large hardwoods.

3. Results

3.1. Distribution of Basal Area and Rates of Disturbances across Ecoregions and Ownership Areas

Table 2 shows the amount of study area within each ecoregion, as well as the amount of disturbance within each ecoregion. Over 40% of the basal area of the eight species was located in the Klamath ecoregion, with the Sierra Nevada (18.3%), California Coast (17%), and California Coast Range (16.6%) ecoregions also being important. Other regions, including the Cascades (3.4%) and Oregon Coast Range (2.8%), had much less basal area of the species of interest. Basal area for each species across the ecoregions during time 2 is included in the Supplementary Materials (Table S1). Tanoak was the most dominant hardwood species in the coastal ecoregions, while black oak and canyon live oak were dominant in the Sierra Nevada. Other ecoregions were less dominated by one or two species, but tanoak, canyon live oak, black oak, and madrone were much greater contributors to the hardwood basal area (each contributing 13%–30% of combined basal area for the eight species) than the remaining four (2%–6% of combined basal area).

Within the study area, 2.5% of forestland was disturbed each year, with cutting being the most prevalent disturbance, followed by fire (Table 2). The highest rate of fire disturbance in the selected forestlands was in the Klamath ecoregion at 1.6%/year, while the highest rate of cutting disturbance was in the Cascades ecoregion at 2.6%/year (Table 2). Over half of the study area (51.8%) was managed by the NFS, 10.6% by other public agencies, and the remaining 37.6% was under private ownership. Sixteen percent of the forested area was on public lands "reserved" from management for timber production.

Table 2. Total area and annual rates of disturbance on forestlands in the study area between times 2 and 3, by ecoregion

Ecoregion	Total Area		Fire		Cut		Fire + Cut		Other		Total Disturbed	
	ha	%	ha/year	%	ha/year	%	ha/year		ha/year		ha/year	%
Klamath	3,037,602	33	47,582	1.6	26,982	0.9	1928	0.1	2455	0.1	78,947	2.6
CA Coast	743,446	8	766	0.1	10,205	1.4	484	0.1	592	0.1	12,047	1.6
CA Coast Range	865,360	9	11,645	1.3	3043	0.4	0.0	0.0	1090	0.1	15,779	1.8
OR Coast Range	178,611	2		0.0	2687	1.5	445	0.2	552	0.3	3684	2.1
Sierra	2,766,788	30	26,437	1.0	31,994	1.2	4252	0.2	2511	0.1	65,194	2.4
Cascades	1,559,874	17	5276	0.3	39,948	2.6	2626	0.2	3649	0.2	51,499	3.3
Totals	9,151,682	100	91,706	1.0	114,859	1.3	9735	0.1	10,849	0.1	227,149	2.5

3.2. Overall Changes in Basal Area and Associated Causes of Mortality

Most of the species appeared to increase modestly in basal area over the three time periods, except for slight, suggestive declines across periods for black oak and substantial increases for tanoak, canyon live oak and laurel from time 1 to time 2 (Figure 5). However, none of the changes were statistically significant because of wide variance, which in part reflects the smaller sample size and the lack of remeasured trees in the three-period analysis.

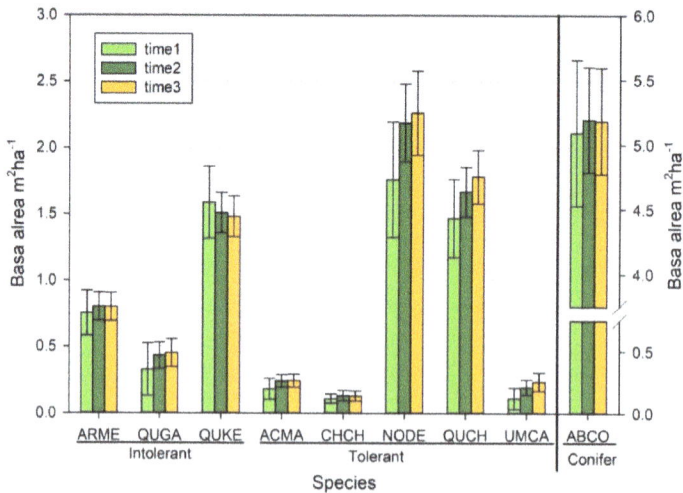

Figure 5. Species basal area across three time periods, National Forest System (NFS), and private lands only. Species codes are first two letters of genus and species, order of shade-intolerants on the left and shade-tolerants on the right used in all figures; white fir (a conifer) is shown for comparison using a second y-axis on the right. Error bars represent 95% confidence intervals.

The trends among remeasured trees across all ownerships from time 2 to time 3 were similar to those seen in the three-period analysis, with growth generally compensating for mortality (Figure 6, Supplemental Table S2). However, several species experienced statistically significant increases, including white oak (+6.3%, $p = 0.003$), tanoak (+3.75%, $p = 0.041$), canyon live oak (+6.16%, $p = 0.002$), and laurel (+20.7%, $p < 0.001$). Giant chinquapin registered a slight but statistically insignificant decline, while madrone and maple had small and statistically insignificant increases. White fir, a common conifer species included for comparison, showed a negligible increase (0.34%, $p = 0.45$). On the other hand, black oak experienced a relatively modest (1.54%) and suggestive ($p = 0.060$) decline.

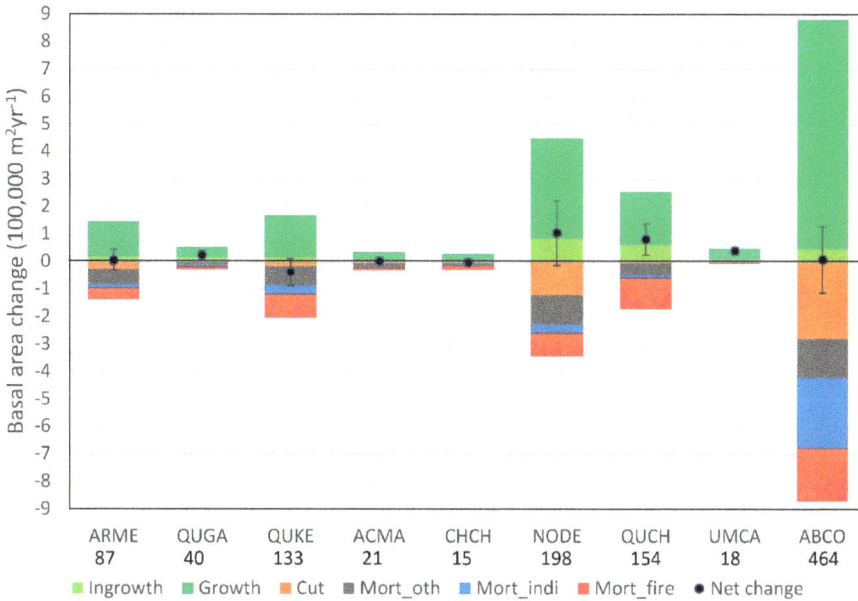

Figure 6. Overall change in basal area, and changes associated with particular causes of mortality (abbreviated terms in legend are defined in Figure 4), from time 2 to time 3 for eight hardwood species and white fir (ABCO). Number under species code is live basal area ($\times 100,000$ m^2) at beginning of time 2, and error bars on the net change values represent the 95% confidence interval.

Mortality was attributable to fire, cutting, insects/disease, and other causes for all species, although causes did vary somewhat by species. For example, mortality of individual black oak and canyon live oak trees was most commonly attributed to fire (Figure 6). By contrast, insects/disease and cutting were more important than fire for the common conifer, white fir (ABCO in Figure 6). Tanoak was the only hardwood species for which cutting was identified as the predominant cause of mortality. The rate of mortality associated with cutting was twice as high for tanoak (0.6%/year) as for any other species. However, the data did not indicate whether the cut tanoak were already dead or infected by *P. ramorum*. Although tanoak had high rates of cutting and natural mortality, its high rate of growth more than compensated. Within plots that were undisturbed, we found a significant loss of basal area for understory black oak, while overstory black oaks significantly increased in basal area; the basal area in overstory trees also increased for the other species, while the basal area in understory trees was maintained, or increased in the case of tanoak, canyon live oak, and laurel (Supplemental Figure S3).

3.3. Changes in Basal Area within Size Classes

When examining the results within particular size classes, the three-period comparison also suggested a steady decline in basal area of very large black oak, while very large madrone and tanoak exhibited substantial increases from time 1 to time 2 (Figure 7). However, in the remeasured data from time 2 to time 3, changes in the largest diameter classes were not apparent. The confidence intervals for net change for most of those estimates included zero change, although there was a significant net loss (−0.91%) of small black oak (13–28 cm DBH).

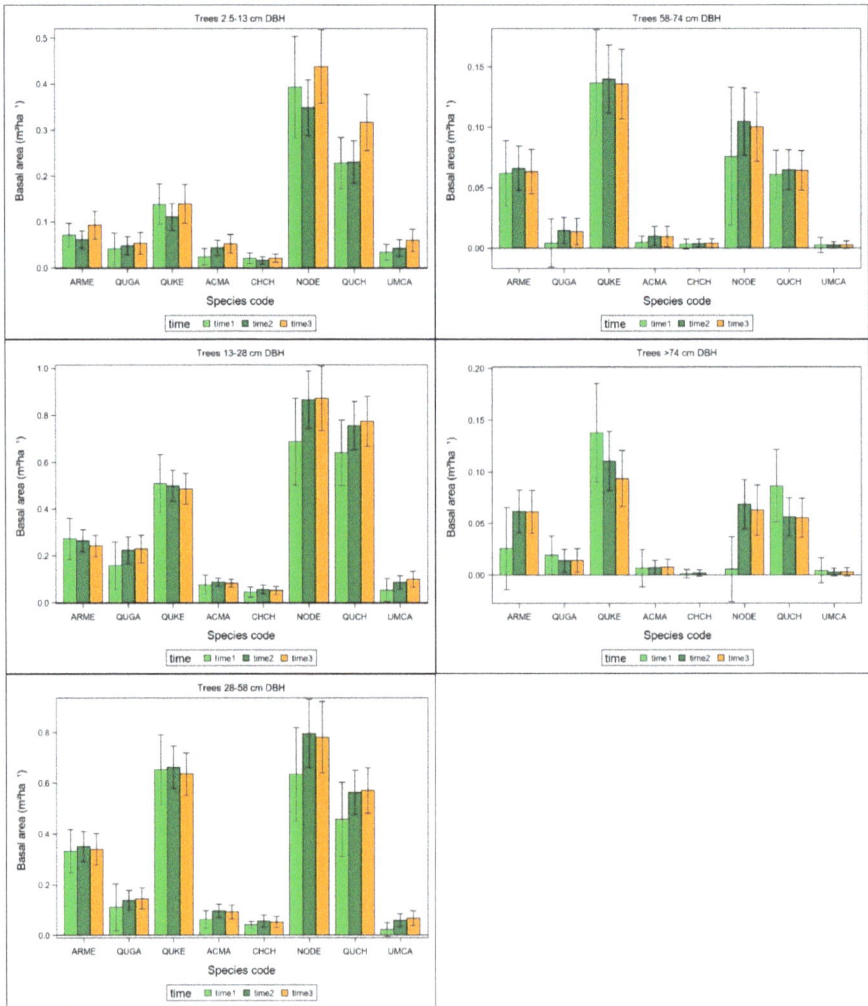

Figure 7. Species basal area by diameter class for three times with 95% confidence intervals (NFS and private lands only).

3.4. Changes in Basal Area across Ecoregions and Ownership

Within ecoregions, changes in basal area between times 2 and 3 (Figure 8) were generally consistent with those from the three-period analysis (Supplemental Figure S1). We focus on the two-period analysis, since several results were statistically significant, including increases in canyon live oak and laurel in the California Coast and California Coast Range, increases in tanoak in the California Coast Range, and increases in white oak in the Cascades. While not statistically significant, the average black oak basal area declined in the California Coast Range, Sierra Nevada, and Klamath, and much of those mortality in those areas was attributed to fire. Meanwhile, the mean basal area of black oak increased in the Cascades, where the fire mortality was relatively low. Other regions contributed little to the basal area of black oak. Tanoak basal area had a non-significant decline in the Oregon Coast, but generally was stable or increased, including a significant increase in the California Coast Range. Mortality of tanoak in the coastal regions was dominated by cutting rather than by disease or fire.

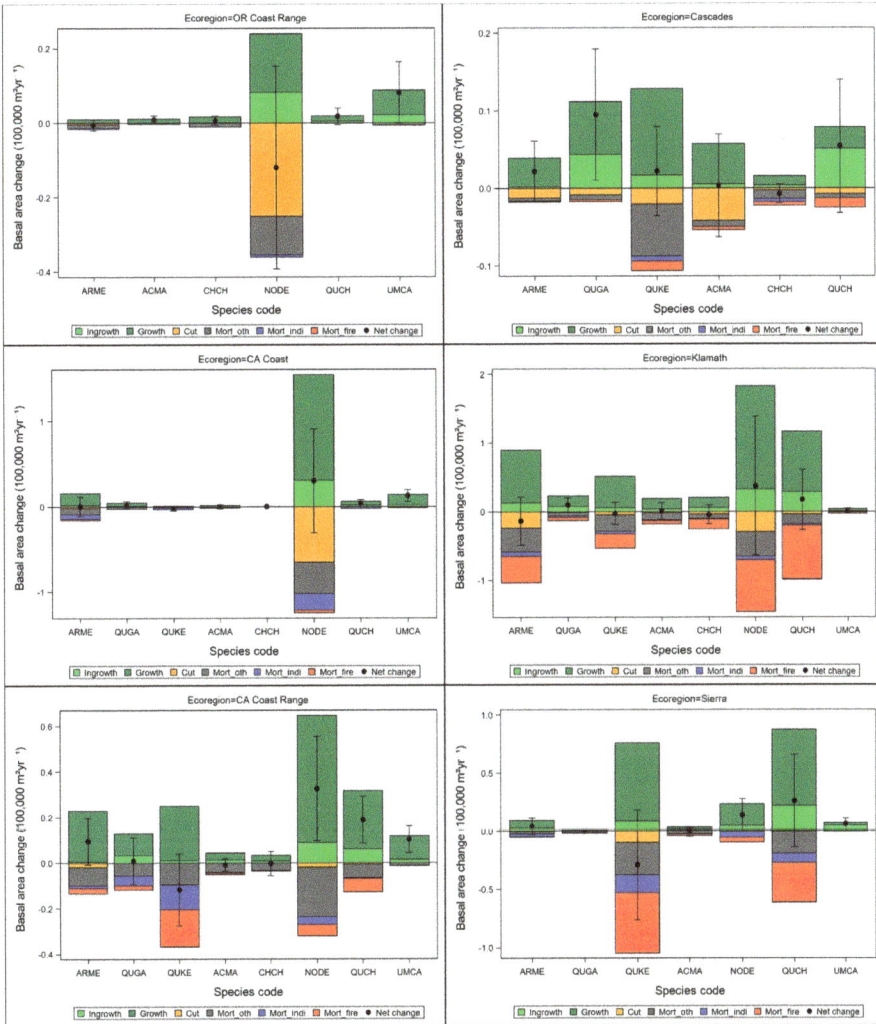

Figure 8. Basal area change by species by ecoregions with net changes shown by points with 95% confidence interval bars.

The analyses by ownership provides additional insights into trends. The three-period analysis did not reveal statistically significant trends and suggested overall stability, although shade-tolerant tanoak and canyon live oak tended to increase in NFS and private unreserved lands, while the shade-intolerant black oak and madrone tended to decline in reserved lands (Supplemental Figure S2). The two-period analysis (Figure 9) revealed similar patterns, although some trends were statistically significant, including a decline in madrone on NFS reserved lands, increases in white oak on non-federal public unreserved lands, and increases in white oak, canyon live oak and laurel on private lands. Results from that analysis also highlight variation in causes of mortality (Figure 9). On public lands, fire was a major cause of mortality, while cutting was fairly minor; the opposite relationship held on private lands. Six of the species (all but tanoak and laurel) tended to decline within NFS-reserved areas (although only the decline in madrone was statistically significant). Mortality in those areas was strongly associated with fire. Increases in hardwood basal area tended to be greater on NFS unreserved lands than on private lands, except for laurel. Black oak basal area tended to decline on NFS lands, in both reserved and nonreserved areas, but it was stable on other public and private lands.

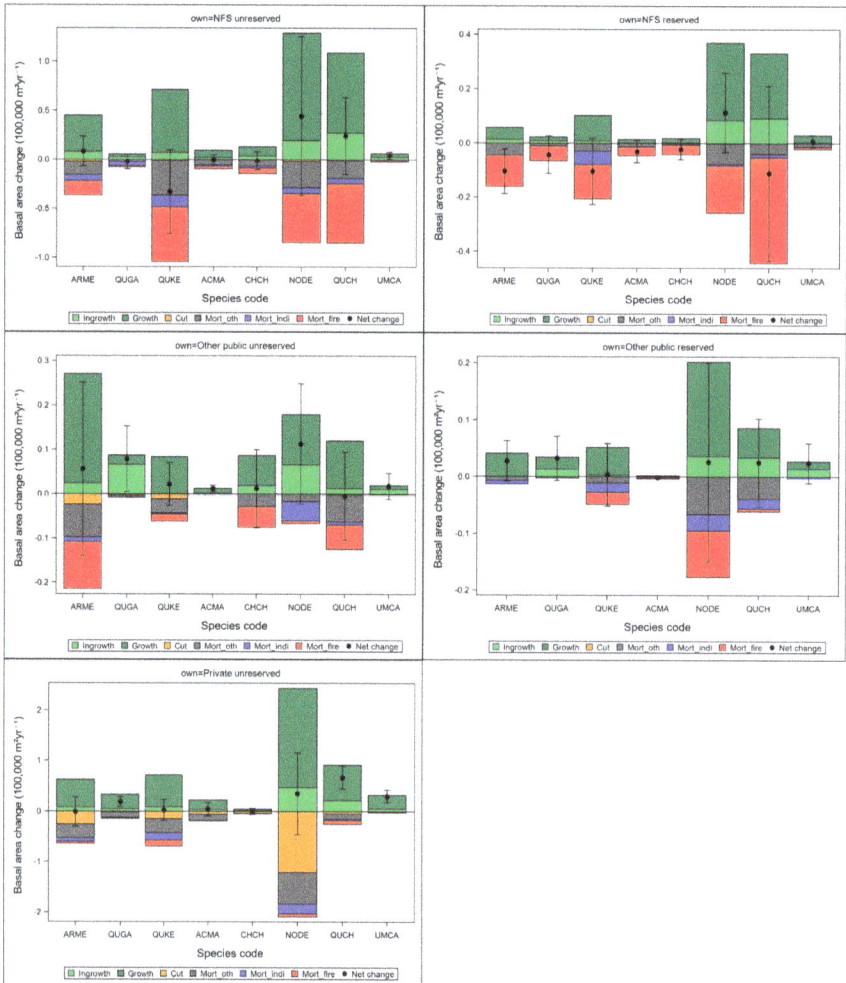

Figure 9. Basal area change by ownership and reserve status with net change shown by points with 95% confidence interval bars.

3.5. Changes in Basal Area by Crown Condition Class

Combined basal area of medium-sized and larger hardwoods (>28 cm DBH) that were classified as either open-grown or dominant declined across all eight species (Figure 10). These declines were generally offset by corresponding increases in the co-dominant crown condition class, except in the case of black oak and maple. This shift was also evident for white fir, a dominant conifer, but it was not as pronounced as for the hardwoods.

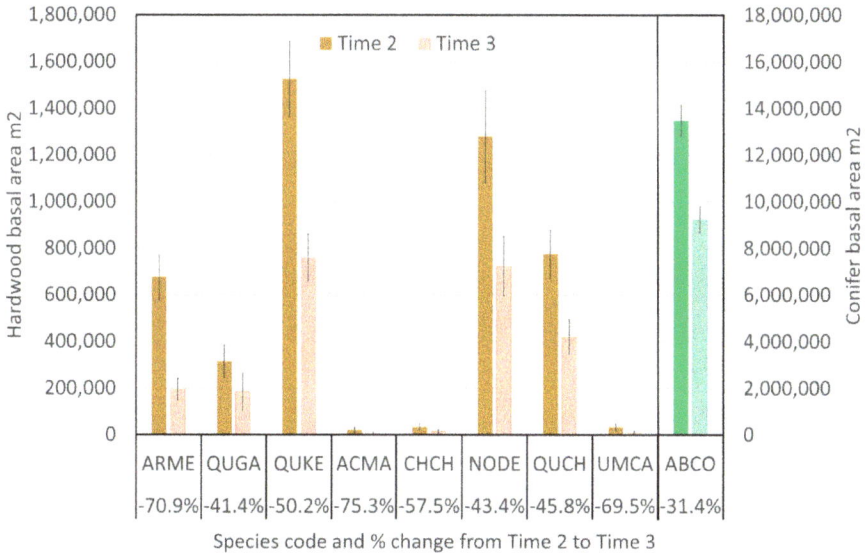

Figure 10. Basal area of trees >28 cm DBH in the open-grown and dominant crown condition categories at times 2 and 3 for eight hardwood species and white fir (on the second Y-axis on the right), with percent reductions labeled (bars are 95% confidence intervals).

4. Discussion

4.1. Overall Trends

The eight species included in this analysis mostly exhibited stable or increased basal area, contrary to our hypothesis that they might be declining in the study region. We also did not find declines within the California Coast region, where *P. ramorum* has been causing hardwood mortality. Even though the FIA database is the best available information for determining trends over a large area, the power of these analyses is still limited, especially for species that are often minor forest components. Furthermore, by confining the analysis plots to frequent fire regime areas within suitable fisher habitat, important trends for particular species across their entire ranges may be missed. For example, impacts of *P. ramorum* will be more evident at lower elevations and latitudes along the coast compared to the areas of focus in this analysis.

Black oak stood out from the other seven species for experiencing downward trends in overall the basal area in the two-period comparison, and in the basal area of very large trees in the three-period comparison, although neither result was statistically significant, based upon a 95% confidence level. Statistically significant declines were evident in small black oaks and those classified as understory trees. Our results also indicated that declines in black oak were more pronounced on NFS lands than on other ownerships, and fire was a major cause of that mortality. This finding was consistent with an earlier finding [59] of a modest but statistically significant recent decline in black oak basal area (particularly in both small and very large trees) on national forest lands throughout the entire range of the species.

Our hypothesis that shade tolerance would explain trends was consistent with the finding that three shade-tolerant species increased in basal area (Figure 6). However, the two shade-intolerant species other than black oak, white oak, and madrone did not show similar declines. This result suggests that characteristics other than shade intolerance are influencing these trends. One possibility is that fires have affected areas where black oak is relatively more dominant than other species, particularly the Sierra Nevada, Klamath, and California Coast Range ecoregions, which altogether contain 92% of the basal area of the species (Supplemental Table S1). These three regions had the highest rates of fire mortality (over 1%/year; Table 2). Miller and Safford [79] reported significant increases in area burned by high severity fires within the Sierra Nevada, Modoc Plateau and southern Cascades region in yellow pine and mixed-conifer forests from 1984 to 2010. A similar study in northwestern California (including parts of the Klamath, the northern California Coast Range, and part of the southern Cascade Range ecoregions) reported that total burned area strongly increased from 1987 to 2008; even though the percentage burned at high severity did not increase, their results indicate increases in total area burned at high severity [80]. By contrast, a third recent study in Oregon (including the Oregon Coast Range and Western Cascades) [81] found only a small increase in total area burned and no increase in proportion burned at high severity from 1985 to 2010. Those results suggest that increases in wildfire-caused mortality are presently a greater threat to hardwoods in California. However, vulnerability varies by species, even within those regions. For example, canyon live oak has experienced a net increase, despite the fact that it is more sensitive to fire, and it predominates in inland California ecoregions that did experience considerable fire mortality. That species may be able to compensate for fire-related losses by growing in more shaded conditions and on harsher sites that have not burned severely in recent wildfires.

Our results demonstrate the value of indicators beyond overall basal area, tree densities, or species distributions to inform forest planning and monitoring. For example, the three-period analysis suggested a decline in the basal area of very large black oaks, while the overall basal area was relatively stable. Although the previous habitat modeling for fisher has relied heavily on overall basal areas, large and very large trees appear disproportionately important. In addition, because crown condition is an important driver of fruit production, increases in the basal area of trees that are not in dominant crown conditions may not necessarily result in increased productivity. Consequently, measures of crown condition, especially for large trees, may be important leading indicators. The declines in the more open crown condition classes across species suggest that these forests are widely becoming more densely canopied, which is consistent with expectations that hardwoods might decline [3]. However, the decline in basal area among open-grown and dominant trees might also reflect changes in how field crews categorized dominant and co-dominant classes between sampling periods. That more trees were being classified as co-dominant would explain why basal area in "overstory" black oaks appeared to increase from time 2 to time 3, even while full-crowned black oaks declined. However, the decline in full-crown trees was more pronounced for all of the hardwood species than it was for white fir, which suggests that there may be an ecological dimension to this trend. Evaluation of the next panel of inventory data would test whether this a real and continuing ecological trend, or a transient artifact of shifting field protocols.

As FIA sampling extends into the next period of tree remeasurement, trends may become clearer, especially as the effects of recent very large wildfires become more evident. On the other hand, broad-scale drought and bark beetle epidemics, particularly in the southern Sierra Nevada, have tended to reduce conifer trees much more than black oak [82]. Where such conifer mortality unfolds, hardwoods might enjoy gains [83] that offset declines. This complicated interplay between species' interactions and disturbances make it difficult to predict net trends during rapid climatic change.

4.2. Strategies and Tactics Conserving Large Hardwood Trees

Conservation efforts for hardwoods have already been initiated on national forests. For example, in the Sierra Nevada, planting of conifer trees is restricted to within 6 m of hardwood crowns, under the Sierra Nevada Forest Plan amendment [8]. In Oregon, treatments have been implemented to clear conifers from the crowns of white oaks [63]. However, in isolated cases, such as the southernmost stand of Pacific madrone, managers have also recommended retaining a conifer overstory to sustain favorable micro-site conditions [84].

Many of the strategies and tactics to promote large hardwoods have been guided by tribal traditional knowledge to promote food, spiritual values, and habitat for valued wildlife [6]. In pre-colonial times, tribal agroforestry systems across the study region facilitated and fostered site conditions that promoted large trees with full, mushroom-shaped crowns, which were more beneficial for tribal harvesters because they had low branches to facilitate harvesting, as well as increased fruit production [17]. Traditional practices included burning forest patches to promote acorn production and other food plants, as well as to enhance habitat for game animals such as deer and elk. To address current tribal goals, the cultivation and conservation of large trees may be prioritized in specially managed stewardship areas that center on former gathering areas and remnant groves of large hardwoods [59] (Figure 1).

Frequent burning may be an important strategy for promoting large trees and basal cavities [8], especially if precautions can be taken to avoid the losses of legacy trees. Due to the sensitivity of hardwoods to fire, promoting large hardwoods in the near term may depend upon removing or girdling competing conifers and reducing fuel loads mechanically. Accumulations of fuels (especially conifer litter and small trees) allow flames to scorch trees higher on their stems and enter existing stem cavities (Figure 11). Such damage often results in top-kill, and it may reduce the potential for forming and sustaining enclosed cavities [8]. In addition to removing conifers, tactics to protect legacy hardwood trees from wildfires and prescribed burning include raking fuels away from boles of large legacy trees [8], cutting lower limbs to reduce potential for crown fire, burning off lichens hanging on large hardwood branches prior to prescribed burning, and using tree-centered firing [85]. Because shrubs in the understory can intensify fires in white oak woodlands [57] and potentially other hardwood stands, site preparation to reduce oak mortality could include treatments to reduce shrub continuity and to promote grasses or other herbaceous species that facilitate low-intensity fire spread [11]. Collectively, such measures could conserve legacy hardwoods by reducing fire scorching and mortality.

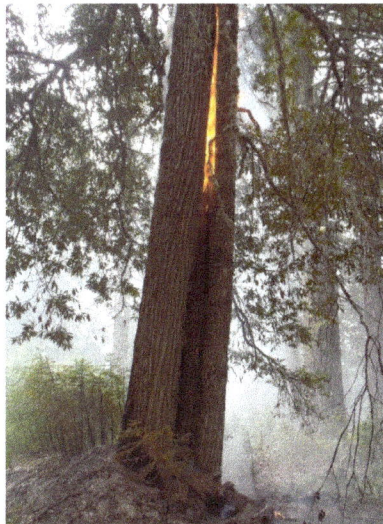

Figure 11. During a prescribed burn in June 2017, fire entered the cavity of this large chinquapin tree, which compromised its integrity (Photo credit: Frank K. Lake).

5. Conclusions

Our results demonstrate how forest inventory data can be used to evaluate trends, the impact of interventions, and potential effects on benefits that these legacy trees have long provided. Our findings of current trends from recent FIA data did not find that the basal areas of eight of the most important large hardwood species, and large trees in particular, were declining overall, with a large region of southwestern Oregon and California, including the California Coast ecoregion where *P. ramorum* is killing many of these trees. However, some indicators suggested modest downward trends in black oak, particularly on national forest lands in inland parts of California, where fires have been a major cause of mortality. Because black oak appears to be the most vulnerable of these important hardwoods, and because it ranges across much of California, it could be a particularly useful and sensitive indicator of trends across the state. We also found significant declines in large full-crowned hardwoods, which are particularly important for fruit production and wildlife habitat quality. However, that finding may have reflected changes in protocols that are used to classify trees as dominant versus co-dominant. This kind of analysis will become more powerful as more inventory panels are completed, and it will be important to effectively monitor trends across the region. This information can help land managers and tribes in the region to conserve, restore, and monitor large, centuries-old hardwood trees with full crowns. Strategies to conserve these legacy trees may be particularly important within interior regions of California to offset losses from wildfires.

Supplementary Materials: The following are available online at http://www.mdpi.com/1999-4907/9/10/651/s1, Figure S1: Three-period basal area change by ecoregion (NFS and private lands). Figure S2: Three-period basal area change by ownership. Figure S3: Basal area changes on undisturbed plots by crown class group. Table S1: Basal area by ecoregion section at time 2. Table S2: Annual rate of basal area change (m^2 year^{-1}) on disturbed plots by disturbance type.

Author Contributions: Conceptualization, J.W.L.; Formal analysis, J.W.L. and A.G.; Methodology, J.W.L. and A.G.; Visualization, J.W.L., A.G. and F.K.L.; Writing—original draft, J.W.L. and A.G.; Writing—review & editing, J.W.L., A.G. and F.K.L.

Funding: This research received no external funding.

Acknowledgments: This analysis would not have been possible without the efforts of many dedicated crew members who collected data in the field, and people who compiled and reported data to make the FIA datasets available. We also thank Carl Skinner, research geographer emeritus, for providing a constructive review of an early draft of the manuscript and other insights, and Jim Steed, with the Forest Inventory and Analysis program, for insights into field measurements.

Conflicts of Interest: The authors declare no conflict of interest.

References

1. Taylor, A.H.; Skinner, C.N. Spatial patterns and controls on historical fire regimes and forest structure in the Klamath Mountains. *Ecol. Appl.* **2003**, *13*, 704–719. [CrossRef]
2. McShea, W.J.; Healy, W.M.; Devers, P.; Fearer, T.; Koch, F.H.; Stauffer, D.; Waldon, J. Forestry Matters: Decline of Oaks Will Impact Wildlife in Hardwood Forests. *J. Wildl. Manag.* **2007**, *71*, 1717–1728. [CrossRef]
3. McComb, B.C.; Spies, T.A.; Olsen, K.A. Sustaining biodiversity in the Oregon Coast Range: Potential effects of forest policies in a multi-ownership province. *Ecol. Soc.* **2007**, *12*, art29. [CrossRef]
4. Hagar, J.C. Wildlife species associated with non-coniferous vegetation in Pacific Northwest conifer forests: A review. *For. Ecol. Manag.* **2007**, *246*, 108–122. [CrossRef]
5. Long, J.W.; Lake, F.K. Escaping a socioecological trap through tribal stewardship on national forest lands in the Pacific Northwest, USA. *Ecol. Soc.* **2018**, *23*, art10. [CrossRef]
6. Bowcutt, F. Tanoak landscapes: Tending a Native American nut tree. *Madrono* **2013**, *60*, 64–86. [CrossRef]
7. McGregor, R.R.; Sakalidis, M.L.; Hamelin, R.C. *Neofusicoccum arbuti*: A hidden threat to *Arbutus menziesii* characterized by widespread latent infections and a broad host range. *Can. J. Plant Pathol.* **2016**, *38*, 70–81. [CrossRef]

8.   Long, J.W.; Anderson, M.K.; Quinn-Davidson, L.N.; Goode, R.W.; Lake, F.K.; Skinner, C.N. *Restoring California Black Oak Ecosystems to Promote Tribal Values and Wildlife*; U.S. Department of Agriculture, Forest Service, Pacific Southwest Research Station: Albany, CA, USA, 2016.

9.   Coops, N.C.; Waring, R.H. Estimating the vulnerability of fifteen tree species under changing climate in Northwest North America. *Ecol. Model.* **2011**, *222*, 2119–2129. [CrossRef]

10. Case, M.J.; Lawler, J.J. Relative vulnerability to climate change of trees in western North America. *Clim. Chang.* **2016**, *136*, 367–379. [CrossRef]

11. Nemens, D.G.; Varner, J.M.; Kidd, K.R.; Wing, B. Do repeated wildfires promote restoration of oak woodlands in mixed-conifer landscapes? *For. Ecol. Manag.* **2018**, *427*, 143–151. [CrossRef]

12. Cocking, M.I.; Varner, J.M.; Sherriff, R.L. California black oak responses to fire severity and native conifer encroachment in the Klamath Mountains. *For. Ecol. Manag.* **2012**, *270*, 25–34. [CrossRef]

13. Hammett, E.J.; Ritchie, M.W.; Berrill, J.-P. Resilience of California black oak experiencing frequent fire: Regeneration following two large wildfires 12 years apart. *Fire Ecol.* **2017**, *13*, 91–103. [CrossRef]

14. Filipe, J.A.; Cobb, R.C.; Meentemeyer, R.K.; Lee, C.A.; Valachovic, Y.S.; Cook, A.R.; Rizzo, D.M.; Gilligan, C.A. Landscape epidemiology and control of pathogens with cryptic and long-distance dispersal: Sudden oak death in northern Californian forests. *PLoS Comp. Biol.* **2012**, *8*, e1002328. [CrossRef] [PubMed]

15. Lofroth, E.; Raley, C.; Higley, J.; Truex, R.; Yaeger, J.; Lewis, J.; Happe, P.; Finley, L.; Naney, R.; Hale, L. *Conservation of Fishers (Martes pennanti) in South-Central British Columbia, Western Washington, Western Oregon, and California–Volume I: Conservation Assessment*; USDI Bureau of Land Management: Denver, CO, USA, 2010.

16. Long, J.W.; Lake, F.K.; Lynn, K.; Viles, C. *Tribal Ecocultural Resources and Engagement*; USDA Forest Service, Pacific Northwest Research Station: Portland, OR, USA, 2018.

17. Anderson, M.K. *Indigenous Uses Management, and Restoration of Oaks of the Far Western United States*; NRCS National Plant Data Center: Washington, DC, USA, 2007.

18. Baumhoff, M.A. California Athabascan groups. *Anthr. Rec.* **1958**, *16*, 157–238.

19. Kniffen, F.B. Achomawi geography. *Univ. Calif. Publ. Am. Archaeol. Ethnol.* **1928**, *23*, 297–332.

20. Kroeber, A.L.; Gifford, E.W. World renewal: A cult system of native northwest California. *Anthr. Rec.* **1949**, *13*, 1–156.

21. Du Bois, C. Wintu ethnography. *Univ. Calif. Publ. Am. Archaeol. Ethnol.* **1935**, *36*, 1–148.

22. Bunnell, F.L.; Kremsater, L.L.; Wind, E. Managing to sustain vertebrate richness in forests of the Pacific Northwest: Relationships within stands. *Environ. Rev.* **1999**, *7*, 97–146. [CrossRef]

23. Wu, J.X.; Siegel, R.B.; Loffland, H.L.; Tingley, M.W.; Stock, S.L.; Roberts, K.N.; Keane, J.J.; Medley, J.R.; Bridgman, R.; Stermer, C. Diversity of great gray owl nest sites and nesting habitats in California. *J. Wildl. Manag.* **2015**, *79*, 937–947. [CrossRef]

24. Lindenmayer, D.B. Integrating forest biodiversity conservation and restoration ecology principles to recover natural forest ecosystems. *New For.* **2018**. [CrossRef]

25. Wasyl, J. Uncle oak: The giant of Palomar Mountain. *Int. Oaks* **2011**, *22*, 100–103.

26. Duckworth, J. *Remembering Fallen Champion Trees*; American Forests: Washington, DC, USA, 2016. Available online: http://www.americanforests.org/blog/remembering-fallen-champion-trees/ (accessed on 1 April 2018).

27. Cocking, M.I.; Varner, J.M.; Knapp, E.E. Long-term effects of fire severity on oak-conifer dynamics in the southern Cascades. *Ecol. Appl.* **2014**, *24*, 94–107. [CrossRef] [PubMed]

28. Baker, W.L. Historical forest structure and fire in Sierran mixed-conifer forests reconstructed from General Land Office survey data. *Ecosphere* **2014**, *5*, 1–70. [CrossRef]

29. Cocking, M.I.; Varner, J.M.; Engber, E.A. *Conifer Encroachment in California Oak Woodlands*; U.S. Department of Agriculture, Forest Service, Pacific Southwest Research Station: Albany, CA, USA, 2015.

30. Crawford, J.N.; Mensing, S.A.; Lake, F.K.; Zimmerman, S.R. Late Holocene fire and vegetation reconstruction from the western Klamath Mountains, California, USA: A multi-disciplinary approach for examining potential human land-use impacts. *Holocene* **2015**, *25*, 1341–1357. [CrossRef]

31. Taylor, A.H.; Trouet, V.; Skinner, C.N.; Stephens, S. Socioecological transitions trigger fire regime shifts and modulate fire–climate interactions in the Sierra Nevada, USA, 1600–2015 CE. *Proc. Natl. Acad. Sci. USA* **2016**, *113*, 13684–13689. [CrossRef] [PubMed]

32. Anderson, K. *Tending the Wild: Native American Knowledge and the Management of California's Natural Resources*; University of California Press: Berkeley, CA, USA, 2005.

33. Abrams, M.D.; Nowacki, G.J. Native Americans as active and passive promoters of mast and fruit trees in the eastern USA. *Holocene* **2008**, *18*, 1123–1137. [CrossRef]
34. Chamberlain, J.L.; Emery, M.R.; Patel-Weynand, T. *Assessment of Nontimber Forest Products in the United States under Changing Conditions*; USDA Forest Service Southern Research Station: Asheville, NC, USA, 2018.
35. Chamberlain, J.L.; Prisley, S.; McGuffin, M. Understanding the relationships between American ginseng harvest and hardwood forests inventory and timber harvest to improve co-management of the forests of eastern United States. *J. Sustain. For.* **2013**, *32*, 605–624. [CrossRef]
36. Emery, M.R.; Wrobel, A.; Hansen, M.H.; Dockry, M.; Moser, W.K.; Stark, K.J.; Gilbert, J.H. Using traditional ecological knowledge as a basis for targeted forest inventories: Paper birch (*Betula papyrifera*) in the US Great Lakes region. *J. For.* **2014**, *112*, 207–214. [CrossRef]
37. Vance, N.; Gray, A.; Haberman, B. *Assessment of Western Oregon Forest Inventory for Evaluating Commercially Important Understory Plants*; USDA Forest Service Pacific Northwest Research Station: Portland, OR, USA, 2002.
38. Christensen, G.A.; Waddell, K.L.; Stanton, S.M.; Kuegler, O. *California's Forest Resources: Forest Inventory and Analysis, 2001-2010*; USDA Forest Service, Pacific Northwest Research Station: Portland, OR, USA, 2016.
39. McIntyre, P.J.; Thorne, J.H.; Dolanc, C.R.; Flint, A.L.; Flint, L.E.; Kelly, M.; Ackerly, D.D. Twentieth-century shifts in forest structure in California: Denser forests, smaller trees, and increased dominance of oaks. *Proc. Natl. Acad. Sci. USA* **2015**, *112*, 1458–1463. [CrossRef] [PubMed]
40. Dolanc, C.R.; Safford, H.D.; Thorne, J.H.; Dobrowski, S.Z. Changing forest structure across the landscape of the Sierra Nevada, CA, USA, since the 1930s. *Ecosphere* **2014**, *5*, art101. [CrossRef]
41. Lutz, J.A.; van Wagtendonk, J.W.; Franklin, J.F. Twentieth-century decline of large-diameter trees in Yosemite National Park, California, USA. *For. Ecol. Manag.* **2009**, *257*, 2296–2307. [CrossRef]
42. Zielinski, W.J.; Gray, A.N. Using routinely collected regional forest inventory data to conclude that resting habitat for the fisher (*Pekania pennanti*) in California is stable over ~20 years. *For. Ecol. Manag.* **2018**, *409*, 899–908. [CrossRef]
43. Zielinski, W.J.; Truex, R.L.; Dunk, J.R.; Gaman, T. Using forest inventory data to assess fisher resting habitat suitability in California. *Ecol. Appl.* **2006**, *16*, 1010–1025. [CrossRef]
44. Zielinski, W.J.; Dunk, J.R.; Gray, A.N. Estimating habitat value using forest inventory data: The fisher (*Martes pennanti*) in northwestern California. *For. Ecol. Manag.* **2012**, *275*, 35–42. [CrossRef]
45. Waddell, K.L.; Barrett, T.M. *Oak Woodlands and other Hardwood Forests of California, 1990s*; USDA Forest Service, Pacific Northwest Research Station: Portland, OR, USA, 2005.
46. USDA Forest Service. *Draft Revised Land Management Plan for the Sierra National Forest*; USDA Forest Service, Pacific Southwest Region, Sierra National Forest: Clovis, CA, USA, 2016.
47. McDonald, P.M.; Tappeiner, J.C. *California's Hardwood Resource: Seeds, Seedlings, and Sprouts of Three Important Forest-Zone Species*; U.S. Department of Agriculture, Forest Service, Pacific Southwest Research Station: Albany, CA, USA, 2002.
48. Gucker, C.L. *Quercus garryana*; USDA Forest Service, Rocky Mountain Research Station: Fort Collins, CO, USA. Available online: https://www.fs.fed.us/database/feis/plants/tree/quegar/all.html (accessed on 30 July 2018).
49. Tollefson, J.E. *Quercus chrysolepis*; USDA Forest Service, Rocky Mountain Research Station: Fort Collins, CO, USA. Available online: https://www.fs.fed.us/database/feis/plants/tree/quechr/all.html (accessed on 1 May 2018).
50. Meyer, R. *Chrysolepis chrysophylla*; USDA Forest Service, Rocky Mountain Research Station: Fort Collins, CO, USA. Available online: https://www.fs.fed.us/database/feis/plants/tree/chrchr/all.html (accessed on 30 July 2018).
51. Reeves, S.L. *Arbutus menziesii*; USDA Forest Service, Rocky Mountain Research Station: Fort Collins, CO, USA. Available online: https://www.fs.fed.us/database/feis/plants/tree/arbmen/all.html (accessed on 30 July 2018).
52. Fryer, J.L. *Acer macrophyllum, bigleaf maple*; USDA Forest Service, Rocky Mountain Research Station: Fort Collins, CO, USA. Available online: https://www.fs.fed.us/database/feis/plants/tree/acemac/all.html (accessed on 1 May 2018).

53. Howard, J.L. *Umbellularia californica*; USDA Forest Service Rocky Mountain Research Station: Fort Collins, CO, USA. Available online: https://www.fs.fed.us/database/feis/plants/tree/umbcal/all.html (accessed on 1 May 2018).

54. Niemiec, S.S.; Ahrens, G.R.; Willits, S.; Hibbs, D.E. *Hardwoods of the Pacific Northwest*; Research Contribution; Oregon State University: Corvallis, OR, USA, 1995.

55. Frankel, S. USDA Forest Service, Pacific Southwest Research Station, Albany, CA, USA. Personal communication, 2018.

56. Bolsinger, C.L. *The Hardwoods of California's Timberlands, Woodlands, and Savannas*; USDA Forest Service, Pacific Northwest Research Station: Portland, OR, USA, 1988.

57. Little, E.L., Jr. *Conifers and Important Hardwoods*; Miscellaneous Publication 1186; USDA: Washington DC, USA, 1971.

58. Little, E.L., Jr. *Minor Western Hardwoods*; Miscellaneous Publication 1314; USDA: Washington, DC, USA, 1976.

59. McDonald, P.M. Chrysolepis Hjelmqvist. In *Silvics of North America*; Burns, R.M., Honkala, B.H., Eds.; USDA Forest Service: Washington, DC, USA, 1990; pp. 404–406.

60. Peter, D.; Harrington, C. Site and tree factors in Oregon white oak acorn production in western Washington and Oregon. *Northwest Sci.* **2002**, *76*, 189–201.

61. Devine, W.D.; Harrington, C.A. Restoration release of overtopped Oregon white oak increases 10-year growth and acorn production. *For. Ecol. Manag.* **2013**, *291*, 87–95. [CrossRef]

62. Long, J.W.; Goode, R.W.; Gutteriez, R.J.; Lackey, J.J.; Anderson, M.K. Managing California black oak for tribal ecocultural restoration. *J. For.* **2017**, *115*, 426–434. [CrossRef]

63. Yaeger, J.S. Habitat at Fisher Resting Sites in the Klamath Province of Northern California. Master's Thesis, Humboldt State University, Arcata, CA, USA, May 2005.

64. Aubry, K.B.; Raley, C.M.; Buskirk, S.W.; Zielinski, W.J.; Schwartz, M.K.; Golightly, R.T.; Purcell, K.L.; Weir, R.D.; Yaeger, J.S. Meta-analyses of habitat selection by fishers at resting sites in the Pacific coastal region. *J. Wildl. Manag.* **2013**, *77*, 965–974. [CrossRef]

65. North, M.; Steger, G.; Denton, R.; Eberlein, G.; Munton, T.; Johnson, K. Association of weather and nest-site structure with reproductive success in California spotted owls. *J. Wildl. Manag.* **2000**, *64*, 797–807. [CrossRef]

66. Devine, W.D.; Harrington, C.A. Changes in Oregon white oak (*Quercus garryana* Dougl. ex Hook.) following release from overtopping conifers. *Trees* **2006**, *20*, 747–756. [CrossRef]

67. Arno, S.F. *Fire in Western Forest Ecosystems*; USDA Forest Service Rocky Mountain Research Station: Fort Collins, CO, USA, 2000.

68. Fonda, R. Postfire response of red alder, black cottonwood, and bigleaf maple to the Whatcom Creek fire, Bellingham, Washington. *Northwest Sci.* **2001**, *75*, 25–36.

69. Dale, V.H.; Hemstrom, M.; Franklin, J. Modeling the long-term effects of disturbances on forest succession, Olympic Peninsula, Washington. *Can. J. For. Res.* **1986**, *16*, 56–67. [CrossRef]

70. Cobb, R.C.; Filipe, J.A.N.; Meentemeyer, R.K.; Gilligan, C.A.; Rizzo, D.M. Ecosystem transformation by emerging infectious disease: Loss of large tanoak from California forests. *J. Ecol.* **2012**, *100*, 712–722. [CrossRef]

71. Bechtold, W.A.; Patterson, P.L. *The Enhanced Forest Inventory and Analysis Program—National Sampling Design and Estimation Procedures*; USDA Forest Service, Southern Research Station: Asheville, NC, USA, 2005.

72. Eyre, F.H. *Forest Cover Types of the United States and Canada*; Society of American Foresters: Washington, DC, USA, 1980.

73. McNab, W.H.; Cleland, D.T.; Freeouf, J.A.; Keys Jr, J.E.; Nowacki, G.J.; Carpenter, C.A. *Description of Ecological Subregions: Sections of the Conterminous United States*; USDA Forest Service: Washington, DC, USA, 2007.

74. LANDFIRE. Fire Regime Groups. USDI, Geological Survey. Available online: https://landfire.gov/frg.php (accessed on 31 August 2018).

75. Spencer, W.D.; Sawyer, S.C.; Romsos, H.L.; Zielinski, W.J.; Sweitzer, R.A.; Thompson, C.M.; Purcell, K.L.; Clifford, D.L.; Cline, L.; Safford, H.D.; et al. *Southern Sierra Nevada Fisher Conservation Assessment*; Conservation Biology Institute: Portland, OR, USA, 2015.

76. USDA Forest Service. Inventory Data. Available online: https://www.fs.fed.us/pnw/rma/fia-topics/inventory-data/index.php (accessed on 31 August 2018).

77. Scott, C.T.; Bechtold, W.A.; Reams, G.A.; Smith, W.D.; Westfall, J.A.; Hansen, M.H.; Moisen, G.G. *Sample-Based Estimators Used by the Forest Inventory and Analysis National Information Management System*; USDA Forest Service, Southern Research Station: Asheville, NC, USA, 2005.

78. Tappeiner, J.; McDonald, P.M.; Roy, D.F. Lithocarpus densiflorus (Hook. & Arn.) Rehd. Tanoak. In *Silvics of North America*; Burns, R.M., Honkala, B.H., Eds.; USDA: Washington, DC, USA, 1990; pp. 417–425.

79. Miller, J.D.; Safford, H.D. Trends in wildfire severity 1984–2010 in the Sierra Nevada, Modoc Plateau, and southern Cascades, California, USA. *Fire Ecol.* **2012**, *8*, 41–57. [CrossRef]

80. Miller, J.D.; Skinner, C.N.; Safford, H.D.; Knapp, E.E.; Ramirez, C.M. Trends and causes of severity, size, and number of fires in northwestern California, USA. *Ecol. Appl.* **2012**, *22*, 184–203. [CrossRef] [PubMed]

81. Reilly, M.J.; Dunn, C.J.; Meigs, G.W.; Spies, T.A.; Kennedy, R.E.; Bailey, J.D.; Briggs, K. Contemporary patterns of fire extent and severity in forests of the Pacific Northwest, USA (1985–2010). *Ecosphere* **2017**, *8*, e01695. [CrossRef]

82. Paz-Kagan, T.; Brodrick, P.G.; Vaughn, N.R.; Das, A.J.; Stephenson, N.L.; Nydick, K.R.; Asner, G.P. What mediates tree mortality during drought in the southern Sierra Nevada? *Ecol. Appl.* **2017**, *27*, 2443–2457. [CrossRef] [PubMed]

83. Stephens, S.L.; Collins, B.M.; Fettig, C.J.; Finney, M.A.; Hoffman, C.M.; Knapp, E.E.; North, M.P.; Safford, H.; Wayman, R.B. Drought, tree mortality, and wildfire in forests adapted to frequent fire. *Bioscience* **2018**, *68*, 77–88. [CrossRef]

84. USFS. *Rim Fire Restoration Environmental Impact Statement*; Stanislaus National Forest: Sonora, CA, USA, 2016.

85. Weatherspoon, C.P.; Almond, G.A.; Skinner, C.N. Tree-centered spot firing-a technique for prescribed burning beneath standing trees. *West. J. Appl. For.* **1989**, *4*, 29–31.

forests

MDPI

Article

Effect of Predation, Competition, and Facilitation on Tree Survival and Growth in Abandoned Fields: Towards Precision Restoration

Annick St-Denis [1,*], Daniel Kneeshaw [1] and Christian Messier [1,2,*]

[1] Centre for Forest Research (CFR), Department of Biological Sciences, Université du Québec à Montréal, PO Box 8888, Station Centre-Ville, Montréal, QC H3C 3P8, Canada; kneeshaw.daniel@uqam.ca

[2] Institute of Temperate Forest Sciences (ISFORT), Department of Natural Sciences, Université du Québec en Outaouais, 58 Rue Principale, Ripon, QC J0V 1V0, Canada

* Correspondence: st-denis.annick.2@courrier.uqam.ca (A.S.-D.); christian.messier@uqo.ca (C.M.)

Received: 8 September 2018; Accepted: 1 November 2018; Published: 7 November 2018

Abstract: Tree seedlings planted in abandoned agricultural fields interact with herb communities through competition, tolerance, and facilitation. In addition, they are subject to herbivory by small mammals, deer or invertebrates. To increase the success of forest restoration in abandoned fields and reduce management costs, we should determine which species are tolerant to or facilitated by herbaceous vegetation and those which require protection from competition and predation. Eight native tree species were planted in plots covered by herbaceous vegetation, plots where herbaceous vegetation was removed, and plots where seedlings were surrounded by an organic mulch mat. Half of the seedlings were protected against small mammal damage. Results showed that two non-pioneer and moderately shade-tolerant species (yellow birch and red oak) were inhibited by herbaceous vegetation. Birch species were particularly affected by small mammal predation. No effects of predation or herbaceous competition were observed for conifer species. Rather, herbaceous vegetation had a positive effect on the survival and the height growth of tamarack (*Larix laricina*). None of the tested herb communities had a stronger competitive effect on tree growth than another. Restoration of abandoned fields using multi-tree species should be designed at the seedling scale rather than at the site scale to account for different tree responses to predation and competition as well as variable site conditions. An approach resembling precision agriculture is proposed to lower costs and any potential negative impact of more intensive vegetation management interventions.

Keywords: tree plantation; abandoned agricultural field; predation; competition; tolerance; facilitation; precision restoration

1. Introduction

Facilitation has been proposed as a possible restoration tool for woody species [1,2]. It is well known that facilitation effects on growth are generally restricted to less favorable environments [3]. However, facilitation may occur in productive systems such as mesic temperate habitats [2,3]. Indeed, positive effects of herbaceous cover on tree emergence and survival have been previously observed in temperate-zone abandoned fields, whereby the presence of herbaceous vegetation reduced frost heaving, heat and desiccation stresses on tree seedlings [4–7].

Usually, the improvement in emergence, survival, growth, or fitness of young trees (facilitation) is assisted by nurse shrubs and trees rather than herbaceous neighbors [2,8]. Dense herbaceous communities colonizing abandoned agricultural fields may inhibit the establishment and growth of numerous tree species for many years [9,10]. Competition for soil water from herb species is recognized as a primary factor affecting tree survival and growth, although when water is not limited, herbaceous vegetation may have no perceptible effect on tree seedlings (tolerance) [5,11–14].

The relationship between herbaceous plants and tree seedlings is influenced by the functional characteristics of trees and herbs. Pioneer tree species have a greater proportion of deep roots, higher cumulative root length and number of root apices than non-pioneer species, allowing them to explore a larger volume of soil and to be better adapted to water and nutrient limited sites [15,16]. Moreover, pioneer species can rapidly outgrow the vegetation layer due to their faster growth rates, although their establishment may be limited by herbaceous vegetation, whereas moderately to highly shade-tolerant species establish better under herb cover but grow more slowly [17–19]. On the other hand, some herb communities are known to be stronger competitors than others. With their high root/shoot ratio, their clonal growth form, their ability to produce tillers and spread rapidly, grasses are generally stronger competitors than forbs [2,20,21]. However, some forb species that form dense communities (e.g., *Solidago* and *Aster* sp.) may compete with tree seedlings or inhibit them via the production of allelopathic compounds, although allelopathy remains mainly hypothetical [11,22,23].

In addition to its variable direct effect on different species of tree seedlings, herbaceous vegetation cover may increase rates of small mammal predation by sheltering them from larger predators [6,24,25]. Rabbits and voles, as well as deer and invertebrates may cause serious damage to tree seedlings, but the severity of herbivory depends on predator density, tree species, the season, the presence of a vegetative cover, the area of tree plantation, and the distance to neighboring woodlands [6,26–30].

Due to predation and competition, most tree plantation projects on abandoned fields have used protection against predation as well as some form of control of herbaceous vegetation [31–34]. Mechanical control, such as plowing, has been used to reduce herbaceous competition, but this can also slow down site restoration by eliminating natural regeneration, disturbing the soil, and decreasing organic matter, microbial activity, and mycorrhizal diversity [35–37]. Light mechanical treatments, such as mowing and shallow cultivation, have also been used, but their effects fade rapidly [38,39]. Despite their lower efficiency in controlling root competition by herbaceous plants, plastic and weed cloth mulch mats are often used as they decrease competition for light, increase soil temperature and moisture, and are more socially acceptable than herbicides [40–43]. However, using non-biodegradable mulching materials, or plastic spiral protectors and tree-shelters against predation could be costly since it is necessary to return to the planting site after a few years to remove these materials.

The main goal of this research was to test whether the success of restoration of abandoned agricultural fields using multi-tree species could be improved through a greater understanding, at the tree species level, of the effects of predation, competition, and facilitation on tree survival and growth. More specifically, we addressed the following questions: (1) Are tree species or tree functional groups affected differently by small mammal predation and herbaceous competition? and (2) Is growth of various tree species influenced differently by various herb groups and soil moisture? We hypothesized that: (1) survival of moderately to highly shade-tolerant species is not negatively affected by herbaceous vegetation; but that (2) the growth of these non-pioneer species is more strongly affected by competition than are pioneer species; (3) competition effects on tree growth increase with the abundance of *Solidago* and *Aster* species that surround tree seedlings; and (4) conifers are less affected by small mammal predation than hardwoods.

2. Materials and Methods

2.1. Study Site

The study was conducted in a peri-urban area of Montréal (Québec, Canada), in the agricultural zone of Laval (45°40′ N; 73°43′ W). The dominant regional forest is sugar maple—hickory and the climate is humid continental. Average annual temperature, recorded at the Montréal Pierre Elliott Trudeau weather station (45°28′ N; 73°45′ W), is 6.8 °C with monthly means of 21.2 °C in July and −9.7 °C in January, the warmest and the coldest months (means were calculated for the

1981–2010 period) [44]. Annual precipitation is 1000 mm, of which around 20% falls as snow [44]. From 2010 to 2012, the average annual temperature was 8.2 °C and the average annual precipitation was 1134 mm [45]. The experiment was carried out in three abandoned agricultural fields. Field #1 (surface area ≈ 9000 m^2) is separated from field #2 (≈23.000 m^2) by a ditch bordered by eastern cottonwood (*Populus deltoides* Marsh.) trees. These two fields are situated on the south side of a 3.5 ha forest composed of silver maple (*Acer saccharinum* L.), black ash (*Fraxinus nigra* Marsh.), green ash (*Fraxinus pennsylvanica* Marsh.), eastern cottonwood (*Populus deltoides* Marsh.), and eastern white cedar (*Thuja occidentalis* L.) trees. Field #3 (≈9000 m^2) is located on the north side of this forest.

The abandoned fields had a similar past land use with grains and vegetables having been cultivated for more than 25 years until the early 2000s. Following this agricultural period, the sites were colonized by ruderal herbaceous species and were mown once or twice a year, until the fall preceding the experiment (2009). The vegetation was principally dominated by grass species (*Poaceae* spp. and *Cyperaceae* spp.), *Solidago*, *Trifolium*, *Sonchus*, and *Aster* species. *Daucus carota*, *Taraxacum officinale*, *Cirsium* spp., *Arctium* spp., and *Erigeron* spp. were also common. Total ground coverage by all species was around 65% in each field at the time of planting. Mowing prevented tree and shrub establishment, but ash (*Fraxinus* sp.) seedlings colonized the sites as soon as mowing ceased, i.e., in the first summer (2010) following planting.

The surface deposit is a mix of glacial (till) and marine deposits. Soil is an orthic melanic brunisol type [46]. Fields #1 and #2 and the majority of field #3 are covered by a stony clay loam that is moderately well-drained, while the remainder of field #3 is a well-drained clay loam [46]. Topography is mainly flat although there are slight depressions. The experimental design (see below) was included in a restoration project of the abandoned fields using more than 15,000 tree and shrub seedlings (height < 1 m) planted from June to August 2010, at least 2 m from experimental plots.

2.2. Experimental Design

The study took place between the end of May 2010 and the end of September 2012. Trees were also measured in June 2015, but no vegetation treatment (weeding or mowing) was done between 2012 and 2015. Four 40.5 m × 27 m experimental blocks were established at least 25 m from mature trees and roads, along an east-west axis: one in field #1 (surface area of ≈1 ha), two in field #2 (23 ha), and one in field #3 (1 ha). Experimental blocks were divided into six subplots following a split-plot design where the main factor was "protection" (plastic spirals against small mammals vs. no protection) and the subplot factor was "vegetation" (VG, intact herbaceous vegetation; M, mulch mats; BS, bare soil). Eight tree seedlings of eight species were randomly planted (within each subplot) for a total of 192 trees per species. Species that were used in the experiment are native to the area and represent a gradient of growth rates and shade tolerances (Table 1). The chosen species were: paper birch (*Betula papyrifera* Marsh.), yellow birch (*Betula alleghaniensis* Britt.), tamarack (*Larix laricina* (Du Roi) K. Koch), red pine (*Pinus resinosa* Ait.), northern red oak (*Quercus rubra* L.), red ash (*Fraxinus pennsylvanica* Marsh.), red maple (*Acer rubrum* L.) and sugar maple (*Acer saccharum* Marsh.).

Container-produced tree seedlings were obtained from the Berthier nursery (Berthierville, QC, Canada) of the Ministère des Fôrets, de la Faune et des Parcs du Québec and were delivered in cold-storage. All tree seedlings were container-produced (initial sizes are provided in Table 1). Hardwood species were one year old, while conifer species were two years old. Seedlings were kept in a dark cool room (≈15 °C) until manual planting from 28 May to 11 June 2010. Seedlings were planted at 1.5 m spacing and were watered once after planting.

One-third of the tree seedlings were planted directly in the herbaceous vegetation (VG) which was less than 20 cm high at the time of planting. Another third of the tree seedlings were surrounded by a 50 cm × 50 cm × 8−10 mm organic mulch mat (M) made of coconut fiber (Biomat, Multi-formes Inc., La Guadeloupe, QC, Canada) installed immediately following planting. Mats were affixed to the ground surface with four U-nails. Finally, the remaining tree seedlings were planted on bare soil (BS). Before planting, herbaceous vegetation was cut to ground level with a gasoline-powered weed cutter. A herbicide (Roundup® concentrate, glyphosate 143 g/L; 100 mL diluted in two liters of water, 25 L/ha) was applied covering a 50-cm radius at each location where seedlings were to be planted. Tree seedlings were planted one week after herbicide application. From July 2010 to September 2012, the vegetation was regularly hand-weeded to maintain bare soil conditions in a 30- to 35-cm radius around each tree. In rows between bare soil seedlings, vegetation was mowed with the weed cutter to prevent the herbaceous vegetation from exceeding 20 cm in height.

In July 2010, half of the tree seedlings were protected from small mammals (e.g., voles, rabbits, etc.) using plastic spiral protectors (TIMM Enterprises Ltd., Milton, ON, USA) 35 cm in length, affixed to the ground with a U-nail. If trees were shorter than 50 cm, plastic protectors were cut to an appropriate length for seedling size. Seedlings affected by predation before the installation of the protectors (<2%) were excluded from the analyses.

2.2.1. Survival and Growth

Tree survival and signs of mammal predation were evaluated every spring and fall from fall 2010 to fall 2012 as well as in June 2015. Seedlings were identified as live (no sign of predation), damaged live or damaged dead by voles (i.e., gnawed stem, removed bark near the ground, or gnawed roots), damaged live or damaged dead by rabbits (i.e., clean-cut edges), damaged live or damaged dead by deer (i.e., rough-torn edges) [24,27]. Tree mortality due to small mammal predation may also be inferred when mortality rates for unprotected seedlings were greater than for seedlings protected with a plastic spiral protector.

Height and diameter were measured every fall (2010−2012) and in June 2015. Height was measured as the distance between the soil surface and the apical meristem while diameter was measured 5 cm above the soil surface. The relative growth rate (RGRX) of height or diameter was calculated using the formula:

$$RGRX = \frac{\ln(X2) - \ln(X1)}{T2 - T2}$$

where $X1$ corresponds to seedling height or diameter in year $T1$, $X2$, height or diameter in year $T2$.

2.2.2. Foliar Measurements

Foliar predation by invertebrates was estimated for all tree seedlings in September 2011. When more than 30% of leaves showed signs of invertebrate predation (such as sawfly damage), the seedling was classified as being affected by invertebrate predation. Further foliar measurements were made between 15−23 August 2012 on two trees per species randomly chosen in each subplot among those having a minimum of 25 leaves (if not possible, trees having the highest number of leaves were chosen). This minimum number of leaves per individual was required for foliar analyses. Leaves and needles were placed between wet paper towels and were kept moist and cool until laboratory analysis. Specific leaf area (SLA) was only measured for leaves of the following species: paper birch, red maple, yellow birch, red oak and sugar maple. Specific leaf area was measured following the method described by Cornellissen et al. [47]. Ten healthy leaves, or the maximum number of healthy leaves, if there were less than ten leaves, were scanned the day of collection. Leaf area was calculated using the Winfolia software (Régent Instruments, QC, Canada). All leaves and needles were dried at 70 °C for at least 48 h before their mass was measured or before grinding. Specific leaf area was calculated as the total one-sided area of fresh leaves divided by their oven-dried mass and was expressed in $mm^2\ mg^{-1}$. Leaves were finely ground with a vibratory pulverizer (Fritsch, Idar-Oberstein, Germany). Between each sample, the pulverizer was cleaned with a vacuum and rinsed with ethanol (70%). Concentrations of nitrogen (leaf N) were analyzed on a Leco CNS−2000 (LECO, St-Joseph, MO, USA) in the laboratory at the Canadian Forest Service's Laurentian Forestry Centre.

Table 1. Characteristics of the eight tree species planted for this restoration experiment.

Common Name	Scientific Name	Mycorrhizal Association	Successional Status	Growth Rate	Shade Tolerance	Initial Mean Height (cm)	Initial Mean Diameter (mm)
Tamarack	*Larix laricina* (Du Roi) K. Koch	EM	Pioneer	Rapid	1	38.7	6.8
Red pine	*Pinus resinosa* Aiton	EM	Pioneer	Rapid	1.9	25.9	5.9
Paper birch	*Betula papyrifera* Marsh.	EM	Pioneer	Rapid	1.5	30.1	4.0
Red ash	*Fraxinus pennsylvanica* Marsh.	AM	Pioneer	Rapid	3.1	33.8	4.8
Red maple	*Acer rubrum* L.	AM	Pioneer	Rapid	3.4	62.3	6.9
Northern red oak	*Quercus rubra* L.	EM	Non-pioneer	Moderate	2.8	41.8	5.7
Yellow birch	*Betula alleghaniensis* Britt.	EM	Non-pioneer	Slow	3.2	37.0	3.8
Sugar maple	*Acer saccharum* Marsh.	AM	Non-pioneer	Slow	4.8	53.0	5.6

AM: arbuscular mycorrhizal species; EM: ectomycorrhizal species; Successional status and growth rate [48,49]; Shade tolerance scales range from 0 (not tolerant) to 5 (maximum tolerance) [50].

2.2.3. Soil Water Content and Light

Environmental measurements were taken on two seedlings per species randomly selected in each subplot. Soil water content (SWC) was measured three times in the summer of 2011 with a TDR-200 probe (Spectrum Technologies Inc., Plainfield, IL, USA), using 12-cm rods. The first measurements were taken on 10 June, two days after a 10-mm rainfall, but the 9 preceding days had been without rain (Environment and Climate Change Canada, 2013b). The second measurements were taken on August 17th, one day after 11 mm of rain and three days after 22 mm of rain. The third measurements were taken on August 23rd, one day after 3 mm of rain and 2 days after 40 mm of rain (Environment and Climate Change Canada, 2013b). Soil water content was also estimated on 21 September 2012, three days after a 21-mm rainfall (Environment and Climate Change Canada, 2013b). Photosynthetic photon flux density (%PPFD) was measured using point quantum sensors (LI-COR Inc., Lincoln, NE, USA) at a height of 30 cm on the south side of each seedling. Measurements were taken at the end of July 2011, on cloudy days, following the method described in Messier and Puttonen [51].

2.2.4. Herbaceous Vegetation

On 17 August 2011, herbaceous biomass was measured around one randomly chosen seedling per species in each herbaceous vegetation and mulch mat subplot. After the first year, some herb plants began to pierce the organic mulch mats or grow through the crack in the middle of the mulch mats. All plants inside a 50 × 25 cm plot installed on the east side of the selected seedling were cut to ground level and placed in a paper bag. Samples were air dried at 25 °C until there was no further loss of mass due to humidity and then weighed to estimate above-ground biomass of the surrounding vegetation. On 23 July 2012, in herbaceous vegetation subplots, an inventory was taken of herbaceous vegetation beside the same two seedlings randomly chosen for SWC. All herbaceous species in a 50 × 25 cm plot were identified and their percent cover was evaluated. The species were grouped into five herb communities: Graminoids (grasses), *Solidago* spp. such as *Solidago rugosa*, *Asteraceae* spp. such as *Sonchus arvensis*, *Fabaceae species* (legumes) such as *Trifolium repens*, and *Apiaceae* mainly dominated by *Daucus carota*. More than 95% of the species identified fit into these groups.

2.3. Statistical Analysis

Seedling mortality was evaluated using Chi-squared tests. We first compared mortality rates between experimental factors (protection, vegetation, and protection × vegetation) from fall 2010 to fall 2012 and in the spring of 2015. We then compared mortality rates due to predation by small mammals (data combined from 2010 to 2012) between the vegetation treatments. Repeated measures of analyses of variance (ANOVAR) and univariate analyses of variance (ANOVA), based on the two-way split-plot design, were performed for each species to evaluate the effects of protection and vegetation treatments on seedling height and diameter, leaf nitrogen (N), SLA, and SWC. Species were analyzed individually because there were significant ($p < 0{,}05$) treatment × species interactions (results not shown), meaning that species reacted differently to the treatments. Herbaceous biomass and light (%PPFD) were only compared between herbaceous vegetation and mulch mat subplots with ANOVAs. Student and Tukey tests were used as post-hoc tests. Log transformations were used when data were not normal. A multiple regression was used to estimate the effects of SWC, average herbaceous cover height, and percentage cover of each herb community on the height and diameter relative growth rates (RGR) of each tree species. All statistical analyses were performed using JMP 10.0 (software from SAS).

3. Results

3.1. General

With the exception of paper birch, seedlings of the pioneer species had less mortality than did non-pioneer species. After five seasons of growth (spring 2015), mortality rates varied from 3% for red ash to 61% for yellow birch seedlings, all treatments combined (Figure 1). Pioneer species such as tamarack, paper birch, and red ash had the highest height and diameter growth between the fall of 2010 and spring of 2015, whereas the non-pioneer species red oak had the lowest growth (Figure 2).

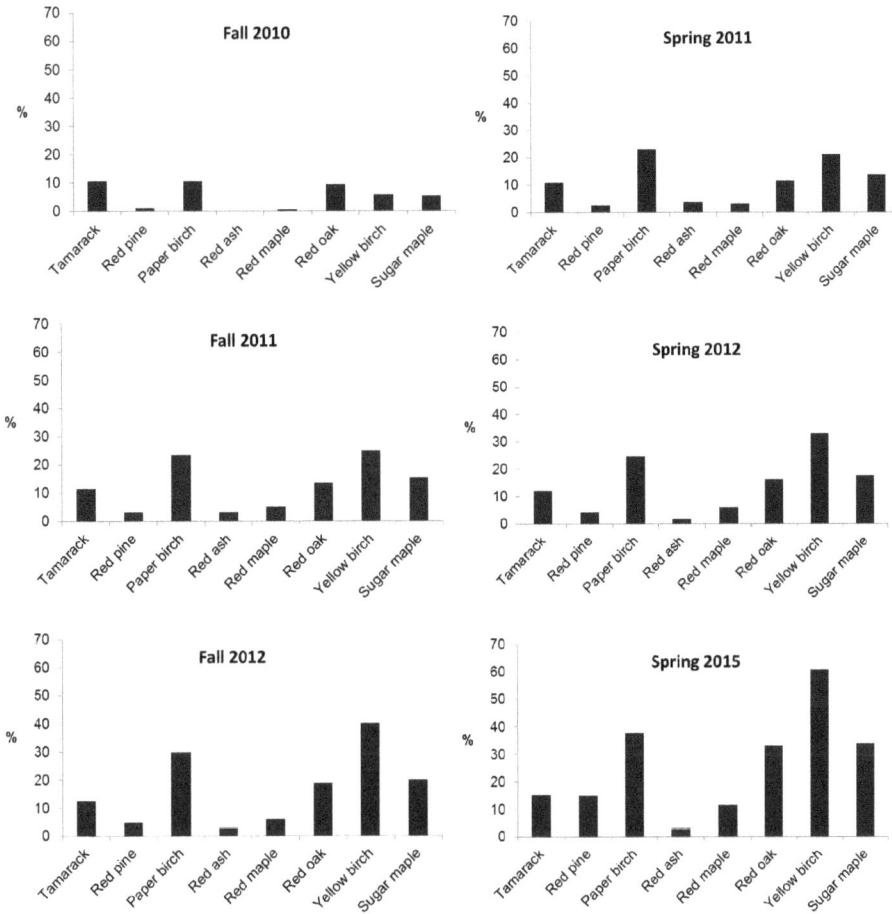

Figure 1. Rates of seedling mortality (all treatments combined) as a function of tree species in fall 2010, spring 2011, fall 2011, spring 2012, fall 2012, and spring 2015.

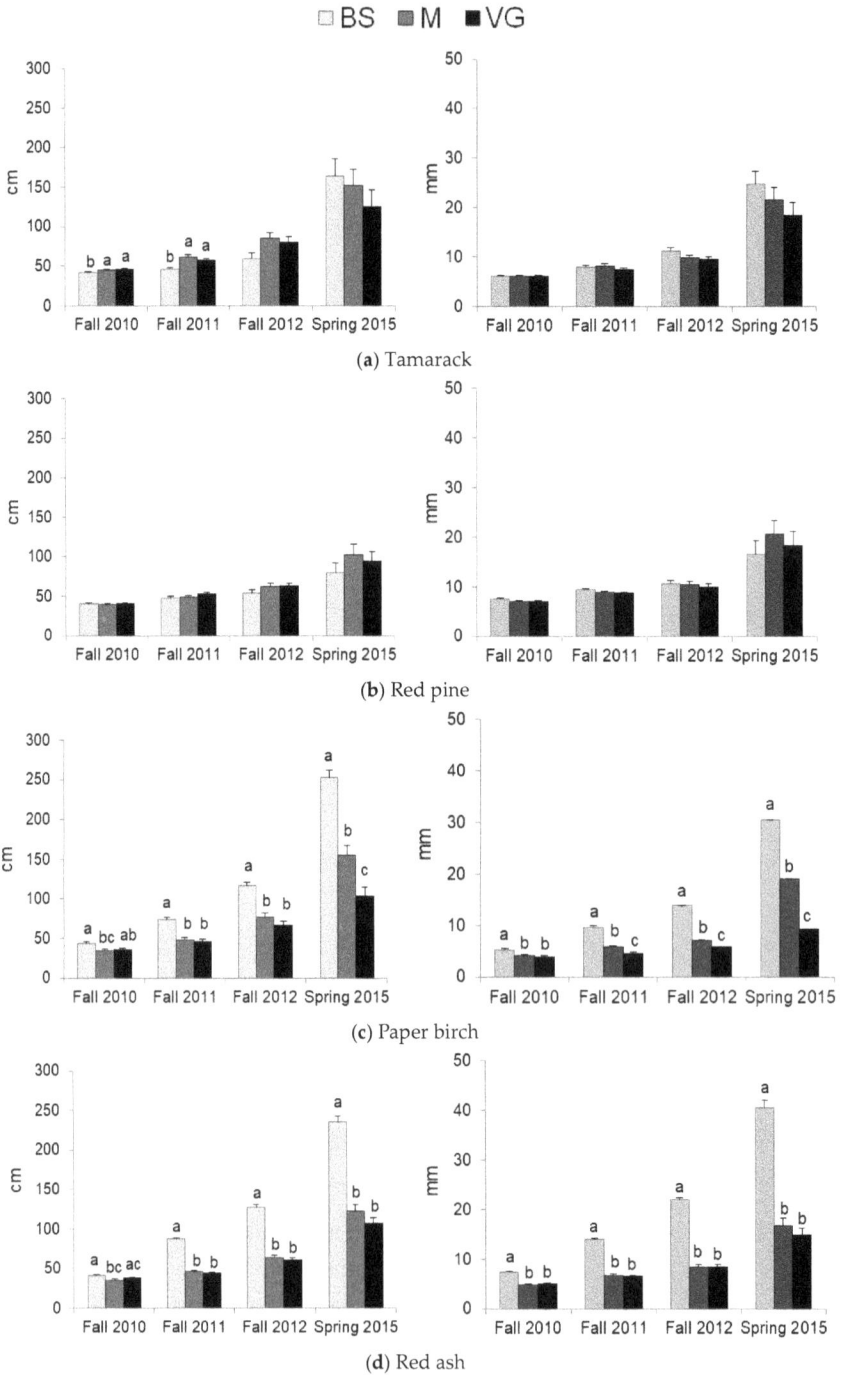

(**a**) Tamarack

(**b**) Red pine

(**c**) Paper birch

(**d**) Red ash

Figure 2. *Cont.*

(e) Red maple

(f) Red oak

(g) Yellow birch

(h) Red oak

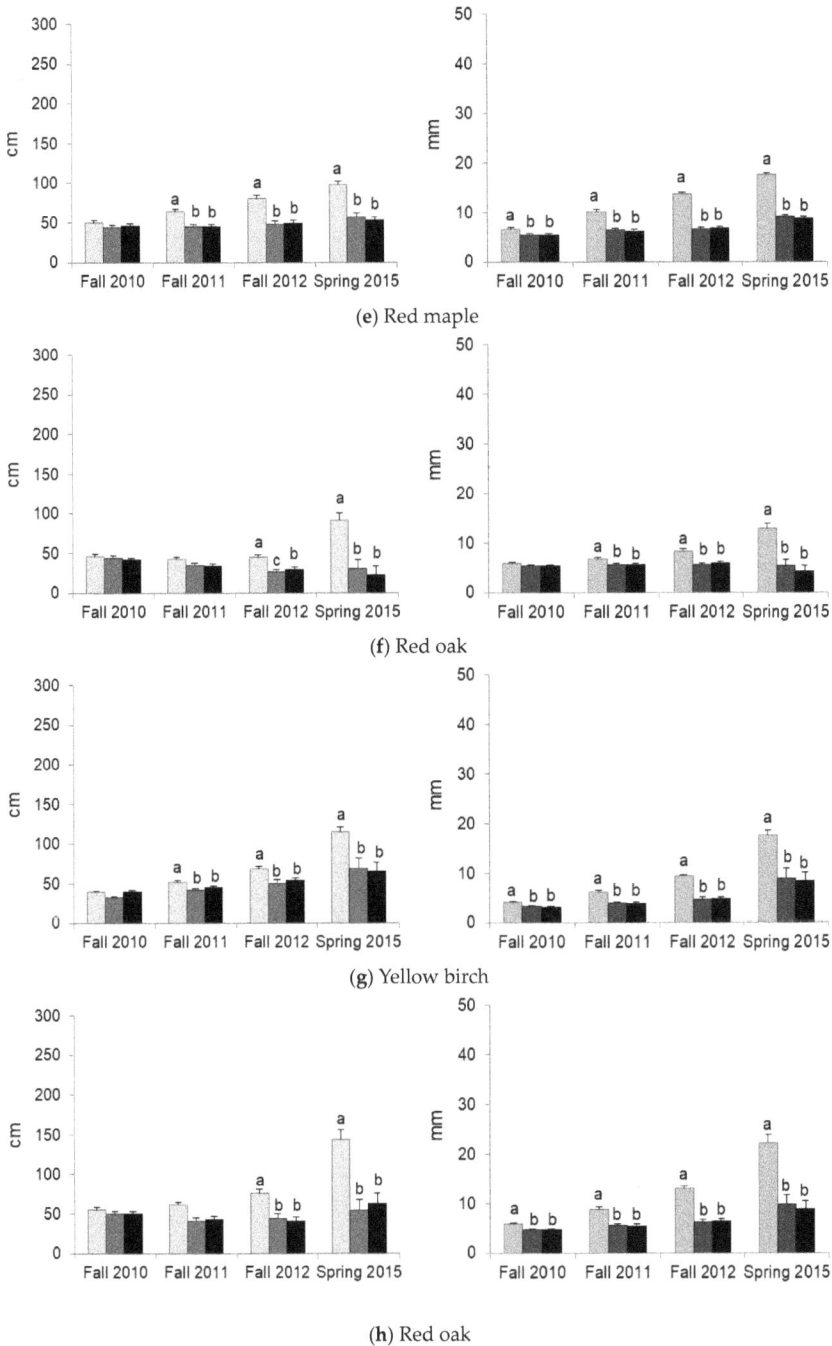

Figure 2. Height and diameter of seedlings (least-squares means and SE) growing in bare soil subplots (BS; light grey), surrounded by mulch mat (M; medium grey) or in vegetation subplots (VG; black), from autumn 2010−2012 and in spring 2015.

3.2. Predation and Herbaceous Vegetation Effects on Tree Survival and Growth

Mortality due to predation was generally higher for non-pioneer species than for pioneer species, with the exception of paper birch (Tables 2 and 3). In fact, both birch species had mortality rates double that of the next most vulnerable species (Figure 1, Table 3). Many birch seedlings were dead after the first winter with their mortality being principally related to predation, although competition also affected yellow birch survival (Table 2). Yellow birch (17%), paper birch (15%), red oak (7%), and sugar maple (6%) seedlings showed more signs of lethal small mammal damage than other species (Table 3). The vegetation control treatments also had a significant impact on predation rates in the most vulnerable species. Seedlings of both species in the birch genus surrounded by a mulch mat were more affected by small mammal damage than seedlings growing in vegetation (Table 3; $\chi^2 = 23.153$, $p \le 0.0001$). However, birch seedlings in both of these treatments were more susceptible to predation than bare soil seedlings ($\chi^2 = 15.634$, $p = 0.0004$).

Mortality due to predation of the non-pioneer species, red oak, and sugar maple, was also inferior or did not occur in bare soil subplots (Table 3; $\chi^2 = 11.544$, $p = 0.0031$; $\chi^2 = 10.050$, $p = 0.0066$, respectively). Effects of predation on red maple mortality were observed in the spring of 2015 survey (Table 2; $\chi^2 = 8.202$, $p = 0.0042$), but between 2010 and 2012 only 2% of dead seedlings showed small mammal damage (Table 3). The use of plastic spiral protectors did not have much of an influence on tree growth with the exception of a positive effect on red oak height from the second growing season onward ($p < 0.03$). The size of unprotected red oak seedlings significantly decreased due to leader mortality or the appearance of a new stem following small mammal damage. In fact, this decline in size was noted in more than 50% of red oak seedlings when all treatments were combined.

In general, predation by rabbits caused two times more seedling mortality than predation by voles (4.2% and 2.0% of seedling mortality, respectively). Both species of birch seedlings (14.3% for yellow birch and 11.5% for paper birch) were most affected by rabbit predation while red oak was the species preferred by voles (5.9%). Mortality due to small mammal damage was observed on some protected seedlings of yellow birch (5.5%), paper birch (2%), and red oak (1%). For these individuals, voles cut roots under the spiral protector while rabbits cut stems above the protector. Some red ash seedlings (3%) snipped by rabbits or voles produced new stems the year following predation, explaining the decrease in mortality rates in 2012 (Figure 1).

Herbivory by deer was negligible, only five seedlings (0.3%) died due to deer predation. Foliar herbivory by invertebrates was also minor (4% of tree seedlings, in total) and did not vary between treatments. Conifer species were not affected by any kind of herbivory.

Table 2. Predation, competition, and facilitation effects estimated from the results obtained from the different experimental treatments on tree seedling mortality from fall of 2010 to spring of 2015.

Species	Factor	Fall 2010	Spring 2011	Fall 2011	Spring 2012	Fall 2012	Spring 2015
Tamarack	Protection						
	Vegetation	Facilitation *	Facilitation *	Facilitation *	Facilitation *	Facilitation **	Facilitation **
Red pine	Protection						
	Vegetation						
Paper birch	Protection		Predation in VG*		Predation **	Predation **	
	Vegetation					M < BS = VG *	Competition ***
Red ash	Protection						
	Vegetation						
Red maple	Protection						Predation **
	Vegetation						Competition **
Red oak	Protection	Predation **			Predation **	Predation **	
	Vegetation		Competition **	Competition **	Competition ***	Competition ***	Competition ***
Yellow birch	Protection		Predation ***	Predation ***	Predation ***	Predation ***	
	Vegetation		Competition **	Competition **	Competition **	Competition ***	Competition ***
Sugar maple	Protection			Predation in VG *	M < BS = VG *		M < BS = VG **
	Vegetation	Facilitation *		Facilitation if Protection * M < BS = VG if No Protection *			

BS: tree seedlings growing in bare soil subplots, M: surrounded by organic mulch mat, VG: or in intact herbaceous vegetation. Blank spaces mean mortality did not differ between treatments. Predation means mortality is higher for seedlings unprotected against small mammals. Competition means mortality is lower when trees are growing in bare soil (BS > M = VG or BS > VG > M). Facilitation means mortality is lower when trees are growing in herbaceous vegetation (BS = M < VG or BS < M < VG). p-value: * < 0.05; ** < 0.01; *** < 0.001.

Table 3. Tree mortality (%) due to predation compared between vegetation treatments (evaluated using signs of lethal small mammal damage on unprotected and protected seedlings, data compiled from Spring 2010 to Autumn 2012), tree mortality in Autumn 2012 compared between protection treatments, and tree mortality in Spring 2015 (total).

Species	Mortality Due to Predation (2010–2012)				% Dead Seedlings (Fall 2012)		% dead Seedlings (Spring 2015)
	Bare Soil	Mulch Mat	Herbaceous Vegetation	Total	Unprotected	Protected	
Tamarack	0	0	0	0	6.8	5.8	15.2
Red pine	0	0	0	0	0.5	4.2	15.1
Paper birch	1.1 a	8.9 b	4.7 c	14.7	20.4 a	9.4 b	37.7
Red ash	0	2.1	0	2.1	2.1	0.5	3.1
Red maple	0	0.5	1.6	2.1	4.2	1.6	11.5
Red oak	0 a	2.6 b	4.2 b	6.8	13.6 a	5.2 b	33.2
Yellow birch	0.5 a	10.1 b	6.4 c	17.0	27.5 a	12.7 b	60.8
Sugar maple	0 a	3.7 b	2.1 b	5.8	12.1	7.9	33.5

Bare soil: seedlings growing in a bare soil, Mulch mats: surrounded by organic mulch mat, Herbaceous vegetation: or in intact herbaceous vegetation. Means for each species were tested using ANOVA and then compared using Tukey tests following ANOVA. Means of each row followed by different letters are significantly different at $p < 0.05$.

In general, the survival of pioneer species was not affected by herbaceous competition except at the end of the experiment for paper birch and red maple seedlings (Table 2). Nevertheless, competition effects were seen on the growth of hardwood pioneer species which was usually higher in bare soil (Figure 2). The conifer species were either tolerant of or facilitated by the presence of the vegetation layer. Red pine mortality (<5% within the first 3 years), height, and diameter did not vary with any treatment at any time (Table 2, Figure 2b). Mortality of tamarack seedlings occurred principally within the first growing season following planting in bare soil or in mulch mat subplots (Figure 1a, Table 2). No effects of herbaceous vegetation were observed on tamarack diameter, but tamarack seedlings growing in bare soil were smaller than those surrounded by mulch mat or in herbaceous vegetation, after the first (F-test = 9.7673; $p = 0.0118$) and the second (F-test = 13.4052; $p = 0.0061$) growing seasons (Figure 2a). After three years, this relationship was only marginally significant ($p = 0.0922$) and it was not significant by the spring of 2015 ($p = 0.4503$).

Herbaceous vegetation also showed some evidence of facilitation on sugar maple survival (Table 2). After the first growing season, sugar maple mortality was higher in bare soil ($\chi^2 = 6.330$, $p = 0.0422$) whereas after the second growing season, sugar maple mortality was lower in vegetation, but only for seedlings protected against predation ($\chi^2 = 6.344$, $p = 0.0419$). Seedling mortality of the other two non-pioneer species was, on the contrary, lower in bare soil subplots (Table 2). The growth of non-pioneer species was greater in bare soil (Figure 2).

No benefit of the organic mulch mats on tree survival was observed in this experiment. Mulch mats had effects comparable to or more negative than herbaceous vegetation (Table 2, Figure 2). In the second summer, herb biomass (47 g) growing through the organic mulch mats (F-test = 2.7507, $p = 0.1958$) and %PPFD at 30 cm high (60%; F-test = 0.7491, $p = 0.4504$) were not different from herb biomass and %PPFD in intact vegetation (65 g and 69%, respectively). By the end of the experiment (five growing seasons), many mulch mats were almost completely decomposed.

After five years, survival of all pioneer species seedlings growing in herbaceous vegetation was higher than 75%, except for paper birch (Figure 3a). Excluding red maple, they all had a height growth increment of more than 30 cm (Figure 3b). By contrast, non-pioneer species seedlings surrounded by vegetation had a lower survival and height growth rate.

(a)

(b)

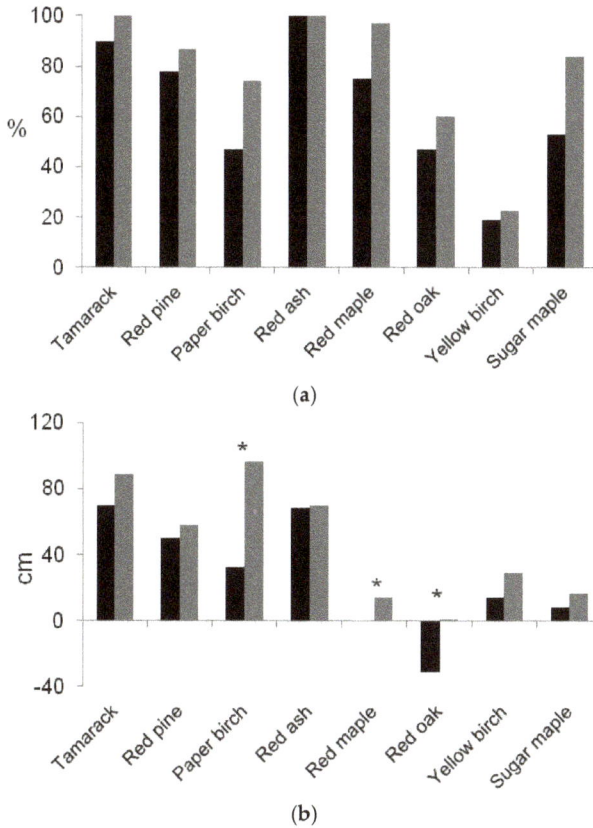

Figure 3. Differences of (**a**) survival rates in spring 2015 and (**b**) height growth increment (2010–2015) for seedlings growing in herbaceous vegetation. Black bars: seedlings unprotected against small mammal predation. Grey bars: seedlings protected. (ª $p < 0.05$, two-sample *t*-test).

3.3. Foliar Attributes, Soil Water Content, and the Influence of Herb Communities

All hardwoods, except yellow birch, had more leaves ($p < 0.01$) and a lower specific leaf area (SLA) ($p < 0.03$) in bare soil subplots (SLA was not measured for red ash and conifers). Nitrogen (N) in conifer needles or in birch leaves was not affected by the presence or absence of herbaceous vegetation (Table 4). Leaf N was, however, higher in bare soil for the three endomycorrhizal (AM) tree species: red ash (F-test = 47.9526; $p = 0.0002$), red maple (F-test = 11.8862; $p = 0.0081$), and sugar maple (F-test = 55.9883; $p = 0.0001$). More seedlings with superficial bare roots were observed in bare soil subplots for all species, except paper birch and sugar maple (Table 4).

Soil water content (SWC) did not vary between vegetation treatments in either 2011 or 2012. Further, it did not influence the relative growth rate (RGR) of seedlings growing in herbaceous vegetation (Table 5). The relative growth rate (RGR) of red maple and conifer species was positively related to the height of the herbaceous layer. In general, no specific herb community influenced tree growth, other than *Solidago* species which had a negative effect on red ash and red maple growth (Table 5). Grasses had a marginally significant positive effect ($p < 0.10$) on tamarack growth (Table 5).

Table 4. Vegetation treatment effects on leaf nitrogen (N) and rooting (%).

Species	Leaf Nitrogen (N)			Superficial Bare Roots		
	BS	M	VG	BS	M	VG
Tamarack	1.91	1.46	1.43	11.6 x	1.7 y	2.3 y
Red pine	1.02	1.07	0.97	12.8 x	1.1 y	2.7 z
Red ash	2.30 a	1.43 b	1.62 b	8.1 x	0 y	3.8 z
Paper birch	2.28	1.84	1.93	3.4	0.7	0.7
Red maple	1.68 a	1.16 b	1.23 b	10.4 x	0.6 y	2.7 y
Yellow birch	1.95	1.86	1.79	6.1 x	0 y	0 y
Red oak	1.68 a	1.11 b	1.24 ab	9.2 x	0.6 y	1.8 y
Sugar maple	1.68 a	1.13 b	1.22 b	1.8	0	2.4

For each variable, means of each row followed by different letters are significantly different at $p < 0.02$.

Table 5. Results of multiple regressions of height and diameter relative growth rates (RGR) (fall 2010–fall 2012) in relation to soil water content, average herb height, percentage cover of grasses, *Solidago*, *Asteraceae*, *Apiaceae*, and *Fabaceae* (legumes) species.

Species		Adjusted R^2	Soil Water Content	Herb Height	Grasses	*Solidago* spp.	*Asteraceae* spp.	*Apiaceae* spp.	*Fabaceae* spp.
Tamarack	RGR_{Height}	0.30		0.03 * (+)	0.09 (+)			0.09 (−)	
	$RGR_{Diameter}$	0.28		0.01 * (+)	0.08 (+)				0.03 * (+)
Red pine	RGR_{Height}	0.39		0.06 (+)					
	$RGR_{Diameter}$	0.14		0.02 * (+)					
Paper birch	RGR_{Height}	−0.07							
	$RGR_{Diameter}$	−0.13							
Red ash	RGR_{Height}	0.11							
	$RGR_{Diameter}$	0.20				0.02 * (−)			
Red maple	RGR_{Height}	0.36		0.004 **(+)		0.04 * (−)			
	$RGR_{Diameter}$	0.31		0.02 * (+)		0.09 (−)			
Red oak	RGR_{Height}	0.40					0.04 * (−)		0.02 * (−)
	$RGR_{Diameter}$	−0.05							
Yellow birch	RGR_{Height}	0.03							
	$RGR_{Diameter}$	−0.26							
Sugar maple	RGR_{Height}	−0.16							
	$RGR_{Diameter}$	−0.11							

Only significant effects ($p < 0.05$) and marginal effects ($p < 0.10$) are presented. Positive (+) or negative (−) effects. *p*-value: * < 0.05; ** < 0.01.

4. Discussion

4.1. Predation

Compared to other studies, predation by deer and small mammals was relatively low, probably due to predation being determined by both the environment and predator densities [26,27,32,52]. Tree species preferences also differed from other studies. For instance, rabbits and voles can potentially inflict damage on pine seedlings in fields or forests [6,52,53], but none of our conifer species exhibited any signs of herbivory. Differences in seedling size, growth rate, density of resin droplets on the stem, the presence of specific monoterpenes in the bark, and palatability are among the factors that have been proposed to explain species and seedling preferences of herbivores [25,26,53]. In our study, birch species were the most vulnerable to small mammal damage, particularly by rabbits. These species were also the smallest at the time of planting (Table 1) and had the smallest diameters, at least in the first two years of growth (Figure 2). By the fifth year of growth, red maple was the only species affected by predation and was amongst the species with the smallest end of experiment diameters

(Table 2, Figure 2). These results suggest that seedlings with small diameters are more vulnerable to small mammal damage.

Predation occurred just as frequently, or even more frequently, with mulch mats than the vegetation layer treatment. Mulch mats appear to offer the same protective benefits to small mammals against their predators as does an herbaceous cover. Mulch mats and the vegetation layer allow voles to feed and move safely beneath cover, whereas for rabbits, the presence of mats may make the young tree seedlings more visible than in the vegetation cover.

4.2. Competition, Tolerance, and Facilitation

Generally, tree mortality was not as affected by herbaceous competition as tree growth. Herbaceous vegetation increased the mortality of four hardwood species (Table 2), whereas growth of all hardwoods was negatively affected by competition (Figure 2).

Contrary to what we expected to see and to what has been observed in the tropics [17–19], the survival of the two moderately shade-tolerant, ectomycorrhizal and non-pioneer species was negatively affected by herbaceous competition. By contrast, slight effects of facilitation were identified for the survival of sugar maple, the species with the highest shade-tolerance [50], consistent with what Berkowitz et al. [7] had observed. Survival and growth of conifer species were not negatively affected by herbaceous vegetation (tolerance), while positive effects were observed on survival and height growth of tamarack seedlings. These positive effects suggest that facilitation does not only occur in harsh conditions such as dry environments [54–56].

The lack of competition effects on survival may be related to the low competition for water. Indeed, no differences in soil water content (SWC) were observed among the different treatments and no effect of SWC on relative growth rates was found for any species. However, herbaceous vegetation may have reduced heat and desiccation stresses in the uppermost few centimeters of soil; reductions that benefited tamarack which has a shallow root system [57]. In the first two growing seasons, few heat wave (>30 °C) events were observed [45]. Tamarack is one of the species (together with red pine) that has a low specific root length (SRL) [58]. It develops roots that are shorter and thicker than any of the hardwoods that were tested. Conifers probably developed a better root system in herbaceous vegetation than in bare soil. The positive effects of herbaceous vegetation on tree survival have been previously observed in other mesic habitats, including facilitation of pine seedlings in temperate-zone abandoned fields, but facilitation effects on growth, such as those that were observed for tamarack, are rare [2,3,5,6]. Similarly, seedling heights of hybrid larch (*Larix* × *marschlinsii* Coaz) planted for a boreal reforestation project were positively related to vegetation cover [59].

While the growth of hardwoods was influenced by competition, only endomycorrhizal (AM?) tree species had a higher leaf N in bare soil subplots. Furthermore, *Solidago* species exerted negative effects on red maple and red ash growth. Burton and Bazzaz [11] also observed lower foliar N for ash seedlings that were growing in *Solidago* patches. In abandoned agricultural fields, forbs and many grass species decrease the rate of N accumulation in the soil [60]. Frequent mowing in the rows of bare soil subplots promoted the legume species, *Trifolium*, from the second growing season onward, which could have increased soil N availability [60]. The explanation may also be related to mycorrhizae. Arbuscular mycorrhizal fungi do not promote tree N acquisition when N availability is low, in contrast to ectomycorrhizal fungi [61,62]. Contrary to what we expected, no herb community seemed to be a stronger competitor than another except for the few negative effects of *Solidago* species on tree growth. The vegetation surrounding the tree seedlings was composed of multiple herbaceous species which may have diluted any allelopathic effects that might occur in pure dense communities of *Solidago* and *Asteraceae* [11,22].

Tree species can be considered to be tolerant to herbaceous vegetation if they have high survival rates, even if growth is reduced, because many of the seedlings will eventually outgrow the vegetation layer [5,63]. Thus, we suggest that conifer species and red ash are tolerant to herbaceous vegetation because they had a high survival rate (>75%) and a height increment of more than 50 cm after five years in vegetation subplots. As expected, pioneer species had a higher growth rate than non-pioneer species, except for the moderately shade-tolerant red maple which had a survival rate higher than 75%. Due to a low survival rate (19% for yellow birch) and a low or negative growth rate (for red oak), we consider these two EM moderately shade-tolerant and non-pioneer species to be inhibited by the herb cover which had both direct and indirect (via predation) effects (Figure 3).

5. Conclusions

In this restoration experiment of an abandoned mesic field using multi-tree species, soil water content did not differ between treatments and did not affect tree growth. However, tree species responded differently to predation and herbaceous vegetation. Therefore, when planting trees in abandoned fields, specific treatments could be used according to the characteristics of the tree species and variable site conditions. Such an approach is inspired by the principles of precision agriculture for increasing efficiency and decreasing environmental and economic costs. Many definitions of precision agriculture have been proposed since its inception in the 1980s, but Gebbers and Adamchuk [64] (2010, p. 828) summarized it as "a way to apply the right treatment in the right place at the right time".

With this in mind, we propose that precision restoration requires the application of the right treatment to the right species at the right place. For instance, on our study sites, conifer species could be planted without controlling for competition and small mammal predation, as they were highly tolerant to or facilitated by herbaceous vegetation and unaffected by herbivory. Nevertheless, herbivory may vary with the environment, predator density, and the diversity of species used. Hardwood pioneer species could also be planted directly into herbaceous vegetation, although paper birch should be protected against small mammal damage. Planting seedlings with a diameter greater than 5 cm could also decrease vulnerability to predation, but this should be further tested. For large-seeded species, such as northern red oak, direct seeding could be a better approach than planting, given that their emergence and survival are not affected by herbaceous competition [65,66], but large-seeded species could be more affected by seed predation [4,67]. A better knowledge of the effect of various factors on different hardwood and conifer species under various site conditions could be used to recommend tree species that require the least amount of protection and vegetation control. Such an approach could be applied to other types of ecosystems, such as forests and industrial wastelands, to reduce costs and improve success rates.

Author Contributions: A.S.-D., D.K., and C.M. conceived and designed the experiment, A.S.-D. collected and analyzed the data, A.S.-D., D.K., and C.M. wrote the paper.

Funding: This study was funded by the National Science and Engineering Research Council in Canada (NSERC, Ottawa, Canada), the Fonds Québécois de la Recherche sur la Nature et les Technologies (FQRNT, Québec, Canada), WSP (GENIVAR Inc.) and the NSERC/Hydro-Québec Research Chair on Tree Growth.

Acknowledgments: We are grateful to the numerous field assistants and occasional contributors to sampling and weeding, particularly: Philippe-Olivier Boucher, Kasia Richer-Juraszek, Sophie Carpentier, Nathan Probst, Christophe Jenkins, Chantal Cloutier, Yann Gauthier, Étienne St-Hilaire, Matthias Schwetterlé, Olivier Lafontaine and Mathieu Messier. We would like to thank also Stéphane Daigle for advice on data analyses. We are thankful to the city of Laval for their collaboration and the use of their land in this research study. We acknowledge the collaboration of Dominic Senecal (WSP) in this study. We are also thankful to three anonymous reviewers for their comments and advice on the manuscript.

Conflicts of Interest: The authors declare no conflict of interest.

References

1. Padilla, F.M.; Pugnaire, F.I. The role of nurse plants in the restoration of degraded environments. *Front. Ecol. Environ.* **2006**, *4*, 196–202. [CrossRef]
2. Gómez-Aparicio, L. The role of plant interactions in the restoration of degraded ecosystems: A meta-analysis across life-forms and ecosystems. *J. Ecol.* **2009**, *97*, 1202–1214. [CrossRef]

3. Goldberg, D.E.; Rajaniemi, T.; Gurevitch, J.; Stewart-Oaten, A. Empirical approaches to quantifying interaction intensity: Competition and facilitation along productivity gradients. *Ecology* **1999**, *80*, 1118–1131. [CrossRef]

4. De Steven, D. Experiments on mechanisms of tree establishment in old-field succession: Seedling emergence. *Ecology* **1991**, *72*, 1066–1075. [CrossRef]

5. De Steven, D. Experiments on mechanisms of tree establishment in old-field succession: Seedling survival and growth. *Ecology* **1991**, *72*, 1076–1088. [CrossRef]

6. Gill, D.S.; Marks, P.L. Tree and shrub seedling colonization of old fields in central New York. *Ecol. Monogr.* **1991**, *61*, 183–205. [CrossRef]

7. Berkowitz, A.R.; Canham, C.D.; Kelly, V.R. Competition vs. Facilitation of tree seedling growth and survival in early successional communities. *Ecology* **1995**, *76*, 1156–1168. [CrossRef]

8. Callaway, R.M. Positive interactions in plant communities and the individualistic-continuum concept. *Oecologia* **1997**, *112*, 143–149. [CrossRef] [PubMed]

9. Niering, W.A.; Goodwin, R.H. Creation of relatively stable shrublands with herbicides: Arresting "Succession" on rights-of-way and pastureland. *Ecology* **1974**, *55*, 784–795. [CrossRef]

10. Benjamin, K.; Domon, G.; Bouchard, A. Vegetation composition and succession of abandoned farmland: Effects of ecological, historical and spatial factors. *Landsc. Ecol.* **2005**, *20*, 627–647. [CrossRef]

11. Burton, P.J.; Bazzaz, F.A. Ecophysiological Responses of Tree Seedlings Invading Different Patches of Old-Field Vegetation. *J. Ecol.* **1995**, *83*, 99–112. [CrossRef]

12. Davis, M.A.; Wrage, K.J.; Reich, P.B. Competition between tree seedlings and herbaceous vegetation: Support for a theory of resource supply and demand. *J. Ecol.* **1998**, *86*, 652–661. [CrossRef]

13. Davis, M.A.; Wrage, K.J.; Reich, P.B.; Tjoelker, M.G.; Schaeffer, T.; Muermann, C. Survival, growth, and photosynthesis of tree seedlings competing with herbaceous vegetation along a water-light-nitrogen gradient. *Plant Ecol.* **1999**, *145*, 341–350. [CrossRef]

14. Laliberté, E.; Bouchard, A.; Cogliastro, A. Optimizing hardwood reforestation in old fields: The effects of treeshelters and environmental factors on tree seedling growth and physiology. *Rest. Ecol.* **2008**, *16*, 270–280. [CrossRef]

15. Gale, M.R.; Grigal, D.F. Vertical root distributions of northern tree species in relation to successional status. *Can. J. For. Res.* **1987**, *17*, 829–834. [CrossRef]

16. Coll, L.; Potvin, C.; Messie, C.; Delagrange, S. Root architecture and allocation patterns of eight native tropical species with different successional status used in open-grown mixed plantations in Panama. *Trees* **2008**, *22*, 585–596. [CrossRef]

17. Hooper, E.; Condit, R.; Legendre, P. Responses of 20 native tree species to reforestation strategies for abandoned farmland in Panama. *Ecol. Appl.* **2002**, *12*, 1626–1641. [CrossRef]

18. Doust, S.J.; Erskine, P.D.; Lamb, D. Direct seeding to restore rainforest species: Microsite effects on the early establishment and growth of rainforest tree seedlings on degraded land in the wet tropics of Australia. *For. Ecol. Manag.* **2006**, *234*, 333–343. [CrossRef]

19. Doust, S.J.; Erskine, P.D.; Lamb, D. Restoring rainforest species by direct seeding: Tree seedling establishment and growth performance on degraded land in the wet tropics of Australia. *For. Ecol. Manag.* **2008**, *256*, 1178–1188. [CrossRef]

20. Caldwell, M.M.; Richards, H.J. Competing root systems: Morphology and models of absorption. In *On the Economy of Plant Form and Function*; Givnish, T.J., Ed.; Cambridge University Press: Cambridge, UK, 1986; pp. 251–273.

21. Pywell, R.F.; Bullock, J.M.; Roy, D.B.; Warman, L.I.Z.; Walker, K.J.; Rothery, P. Plant traits as predictors of performance in ecological restoration. *J. Appl. Ecol.* **2003**, *40*, 65–77. [CrossRef]

22. Horsley, S.B. Allelopathic inhibition of black cherry by fern, grass, goldenrod, and aster. *Can. J. For. Res.* **1977**, *7*, 205–216. [CrossRef]

23. De Blois, S.; Brisson, J.; Bouchard, A. Herbaceous covers to control tree invasion in rights-of-way: Ecological concepts and applications. *Environ. Manag.* **2004**, *33*, 606–619. [CrossRef]

24. Ostfeld, R.S.; Canham, C.D. Effects of meadow vole population density on tree seedling survival in old fields. *Ecology* **1993**, *74*, 1792–1801. [CrossRef]

25. Pusenius, J.; Ostfeld, R.S.; Keesing, F. Patch selection and tree-seedling predation by resident vs. immigrant meadow voles. *Ecology* **2000**, *81*, 2951–2956. [CrossRef]

26. Ostfeld, R.S.; Manson, R.H.; Canham, C.D. Effects of rodents on survival of tree seeds and seedlings invading old fields. *Ecology* **1997**, *78*, 1531–1542. [CrossRef]

27. Stange, E.E.; Shea, K.L. Effects of deer browsing, fabric mats, and tree shelters on *Quercus rubra* seedings. *Rest. Ecol.* **1998**, *16*, 29–34.

28. Moore, N.P.; Hart, J.D.; Langton, S.D. Factors influencing browsing by fallow deer Dama dama in young broad-leaved plantations. *Biol. Conserv.* **1999**, *87*, 255–260. [CrossRef]

29. Moore, N.P.; Hart, J.D.; Kelly, P.F.; Langton, S.D. Browsing by fallow deer (Dama dama) in young broadleaved plantations: Seasonality, and the effects of previous browsing and bud eruption. *Forestry* **2000**, *73*, 437–445. [CrossRef]

30. McPherson, G.R. Effects of herbivory and herb interference on oak establishment in a semi-arid temperate savanna. *J. Veg. Sci.* **1993**, *4*, 687–692. [CrossRef]

31. Ward, J.S.; Gent, M.P.; Stephens, G.R. Effects of planting stock quality and browse protection-type on height growth of northern red oak and eastern white pine. *For. Ecol. Manag.* **2000**, *127*, 205–216. [CrossRef]

32. Sweeney, B.W.; Czapka, S.J.; Yerkes, T. Riparian forest restoration: Increasing success by reducing plant competition and herbivory. *Rest. Ecol.* **2002**, *10*, 392–400. [CrossRef]

33. Groninger, J.W. Increasing the impact of bottomland hardwood afforestation. *J. For.* **2005**, *103*, 184–188. [CrossRef]

34. Cogliastro, A.; Benjamin, K.; Bouchard, A. Effects of full and partial clearing, with and without herbicide, on weed cover, light availability, and establishment success of white ash in shrub communities of abandoned pastureland in southwestern Quebec, Canada. *New For.* **2006**, *32*, 197–210. [CrossRef]

35. Sampaio, A.B.; Holl, K.D.; Scariot, A. Does restoration enhance regeneration of seasonal deciduous forests in pastures in central Brazil? *Rest. Ecol.* **2007**, *15*, 462–471. [CrossRef]

36. Barea, J.M.; Azcon-Aguilar, C.; Azcon, R. Interactions between mycorrhizal fungi and rhisosphere micro-organisms within the context of sustainable soil-plant systems. In *Multitrophic Interactions in Terrestrial Systems*; Gange, A.C., Brown, V.K., Eds.; Blackwell Science: Cambridge, MA, USA, 1997; pp. 65–78.

37. Alguacil, M.M.; Lumini, E.; Roldán, A.; Salinas-García, J.R.; Bonfante, P.; Bianciotto, V. The impact of tillage practices on arbuscular mycorrhizal fungal diversity in subtropical crops. *Ecol. Appl.* **2008**, *18*, 527–536. [CrossRef] [PubMed]

38. Cogliastro, A.; Gagnon, D.; Coderre, D.; Bhereur, P. Responses of seven hardwood tree species to herbicide, rototilling, and legume cover at two southern Quebec plantation sites. *Can. J. For. Res.* **1990**, *20*, 1172–1182. [CrossRef]

39. Coll, L.; Messier, C.; Delagrange, S.; Berninger, F. Growth, allocation and leaf gas exchanges of hybrid poplar plants in their establishment phase on previously forested sites: Effect of different vegetation management techniques. *Ann. For. Sci.* **2007**, *64*, 275–285. [CrossRef]

40. Davies, R.J. The importance of weed control and the use of tree shelters for establishing broadleaved trees on grass-dominated sites in England. *Forestry* **1985**, *58*, 167–180. [CrossRef]

41. Davies, R.J. Sheet mulching as an aid to broadleaved tree establishment II. Comparison of various sizes of black polythene mulch and herbicide treated spot. *Forestry* **1988**, *61*, 107–124. [CrossRef]

42. Truax, B.; Gagnon, D. Effects of straw and black plastic mulching on the initial growth and nutrition of butternut, white ash and bur oak. *For. Ecol. Manag.* **1993**, *57*, 17–27. [CrossRef]

43. Lambert, F.; Truax, B.; Gagnon, D.; Chevrier, N. Growth and N nutrition, monitored by enzyme assays, in a hardwood plantation: Effects of mulching materials and glyphosate application. *For. Ecol. Manag.* **1994**, *70*, 231–244. [CrossRef]

44. Environment and Climate Change Canada. Canadian Climate Normals 1981–2010. Montréal/Pierre Elliott Trudeau Intl A, Québec. Available online: http://climate.weather.gc.ca/climate_normals/ (accessed on 22 January 2013).

45. Environment and Climate Change Canada. Monthly data report for 2010, 2011, 2012. Montréal/Pierre Elliott Trudeau Intl A, Québec. Available online: http://climate.weather.gc.ca/index_f.html (accessed on 22 January 2013).

46. Institut de recherche et de développement en agroenvironnement (IRDA). Études pédologiques. Available online: http://www.irda.qc.ca/fr/outils-et-services/informations-sur-les-sols/etudes-pedologiques/ (accessed on 12 december 2012).

47.   Cornelissen, J.H.C.; Lavorel, S.; Garnier, E.; Díaz, S.; Buchmann, N.; Gurvich, D.E.; Reich, P.B.; Ter Steege, H.; Morgan, H.D.; van der Heijden, M.G.A.; et al. A handbook of protocols for standardised and easy measurement of plant functional traits worldwide. *Aust. J. Bot.* **2003**, *51*, 335–380. [CrossRef]

48.   U.S. Department of Agriculture. The PLANTS Database. Available online: http://plants.usda.gov (accessed on 11 February 2013).

49.   Burns, R.M.; Honkala, B.H. *Silvics of North America: 1. Conifers; 2. Hardwoods*; U.S. Department of Agriculture, Forest Service: Washington, DC, USA, 1990; Volume 2, p. 877.

50.   Niinemets, Ü.; Valladares, F. Tolerance to shade, drought, and waterlogging of temperate Northern Hemisphere trees and shrubs. *Ecol. Monogr.* **2006**, *76*, 521–547. [CrossRef]

51.   Messier, C.; Puttonen, P. Spatial and temporal variation in the light environment of developing scots pine stands: The basis for a quick and efficient method of characterizing light. *Can. J. For. Res.* **1995**, *25*, 343–354. [CrossRef]

52.   Bergman, M.; Iason, G.R.; Hester, A.J. Feeding patterns by roe deer and rabbits on pine, willow and birch in relation to spatial arrangement. *Oikos* **2005**, *109*, 513–520. [CrossRef]

53.   Bucyanayandi, J.D.; Bergeron, J.M.; Menard, H. Preference of meadow voles (*Microtus pennsylvanicus*) for conifer seedlings: Chemical components and nutritional quality of bark of damaged and undamaged trees. *J. Chem. Ecol.* **1990**, *16*, 2569–2579. [CrossRef] [PubMed]

54.   Bertness, M.D.; Callaway, R. Positive interactions in communities. *Trends Ecol. Evol.* **1994**, *9*, 191–193. [CrossRef]

55.   Holmgren, M.; Scheffer, M.; Huston, M.A. The interplay of facilitation and competition in plant communities. *Ecology* **1997**, *78*, 1966–1975. [CrossRef]

56.   Callaway, R.M.; Walker, L.R. Competition and facilitation: A synthetic approach to interactions in plant communities. *Ecology* **1997**, *78*, 1958–1965. [CrossRef]

57.   Strong, W.L.; Roi, G.L. Root-system morphology of common boreal forest trees in Alberta, Canada. *Can. J. For. Res.* **1983**, *13*, 1164–1173. [CrossRef]

58.   Tobner, C.M.; Paquette, A.; Messier, C. Interspecific coordination and intraspecific plasticity of fine root traits in North American temperate tree species. *Front. Plant Sci.* **2013**, *4*, 242. [CrossRef] [PubMed]

59.   Buitrago, M.; Paquette, A.; Thiffault, N.; Bélanger, N.; Messier, C. Early performance of planted hybrid larch: Effects of mechanical site preparation and planting depth. *New For.* **2014**, *46*, 319–337. [CrossRef]

60.   Knops, J.M.; Tilman, D. Dynamics of soil nitrogen and carbon accumulation for 61 years after agricultural abandonment. *Ecology* **2000**, *81*, 88–98. [CrossRef]

61.   Reynolds, H.L.; Hartley, A.E.; Vogelsang, K.M.; Bever, J.D.; Schultz, P.A. Arbuscular mycorrhizal fungi do not enhance nitrogen acquisition and growth of old-field perennials under low nitrogen supply in glasshouse culture. *New Phytol.* **2005**, *167*, 869–880. [CrossRef] [PubMed]

62.   Smith, S.E.; Read, D.J. *Mycorrhizal Symbiosis*, 3rd ed.; Academic Press: London, UK, 2008.

63.   Connell, J.H.; Slatyer, R.O. Mechanisms of succession in natural communities and their role in community stability and organization. *Am. Nat.* **1977**, *111*, 1119–1144. [CrossRef]

64.   Gebbers, R.; Adamchuk, V.I. Precision agriculture and food security. *Science* **2010**, *327*, 828–831. [CrossRef] [PubMed]

65.   Laliberté, E.; Cogliastro, A.; Bouchard, A. Spatiotemporal patterns in seedling emergence and early growth of two oak species direct-seeded on abandoned pastureland. *Ann. For. Sci.* **2008**, *65*, 407. [CrossRef]

66.   St-Denis, A.; Messier, C.; Kneeshaw, D. Seed size, the only factor positively affecting direct seeding success in an abandoned field in Quebec, Canada. *Forests* **2013**, *4*, 500–516. [CrossRef]

67.   Martelletti, S.; Lingua, E.; Meloni, F.; Freppaz, M.; Motta, R.; Nosenzo, A.; Marzano, R. Microsite manipulation in lowland oak forest restoration result in indirect effects on acorn predation. *For. Ecol. Manag.* **2018**, *411*, 27–34. [CrossRef]

MDPI

St. Alban-Anlage 66

4052 Basel

Switzerland

Tel. +41 61 683 77 34

Fax +41 61 302 89 18

www.mdpi.com

Forests Editorial Office

E-mail: forests@mdpi.com

www.mdpi.com/journal/forests

www.ingramcontent.com/pod-product-compliance
Lightning Source LLC
Chambersburg PA
CBHW051314020426
42333CB00028B/3338